AMD-Vol. 104

D0807999

Mechanics of Plastics and Plastic Composites

presented at

THE WINTER ANNUAL MEETING OF
THE AMERICAN SOCIETY OF MECHANICAL ENGINEERS
SAN FRANCISCO, CALIFORNIA
DECEMBER 10–15, 1989

co-sponsored by

THE APPLIED MECHANICS DIVISION,
THE MATERIALS DIVISION, AND
THE PRODUCTION ENGINEERING DIVISION, ASME

edited by

V. K. STOKES
GE CORPORATE RESEARCH AND DEVELOPMENT

THE AMERICAN SOCIETY OF MECHANICAL ENGINEERS

United Engineering Center   345 East 47th Street   New York, N.Y. 10017

ISBN No. 0-7918-0407-0

Library of Congress
Catalog Number 89-046383

FOREWORD

This AMD Volume contains written versions of papers that were presented in a symposium that was jointly sponsored by the Applied Mechanics, Materials, and Production Engineering Divisions at the 1989 Winter Annual Meeting of ASME in San Francisco, California. The main purpose of the symposium was to provide a forum for discussing mechanics issues relating to the use of plastics and plastic composites in load-bearing applications. Much of the work in this area is being done by individuals educated in diverse fields such as mathematics, physics, materials science, mechanics, and other engineering subdisciplines. The Joint AMD/MD Committee on Constitutive Equations, which took the lead in organizing this symposium, felt that the synergy generated by bringing together individuals from different disciplines could stimulate new directions for research.

High-performance, thermoset-polymer-matrix composites have been used effectively in structural applications by the aerospace industry for at least 25 years. While methods for predicting part stiffness in these stiff, high-modulus materials are well understood, failure prediction and cost-effective part manufacturing continue to be problems. On the other hand, efficient processes for the manufacture of neat, particulate-filled, and chopped-fiber-filled resins have been available for quite some time. But procedures for part design are not well understood — especially for highly ductile low-modulus thermoplastics. The increasing use of such materials in the consumer and automotive industries — typified by all-plastic, unfilled (noncomposite) thermoplastic bumpers that are capable of withstanding 5-mph impacts — has highlighted the need for developing accurate methods for predicting part performance. Furthermore, in these materials, part design and processing (manufacture) are very strongly coupled. For example, the blow molding and thermoforming processes result in parts in which the wall thickness can vary by a factor of 20:1, so that part performance cannot be evaluated without using process simulation to predict wall thickness distribution. Similarly, in injection molded parts, local stiffness and strength are controlled by chopped-fiber distribution and orientation, both of which are affected by the molding process. Between these extremes of unfilled resins and high-performance composites, a host of new material systems — such as random glass mat in thermoplastic stampable sheet — are being developed for the consumer and automotive markets. This development of new plastic-based material systems and new processing technologies, and the increasing use of such materials in load-bearing applications, offers new exciting research opportunities for mechanics. This symposium was organized in recognition of the strong link between materials, mechanics, and processing for this class of materials — the common unifying theme being mechanics.

This symposium would not have been possible without the active support and encouragement from many people. First, the authors are to be thanked for having taken the time to prepare the papers in this volume. Second, the help and support provided by Martin A. Eisenberg and George J. Weng (Applied Mechanics Division), Minoru Taya (Materials Division), and Miguel R. Martinez-Heath (Production Engineering Division), and the encouragement from members of the Joint AMD/MD Committee on Constitutive Equations, is greatly appreciated. Finally, the assistance of ASME technical staff in supervising the production of this volume is gratefully acknowledged.

Vijay K. Stokes

CONTENTS

THE EFFECT OF THERMAL HISTORY ON THE MECHANICAL BEHAVIOR OF AMORPHOUS POLYMERS

R. M. Shay, Jr., and J. M. Caruthers
Purdue University
School of Chemical Engineering
West Lafayette, Indiana

ABSTRACT

The effect of thermal history during formation on the mechanical behavior of amorphous polymer solids must be accurately described to enable efficient design of manufacturing processes for polymers. A nonlinear, thermoviscoelastic constitutive equation for amorphous polymers has been derived via the rational thermodynamics framework, where the dissipation is assumed to occur on a material timescale that is controlled by the nonequilibrium entropy. The material properties for the model were determined for poly(vinyl acetate) from equilibrium PVT and linear viscoelastic shear and bulk moduli measurements. Predictions of specific volumes during isobaric cooling from the rubber and nonlinear uniaxial stress-strain curves just below T_g are in good agreement with experimental data. The model predicts that samples cooled below T_g and then isothermally annealed for specified times will exhibit yield stresses that increase with increasing annealing time, as well as the familiar effects of changes in temperature and strain rate on the uniaxial stress-strain and yield behavior. Shortcomings and potential improvements in the model are discussed.

I. INTRODUCTION

The engineering properties of a polymeric component depend not only upon its chemical structure and composition, but also upon the processing history employed during formation of that component. The dependence upon processing conditions arises from the long-time viscoelastic response to the complicated thermal-deformation histories applied during formation from the melt or rubber. Since the processing history has a pronounced effect on the mechanical behavior of the glassy component, a constitutive model of amorphous polymers must not only describe its nonlinear mechanical response, but also the solidification process and its long-term consequences.

There are numerous examples of the effects of thermal history on the linear and nonlinear mechanical properties of glassy polymers. The effect of temperature on the isothermal, linear viscoelastic material properties of amorphous polymers is well-described by time-temperature superposition, at least near T_g and above. In the most idealized situation a polymer melt well above T_g is cooled at a constant rate and pressure through T_g, annealed isothermally below T_g, and finally deformed isothermally. Actual processing operations also include the application of anisotropic stresses/strains throughout this thermal history. Glasses formed via the "simple" thermal history described above exhibit a higher modulus and yield stress for formation histories with higher pressures [1-3], slower cooling rates [4], and longer aging times [4,5].

A general thermoviscoelastic constitutive equation will be required to describe the formation and subsequent nonlinear viscoelastic deformation. Although numerous thermoviscoelastic constitutive equations have been proposed, a key element of these constitutive theories is a material timescale. The material time t^* is related to the laboratory time t by

$$\frac{dt}{dt^*} = a(t^*) \qquad \text{or} \qquad t^* = \int_0^t \frac{d\zeta}{a(\zeta)} \qquad \text{(1-1a,b)}$$

where $a(t^*)$ is a general shift factor that depends on the thermodynamic state of the polymer. The material time t^* was first proposed by Hopkins [6] and Moreland and Lee [7] for nonisothermal *linear* viscoelasticity, where $a = a_T$. This idea was extended by Knauss and Emri [8,9] and Shay and Caruthers [10,11] for nonlinear viscoelasticity by (i) replacing the time argument in the compressible linear or finite viscoelastic constitutive equation with t^* and (ii) allowing the log a shift factor to depend upon volume as well as temperature. The key feature of these constitutive equations is that the deformation-induced dilation effects a nonlinear expansion/contraction of the timescale. These constitutive equations predict a nonlinear stress-strain curve including yield.

Knauss and Emri [8,9] employed the Doolittle expression for the shift factor in terms of a fractional free volume f; specifically,

$$\log a = \frac{B}{2.3}\left[\frac{1}{f} - \frac{1}{f_o}\right] = -\frac{B}{2.3 f_o}\frac{\alpha\Delta T + \delta\Delta_v}{f_o + \alpha\Delta T + \delta\Delta_v} \qquad \text{(1-2a,b)}$$

where B, α, and δ are material constants and f_o denotes the free volume in the reference state. It was assumed that the changes in fractional free volume due to temperature T and mechanically-induced dilation Δ_v are independent. Assuming $\delta = 1$, the material constants for poly (vinyl acetate) were determined as $B = 0.016$, $f_o = 0.01$, and $\alpha = 5.98 \times 10^{-4}\ °C^{-1}$. The constitutive model also required the linear viscoelastic shear G(t) and bulk K(t) moduli. Although both properties could be determined from independent experiments, Knauss and Emri assumed a single relaxation time for the bulk modulus; specifically [8],

$$K(t) = 1.34 + 2.01\exp(-t/\tau) \quad [=]\quad GPa \qquad \text{(1-3)}$$

where $\tau = 4 \times 10^5$ mins. The three parameters in Eqn. 1-3 were determined by fitting the nonlinear tensile viscoelastic relaxation data. The model predicts the nonlinear stress-strain behaviors including yield. However, the rubbery bulk modulus (i.e. 1.34 GPa) and the glassy bulk modulus (i.e. 1.34 + 2.01 = 3.35 GPa) are only 50% of the experimentally measured PVAc bulk modulus of McKinney and Belcher [2].

Shay and Caruthers [10,11] developed a purely mechanical constitutive equation by replacing (i) the time arguments of the viscoelastic functions with t^* and (ii) replacing the infinitesimal strain tensor with a finite strain tensor. Two different relationships for the shift factor have been employed. First, as suggested by Curro, Lagasse, and Simha [12] the fractional free volume in Eqn. 1-2a is identified as the hole fraction in an intuitive extension of the Simha-Somcynsky (SS) equilibrium equation-of-state for polymeric fluids, where the hole fraction is evaluated from the equilibrium equation-of-state at the nonequilibrium pressure-volume-temperature state. However, this free volume model only describes the time-temperature shift factor over a relatively narrow temperature range. Alternatively, the shift factor can depend upon the configurational entropy S as suggested by Adams and Gibbs [13]; specifically,

$$\log a = \frac{B}{2.3}\left[\frac{1}{TS} - \frac{1}{T_o S_o}\right] \qquad \text{(1-4)}$$

where S_o and T_o refer to the entropy and temperature in the reference state. The nonequilibrium entropy is calculated by evaluating the expression for the equilibrium SS entropy at the nonequilibrium pressure, volume, temperature, and calculated hole fraction. Using the nonequilibrium SS entropy, Eqn. 1-4 can describe the time-temperature shift factor over a wider range of temperatures extending above and below the glass transition temperature, where the material parameter B is fit only to equilibrium rubbery shift data [14]. The shear and bulk moduli are the only other material properties and are determined from independent measurements.

The purely mechanical constitutive model is able to predict qualitatively, and quantitatively where comparisons were possible, the isothermal, uniaxial stress-strain and ductile failure of poly(methyl methacrylate), including the strain rate and temperature dependence of the yield stress [10,11,15]. In addition, since a finite strain measure was employed, the model predicts isothermal yield in an experimentally realizable shear deformation. We believe that the purely mechanical constitutive model is the first constitutive equation that can describe isothermal, nonlinear stress-

2

strain and yield in arbitrary three-dimensional deformations from material parameters obtained from independent measurements.

Although the purely mechanical constitutive model has some attractive features, it is unable to describe the time-dependent thermal expansion in the glass-to-rubber transition [25]. Thus, the purely mechanical model is unable to describe the effects of different processing histories on the subsequent deformation of the glassy solid. Because of the inability to incorporate processing history and questions concerning which particular finite stress and strain tensors should be employed, the constitutive model has been reformulated using the rigorous framework of rational thermodynamics [16-18]. In the remainder of this paper we will discuss this new constitutive equation and present some initial predictions for the complicated thermal/deformation histories that are important in processing of polymers.

II. THERMODYNAMIC CONSTITUTIVE EQUATION

A thermodynamic constitutive equation for isotropic amorphous polymer solids will now be presented [19-21]. The approach employed in the constitutive model is the rational thermodynamics framework originally developed by Coleman [17], but where the deformation is assumed to occur on a material timescale that is natural for the polymer. If the free energy functional is assumed to depend only on the deformation history and the current temperature (except to the extent that the material timescale is affected by the history of the temperature), then application of the rational thermodynamics framework for isotropic solids with fading memory results in the following constitutive equation for the stress:

$$\widetilde{\mathbf{T}}(t)=\widetilde{\mathbf{T}}^{\infty}[\theta,\mathbf{C}]+\int_{-\infty}^{t}\left\{G_{\Delta}(t^{*}-\xi^{*};\theta,\mathbf{C})\frac{d}{d\xi}[\mathbf{C}(\xi)-\frac{1}{3}I tr\mathbf{C}(\xi)] + \frac{1}{2}K_{\Delta}(t^{*}-\xi^{*};\theta,\mathbf{C})\mathbf{I}\frac{d}{d\xi}tr\mathbf{C}(\xi)\right\}d\xi \tag{2-1a}$$

where

$$\widetilde{\mathbf{T}}^{\infty}[\theta,\mathbf{C}] = -(\det\mathbf{C})^{1/6}P^{\infty}(\theta,V) + G_{\infty}[\mathbf{C}-\frac{1}{3}\mathbf{I} tr\mathbf{C}], \tag{2-1b}$$

$\widetilde{\mathbf{T}}$ is the second Piola-Kirchhoff stress tensor, the infinity super/subscript denotes the long-time equilibrium response, θ is the absolute temperature, $\mathbf{C}$ is the right Cauchy-Green deformation tensor, V is the specific volume, $\det\mathbf{C}$ is the determinant of $\mathbf{C}$, and $tr\mathbf{C}$ is the trace of $\mathbf{C}$. The material properties G_{Δ} and K_{Δ} are the time-dependent portion of the linear viscoelastic shear and bulk moduli and vanish at long material times. G_{Δ} and K_{Δ} also depend on the current temperature $\theta = \theta(t^{*})$ and on the invariants of the current deformation $\mathbf{C} = \mathbf{C}(t^{*})$.

In this thermodynamic constitutive equation, ordinary laboratory time t is a dependent variable and is related to the independent material time t^{*} by Eqn. 1-1a. The shift functional includes all variables and histories that affect the rate of viscoelastic relaxation. We postulate that log a is related to the entropy as first proposed by Adam and Gibbs [13] and given by Eqn. 1-4, where the nonequilibrium entropy will now be denoted by η. A constitutive equation for the nonequilibrium entropy can be derived through the rational thermodynamics framework from the same nonequilibrium Helmholtz free energy used to obtain the stress constitutive equation. Specifically,

$$\eta(t) = \eta^{\infty}[\theta,\mathbf{C}] - \frac{V_{r}}{8}\int_{-\infty}^{t}\int_{-\infty}^{t}\frac{\partial K_{\Delta}}{\partial\theta}\frac{d tr\mathbf{C}(\xi)}{d\xi}\frac{d tr\mathbf{C}(\zeta)}{d\zeta}d\xi d\zeta$$

$$- \frac{V_{r}}{4}\int_{-\infty}^{t}\int_{-\infty}^{t}\frac{\partial G_{\Delta}}{\partial\theta}\left[\frac{d\mathbf{C}(\xi)}{d\xi}:\frac{d\mathbf{C}(\zeta)}{d\zeta} - \frac{1}{3}\frac{d tr\mathbf{C}(\xi)}{d\xi}\frac{d tr\mathbf{C}(\zeta)}{d\zeta}\right]d\xi d\zeta \tag{2-2}$$

where V_{r} is the specific volume in the reference state and all other symbols are the same as in Eqn. 2-1. Note that all viscoelastic material functions needed for the entropy are identical to those in the stress constitutive equation. Thus, Eqns. 1-1a, 1-4, 2-1, and 2-2 form a complete, self-consistent nonlinear viscoelastic model which only requires as input the equilibrium thermodynamic equation-of-state and the viscoelastic shear and bulk moduli. Alternatively, the entropy may be

3

identified with the nonequilibrium SS entropy discussed in Section I. Predictions of the model employing both entropy alternatives will be presented in Section IV.

III. MATERIAL PROPERTIES

All of the material properties in the model can be determined from independent measurements. We will compare the predictions of the constitutive equation with experimental data for poly(vinyl acetate), PVAc. Although there is only limited nonlinear stress-strain data on PVAc available in the literature, it is the polymer with the most complete bulk viscoelastic data and its glass transition behavior has been extensively studied.

We assume that the viscoelastic material properties obey pseudo time-shift invariance [21]; specifically,

$$G_\Delta(t^*;\theta,C) = G(t^*;\theta,C) - G_\infty(\theta,C) = [G_o(\theta,C) - G_\infty(\theta,C)]g(t^*) \tag{3-1}$$

where $G_o(G_\infty)$ is the instantaneous (equilibrium) modulus. The function $(G_o - G_\infty)$ is analogous to the vertical shift in traditional time-temperature superposition, and $g(t^*)$ corresponds to the modulus master curve along the log time axis. The normalized function $g(t^*)$ can be represented by

$$g(t^*) = \sum_j g_j \exp\left[-\frac{t^*}{\tau_{G_j}}\right] \qquad \text{where} \qquad \sum_j g_j = 1 \tag{3-2a,b}$$

The discrete exponential relaxation spectrum (i.e., the set of relaxation times τ_{G_j} and associated strengths g_j) employed is for computational convenience. The functions $G_o(\theta,C)$, $G_\infty(\theta,C)$, and $g(t^*)$ must now be determined.

The complex dynamic shear modulus $G^*(\omega;\theta)$ of PVAc has been measured from liquid nitrogen temperatures up to well above the glass transition temperature and the data analyzed by time-temperature superposition [14]. The storage and loss modulus master curves are shown in Fig. 1 and the associated log a_T shift factors are shown in Fig. 2. From the high and low frequency plateaus of the G' master curve, $G_o = 208$ MPa and $G_\infty = 0.25$ MPa, where the θ and C dependence has been neglected. This PVAc material was uncrosslinked and the nonzero G_∞ value actually represents an entanglement plateau. The discrete shear relaxation spectrum was computed by a regularization with quadratic programming [20] from the loss modulus alone. As shown in Fig. 1, both the dynamic storage and loss moduli responses computed from the spectrum are in good agreement with the experimental data.

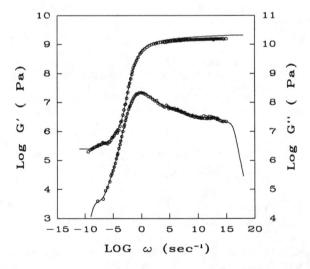

Figure 1. G' and G'' master curves for PVAc. Circles - data [14]; lines - response from shear relaxation spectrum [20].

4

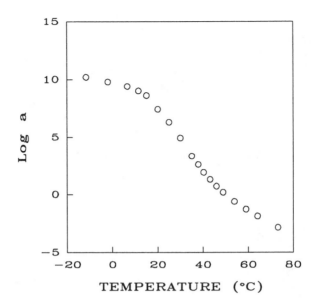

Figure 2.   Time-temperature shift factor for PVAc [14].

The equilibrium pressure $P^\infty(T, V)$ is described by the Tait equation, where the Tait functions $V_o(T)$ and $B(T)$ for PVAc are [22]

$$V_o(T) = 0.82485 + 0.5855 \times 10^{-3} T + 0.282 \times 10^{-6} T^2 \tag{3-3a}$$

$$B(T) = 203.5 \exp(-4.26 \times 10^{-3} T) \tag{3-3b}$$

V_o is in cm^3/g, T is in °C, and B(T) is in MPa. The same PVT data has also been fit to the SS equation-of-state, where the reducing parameters obtained were $P^* = 938.0$ MPa, $V^* = 0.8141$ cm^3/g, and $T^* = 9419$ K for a chain flexibility of s/3c=1 [23]. The SS equation-of-state will be used to calculate (i) the equilibrium entropy in Eqn. 2-2 and (ii) the SS nonequilibrium entropy as described previously in Section I. Both entropy alternatives are equivalent for equilibrium conditions, and the material constant B is fit from time-temperature shift data above T_g and equilibrium V-T data in conjunction with Eqn. 1-5. For PVAc, B = 1.043 $P^* V^*$ [20].

We postulate pseudo time-shift invariance also describes the viscoelastic bulk modulus and again represent the time-dependent relaxation by a spectrum of discrete exponential relaxation times. The functions $\Delta K(\theta, C) = K_o(\theta, C) - K_\infty(\theta, C)$ and $k(t^*)$ must now be determined. The equilibrium bulk modulus K_∞ is easily computed from the Tait equation for the polymer in the rubbery state; specifically, $K_\infty = -V \partial P^\infty / \partial V$.

The instantaneous bulk modulus K_o, which depends upon θ and C, is not the same as the glassy modulus K_g, which depends upon *the formation history* of the glass. K_o was extracted from a single atmospheric volume-temperature curve extending below T_g by employing the concept of an "ideal glass," where an abrupt transition from an equilibrium, rubbery state to the nonequilibrium, glassy state occurs at T_g. Details of the ideal glass analysis are presented elsewhere [20]. K_o along the one atmosphere line is:

$$\Delta K = K_o - K_\infty = 3314.8 - 5.916(T - 30.56) \tag{3-4}$$

where T is in °C and ΔK is in MPa. Although this expression is exact only for temperature-deformation states along the glassy one atmosphere isobar, it is employed in Section IV without any correction for changes in hydrostatic pressure.

The only complete bulk viscoelastic data is the complex bulk compliance data by McKinney and Belcher [2] for PVAc. This material, however, was reported to be slightly swelled, having an abnormally low T_g of 17°C [2] instead of a more typical value near 30°C [24]. McKinney and Belcher normalized their compliance data by the zero and infinite frequency storage compliance plateau values which depended upon temperature and pressure. In order to determine $k(t^*)$ the following procedure was used: (i) the normalized dynamic bulk compliance data [2] was transformed [24] to the dynamic modulus at a reference temperature of 17°C; (ii) the discrete relaxation spectra for $k(t^*)$ was computed from the K'' data [25]; and (iii) the reference temperature for the $k(t^*)$ was changed to 30.5°C without shifting the spectra. The resulting K' and K'' data and predictions of the spectra are shown in Fig. 3. This procedure for determining $k(t^*)$ assumes that the normalized spectra for the unswollen PVAc was identical to the normalized spectra for the swollen PVAc, when the individual T_g's are the reference temperatures. Although this shifting process may introduce artifacts into the $k(t^*)$ response, we believe it is the best way to analyze the only available data suitable for constructing a bulk modulus/compliance master curve.

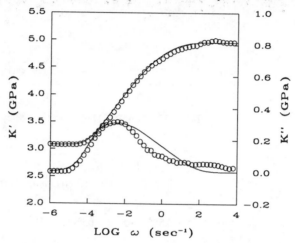

Figure 3. K' and K'' master curves for PVAc. Circles - converted from shifted complex bulk compliance data [2]; lines - response from bulk relaxation spectrum [20].

All of the material properties necessary for the model (P^∞, η^∞ or η_{SS}, G_∞, G_Δ, K_Δ, and B) are now determined from independent measurements on PVAc. The required experimental data included the equilibrium pressure-volume-temperature data, a single glassy volume-temperature isobar, the linear viscoelastic shear and bulk moduli, and the log a_T shift factor. No nonlinear measurements are required to specify the material parameters of the model.

IV. PREDICTIONS

The formation of the nonequilibrium glassy state can effect the subsequent deformation behavior of the glass; thus, the formation of the glass must be accurately described. Since an isobaric cooling process is simply a nonisothermal deformation in which the stress and strain fields are completely specified by the pressure and specific volume, the glass formation is a special case of the thermoviscoelastic constitutive equation. The predictions of our model with the PVAc material parameters given in Section III and both alternatives for the entropy are shown in Fig. 4 for isobaric cooling at 1 bar and a constant cooling rate of 5°C per hour. Also shown are experimental data measured for approximately the same conditions [26] and the equilibrium PVT behavior. The log a shift factor has been computed with both (i) the viscoelastic entropy computed via Eqn. 2-2 and (ii) η_{ss} determined from the nonequilibrium extension of the Simha-Somcynsky equation-of-state. The predicted and measured glass transition temperatures of approximately 30.5°C are in good agreement, and this is very satisfying since T_g was not used to set any of the material parameters. The predicted nonequilibrium specific volumes below T_g are also in good agreement with the data as was expected, since this experimental isobar was used in the ideal glass analysis to obtain ΔK. The two alternatives for the entropy do not produce significant differences in the glassy specific volumes. The model can also predict the effects of pressure and the cooling rate on T_g and the densification that occurs during isobaric, isothermal annealing near T_g [20,25].

6

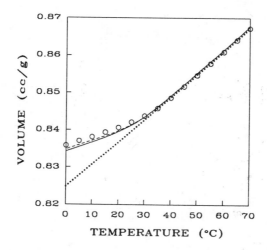

Figure 4. Specific volume versus temperature for cooling at 5°C/hr. Circles - data [26]; solid line
- predictions with nonequilibrium SS entropy; dashed line - predictions with equilibrium viscoelastic entropy; dotted line - equilibrium Tait equation at one atmosphere.

Predictions of the thermodynamic constitutive equation and experimental measurements of
isothermal, uniaxial stress-strain curves for PVAc at temperatures just below the glass transition
temperature are shown in Fig. 5. In the absence of experimental information, the initial state just
prior to deformation is assumed to be the equilibrium state at 1 bar and the temperature of interest.
The predictions for both entropy alternatives are in reasonable agreement with the experimental
curves, although insufficient strain softening is predicted for the 24.3°C data.

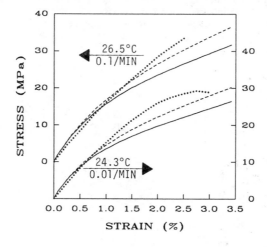

Figure 5. Isothermal, uniaxial stress-strain curves for various temperatures and strain rates. Dotted lines - data [8]; solid lines - predictions with nonequilibrium SS entropy; dashed
lines - predictions with equilibrium viscoelastic entropy.

The effect of strain rate on the predicted stress-strain behavior at 0°C and 20°C is shown in
Fig. 6, where the nonequilibrium SS entropy has been employed. Since T_g is just above 30°C, the
material is still within the glass transition region at 20°C, while at 0°C the material is completely
"frozen" into the glassy state. The predictions at both temperatures exhibit strongly nonlinear

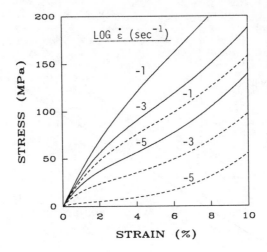

Figure 6. Predicted isothermal, uniaxial stress-strain curves at a variety of strain rates. Solid lines - 0°C; dashed lines - 20°C. Both glasses were formed by cooling from the melt at 1°C/min.

stress-strain behavior at all strain rates, and the magnitudes of the initial Young's modulus and overall stress levels decrease with decreasing strain rate, as is observed experimentally [27]. The predicted stress-strain curves do not, however, exhibit a maximum in stress and post-yield strain softening.

The effect of aging time on the predicted stress-strain behavior is shown in Fig. 7. As the material is physically aged prior to the mechanical deformation, densification occurs which increases both the initial modulus and the yield stress. The difference in the shift factor of the unaged and equilibrium materials just prior to deformation is nearly three orders of magnitude, and this change in shift factor with aging time is in reasonable agreement with experimental data [28].

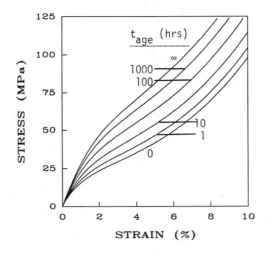

Figure 7. The effect of physical aging on the isothermal stress-strain curves at 20°C. Constant uniaxial strain rate of 10^{-3}/sec. Glass formed by cooling from the melt at 1°C/min.

The bulk modulus data of McKinney and Belcher [2] was for a slightly swollen PVAc material. Since the exact location of the bulk modulus spectra on the log τ axis is subject to question, the effect of shifting the bulk relaxation spectrum has been studied. Specifically, in Fig. 8 predictions are shown (i) when the normalized bulk modulus spectrum is located as described in Section III and (ii) when the spectrum has been shifted to shorter relaxation times by one, two, and four decades. The location of the shear modulus relaxation spectrum was not changed. A shift of -2.8 decades represents the location of the spectrum if the reported bulk compliance data was for an unswelled PVAc and no shifting had been performed to compensate for the change in T_g due to the solvent. A shift of -3.6 decades results in the bulk modulus relaxation spectrum being centered at the same relaxation time as the shear modulus spectrum. Examining the predictions in Fig. 8, a maximum in the stress-strain curve and slight post-yield softening is predicted when the bulk modulus spectrum is centered at a shorter relaxation time than the shear spectrum. Up to the yield point, all of the curves are similar with increasing initial Young's modulus as the bulk spectrum is shifted to shorter times. The predictions in Fig. 8 emphasize the need for reliable bulk modulus data. Specifically the linear viscoelastic shear and bulk relaxation behavior and the stress-strain curves need to be accurately determined on the same material.

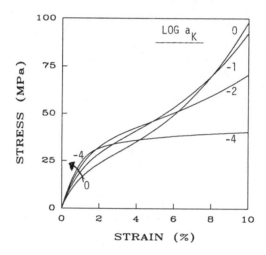

Figure 8.   The effect of the location of the bulk relaxation spectrum on the log τ axis. Unshifted curve is identical to unaged curve in Fig. 7.

V. DISCUSSION

A thermoviscoelastic constitutive equation has been developed to describe the time-dependent viscoelastic behavior of polymer solids including (i) volume relaxation in the glass-to-rubber transition, (ii) the nonlinear stress-strain behavior including yield, and (iii) the effect of complicated thermal histories on the nonlinear mechanical properties. All the thermodynamic properties (i.e., stress, pressure, entropy, etc.) can be computed from the nonequilibrium Helmholtz free energy using the rational thermodynamics framework; thus, we are assured that the proposed constitutive equation is completely self-consistent for all conceivable temperature and deformation histories.

A key feature of the proposed constitutive equation is that it only requires three material properties - i.e. G(t), K(t), and the time-independent equilibrium PVT response. Since these three material properties can be evaluated from independent experiments that are qualitatively different from nonlinear stress-strain and glass transition phenomenon, the predictive capabilities of the theory can be assessed unambiguously. This is in contrast to most constitutive equations, where material parameters are determined from the same nonlinear mechanical properties that the theory is supposed to predict.

The proposed constitutive equation does a good job in predicting the specific volume response upon cooling through the glass transition temperature as well as physical aging below T_g. The model does a reasonable job in describing the nonlinear stress-strain data; however, the model does not predict a definite maximum in the stress-strain curve which would be associated with yield. This problem could be associated with the location of the bulk relaxation spectra on the log time axis. As illustrated in Fig. 8, the shifting of the bulk relaxation spectra to shorter times induces a definite yield, and this shift in the bulk spectra would not be unreasonable considering the manipulations required to correct for the unusual T_g that was reported for this partially swollen PVAc.

The predictions of Knauss and Emri [8,9] are in better agreement with the experimental stress-strain data, but they employed an empirically determined bulk modulus that was only 50% of the experimental bulk modulus data [2]. This difficulty in the Knauss and Emri constitutive model may be a consequence of using the simple free volume form for the log a shift factor. Struik has discussed the problem associated with using simple free volume models [4].

Although the proposed model has some very attractive features, it still needs improvement in order to quantitatively predict the nonlinear stress-strain behavior for complicated thermal/deformation histories. There is a very logical way to improve the constitutive model. Specifically, it was assumed that the nonequilibrium Helmholtz free energy depended on the complete deformation history and the *current value* of the temperature when written on a material timescale. All dependence on the thermal *history* was assumed to be implicitly included in the material time. If instead an explicit dependence on the temperature history is also included, which is the more general assumption, then the new thermodynamic constitutive equation for the stress will contain an additional single-integral term containing the rate of change of the temperature [21]. In addition, the new constitutive equation for the nonequilibrium entropy will contain two single-integral terms, one convolution integral over the volume history and another over the temperature history. An analogy which clearly shows the need for this more general derivation is formed by considering an isotropic deformation where the pressure is a function of the temperature and volume. The total differential of the pressure would be given by

$$dP = \left[\frac{\partial P}{\partial V}\right]_T dV + \left[\frac{\partial P}{\partial T}\right]_V dT \qquad (4\text{-}1)$$

It would be overly restrictive, although allowable, to assume that the pressure is only a function of V and independent of T. The model described in Section II assumed the contribution analagous to the second term on the right was identically zero and the bulk modulus term (the first term on the right) was required to compensate for its absence. Preliminary calculations indicate the new terms will correct the problems described above. While this more general constitutive equation does contain four viscoelastic material functions, they are extractable from the same set of experimental measurements. Thus, the new constitutive equation will still utilize only material properties that can be determined from independent measurements on small quantities of material.

In conclusion, we have shown that a constitutive model rigorously derived to include an entropy-based material timescale and obeying the fundamental postulates of rational mechanical is capable of describing some aspects of nonlinear thermoviscoelasticity, including the effects of processing history on the mechanical response of the glassy solid. A more general model that has been outlined should potentially resolve the shortcomings identified in the current constitutive model.

ACKNOWLEDGEMENTS

This work was supported by the National Science Foundation under grant CD8803017 to the Engineering Research Center for Intelligent Manufacturing Systems.

REFERENCES

1. Moonan, W.K., and Tschoegl, N.W., *J. Polym. Sci.: Polym. Phys., 23,* 623 (1985).

2. McKinney, J.E., and Belcher, H.V., *J. Res. Nat. Bur. Standards, 67A,* 43 (1963).

3. Sauer, J.A., *Polym. Eng. Sci., 17,* 150 (1977).

4. Struik, L.C.E., "Physical Aging in Amorphous Polymers and Other Materials," Elsevier, New York (1978).

5. Matsuoka, S., and Kwei, T.K., in "Macromolecules: An Introduction to Polymer Science," F.A. Bovey and F.H. Winslow, eds., Academic Press, New York (1979).

6. Hopkins, I.M., *J. Polym. Sci., 28,* 631 (1958).

7. Morland, L.W., and Lee, E.H., *Trans. Soc. Rheol., 4,* 233 (1960).

8. Knauss, W.G., and Emri, I.J., *Computers & Structures, 13,* 123 (1981).

9. Knauss, W.G., and Emri, I., *Polym. Eng. Sci., 27,* 86 (1987).

10. Shay, R.M., Jr., and Caruthers, J.M., *J. Rheol., 30,* 781 (1986).

11. Shay, R.M., Jr., and Caruthers, J.M., *Proc. 20th Midwestern Mech. Conf., 14b,* 493, Purdue University (1987).

12. Curro, J.G., Lagasse, R.R., and Simha, R., *J. Appl. Phys., 52,* 5892 (1981).

13. Adam, G., and Gibbs, J.H., *J. Chem. Phys., 43,* 139 (1965).

14. Sedath, R.H., Ph.D. Dissertation, Purdue University, West Lafayette, IN (1987).

15. Shay, R.M., Jr., M.S.E. Thesis, Purdue University, West Lafayette, IN (1986).

16. Coleman, B.D., and Noll, W., *Rev. Mod. Phys., 33,* 239 (1961).

17. Coleman, B.D., *Arch. Rat. Mech. Anal., 17,* 1 & 230 (1964).

18. Truesdell, C., and Noll, W., "The Non-Linear Field Theories of Mechanics," in Encyclopedia of Physics, Vol. III/3, S. Flugge, ed., Springer-Verlag, New York (1965).

19. Lustig, S.R., and Caruthers, J.M., *Proc. Xth Intl. Congr. Rheol.,* Sydney (1988).

20. Shay, R.M., Jr., Ph.D. Dissertation, Purdue University, West Lafayette, IN (1989).

21. Lustig, S.R., Shay, R.M., Jr., and Caruthers, J.M., submitted.

22. McKinney, J.E., and Simha, R., *Macromolecules, 7,* 894 (1974).

23. McKinney, R.E., and Simha, R., *Macromolecules, 9,* 430 (1976).

24. Ferry, R.D., "Viscoelastic Properties of Polymers," 3rd Edition, Wiley, New York (1980).

25. Shay, R.M., Jr., and Caruthers, J.M., *Proc. North Amer. Thermal Anal. Soc.,* Orlando (1988).

26. McKinney, J.E., and Goldstein, M., *J. Res. Nat. Bur. Standards, 78A,* 331 (1974).

27. Bowden, P.B., in The Physics of Glassy Polymers, R.N. Haward, ed., Wiley, New York (1973).

28. Lagasse, R.R., and Curro, J.G., *Macromolecules, 15,* 1559 (1982).

YIELDING, ANISOTROPY AND DEFORMATION PROCESSING OF POLYMERS

S. D. Batterman and J. L. Bassani
Department of Mechanical Engineering and Applied Mechanics
University of Pennsylvania
Philadelphia, Pennsylvania

ABSTRACT

Constitutive equations that characterize the deformation induced anisotropy in polymers at large strains are developed for both flow and deformation (nonlinear elastic) theories of plasticity. These equations are used first to predict the evolution of yield surfaces for anisotropies that develop under simple plane strain deformations. Then, non-uniform deformations are considered including neck propagation and extrusion. Finite element solutions are based on flow theory version of the constitutive models. Neck propagation loads and draw ratios are found to be sensitive to the degree of anisotropy and pressure sensitivity, while extrusion forces are less so. The results are compared to solutions for an isotropic material with the same effective stress - strain behavior.

INTRODUCTION

Polymers exhibit a complex behavior at large strains that arises from the stretch and orientation of molecular chains and, in semi-crystalline polymers, from orientation of crystallites as well. Generally, the oriented polymer is much stiffer than in the unoriented state which leads to the phenomenon of neck propagation. As the microstructural orientation evolves so does the anisotropic variation in continuum properties such as stiffness and yield point. Recently, neck propagation has been studied by several investigators, (Hutchinson and Neale 1983, Fager and Bassani 1986, and Tugcu and Neale 1987), where isotropic constitutive descriptions were incorporated. In this paper constitutive equations, strain-induced anisotropy, and pressure sensitivity are developed and then incorporated into analyses of neck propagation and extrusion. A brief discussion of the kinematics of deformation at large strain follows.

Under a general three-dimensional deformation the three principal stretches λ_i, i=1,3 are determined from an eigenvalue problem, where λ_i^2, i = 1,3, are the eigenvalues of $\underset{\sim}{F}^T \cdot \underset{\sim}{F}$, where $\underset{\sim}{F}$ is the deformation gradient and superscript T denotes transpose (Malvern, 1969). The eigenvectors give the principal (orthogonal) directions of stretch. In the case of incompressible deformation $\lambda_1 \lambda_2 \lambda_3 = 1$ (i.e. the volume dilatation is zero). In what follows we will adopt the convention $\lambda_1 > \lambda_2 > \lambda_3$. Therefore, for plane strain tension of an incompressible material, which will be considered in detail below, $\lambda_2 = 1$ and $\lambda_3 = 1/\lambda_1$.

For the anisotropic constitutive model developed below we assume that the principal directions of stretch correlate with the molecular orientation or "texture" of the polymer, and the principal axes of stretch are taken as the principal (orthogonal) axes of anisotropy. Therefore, under arbitrary histories of deformation each material point of the deforming polymer will possess orthotropic symmetries about the principal axes of stretch. For example, the elastic strain energy function and the criterion for yielding should be orthotropic. Measures of stiffness or yield stress are taken as functions of the stretch λ in the constitutive equations developed in this paper. Hill's quadratic orthotropic yield function (Hill, 1979) is taken as the stress potential for both the (nonlinear) elastic and elastic-plastic constitutive descriptions. Experimental observations that have motivated this study are reviewed in (Bassani and Batterman, 1987).

ANISOTROPIC STRESS-STRAIN RELATIONS

In this section, and several that follow simple deformations are considered where the principal axes of stress and anisotropy coincide. Shear terms that arise in general deformations are included in the finite element analysis. Let σ_e and ε_e be effective stress and strain measures for multiaxial deformations where $\sigma_e = \sigma$ and $\varepsilon_e = \varepsilon$ in uniaxial tension. These measures are defined using the Hill orthotropic yield criterion, Eq. (1), and the notion of work equivalent deformations (Bassani and Batterman, 1987). Consider cartesian axes $\underset{\sim}{e}_i$ that are aligned with the principal stretch directions and $\lambda_1 > \lambda_2 > \lambda_3$.

The Hill yield criterion (Hill, 1979) written to define the effective stress is:

$$\sigma_e^2 = f(\sigma_2 - \sigma_3)^2 + g(\sigma_3 - \sigma_1)^2 + h(\sigma_1 - \sigma_2)^2 \tag{1}$$

where

$$2f = (\sigma_{y1}/\sigma_{y2})^2 + (\sigma_{y1}/\sigma_{y3})^2 - 1$$

$$2g = (\sigma_{y1}/\sigma_{y3})^2 - (\sigma_{y1}/\sigma_{y2})^2 + 1 \tag{2}$$

$$2h = (\sigma_{y1}/\sigma_{y2})^2 - (\sigma_{y1}/\sigma_{y3})^2 + 1$$

and, σ_{yi} denotes the uniaxial yield stress in the direction ε_i. Let $\underset{\sim}{\varepsilon}$ denote the plastic strain-rate (or increment) for flow theories of plasticity or let $\underset{\sim}{\varepsilon}$ denote the total (logarithmic) strain for deformation theories (nonlinear elasticity). Hill's notion of work equivalent deformations under multiaxial states (Hill, 1979) requires that $\sigma_{ij}\varepsilon_{ij} = \sigma_e\varepsilon_e$.

Since σ_e in Eq. (1) is homogeneous of degree one in stress, from Euler's theorem $\sigma_{ij}(\partial\sigma_e/\partial\sigma_{ij}) = \sigma_e$. Therefore, with σ_e taken as the potential for $\underset{\sim}{\varepsilon}(\underset{\sim}{\sigma})$ it follows that

$$\varepsilon_{ij} = \varepsilon_e \partial\sigma_e/\partial\sigma_{ij} \qquad (3)$$

From Eq. (1) with Eq. (3)

$$\varepsilon_1 = \varepsilon_e/\sigma_e \ [-g(\sigma_3-\sigma_1) + h(\sigma_1-\sigma_2)]$$

$$\varepsilon_2 = \varepsilon_e/\sigma_e \ [\ f(\sigma_2-\sigma_3) - h(\sigma_1-\sigma_2)] \qquad (4)$$

$$\varepsilon_3 = \varepsilon_e/\sigma_e \ [-f(\sigma_2-\sigma_3) + g(\sigma_3-\sigma_1)]$$

$$\varepsilon_e^2 = (f\varepsilon_1^2 + g\varepsilon_2^2 + h\varepsilon_3^2) \ / \ (fg + gh + hf) \qquad (5)$$

Finally, we consider plane strain tension ($\varepsilon_2 = 0$) and neglect elastic strains so that from Eq. (4), which is considered below in detail, from Eq. (4)

$$\sigma_2 = \sigma_1 h/(f+h) \qquad (6)$$

Therefore, with σ_1 and σ_2 required to have the same sign, h must be non-negative. This restriction is imposed on the analytical description for the evolution of the yield stresses that is presented in the next section. From Eq. (1) with (5) and (6) it follows that

$$\sigma_e = A\sigma_1 \ , \ \varepsilon_e = \varepsilon_1/A \qquad (7)$$

$$A^2 = 1 - h^2/(f+h) \qquad (8)$$

Evolution Of Yield Stresses

To complete the constitutive description the evolution of the yield stresses is specified. These will be taken as a function of the stretches, i.e., $\sigma_{yi}(\underset{\sim}{\lambda})$ and $\tau_{yi}(\underset{\sim}{\lambda})$. For example, for tension in the $\underset{\sim}{e}_1$ direction $\sigma_{y1} > \sigma_{y2}$ and $\sigma_{y1} > \sigma_{y3}$ (Bassani and Batterman, 1987).

In uniaxial tension the stress – strain relation is taken to be (Hutchinson and Neale, 1983, Fager and Bassani, 1986, Neale and Tugcu, 1987)

$$\sigma/\sigma_0 = \begin{cases} \varepsilon/\varepsilon_y & ; \ 0 \leqslant \varepsilon \leqslant \varepsilon_y \\ \alpha\varepsilon^N & ; \ \varepsilon_y < \varepsilon \leqslant \varepsilon_L \\ \beta e^{M\varepsilon^2} & ; \ \varepsilon_L > \varepsilon \end{cases} \tag{9}$$

where $\varepsilon = \ell n(\lambda)$. With σ_0, ε_y, ε_L and N given, the remaining constants α, β, and M are determined by requiring continuity of the stress - strain behavior everywhere as well as continuity of slope at ε_L. The constants in Eq. (9) used in the present analysis are $\varepsilon_y = 0.05$, $\varepsilon_L = 0.4$, $N = 0.2$, $\alpha = 1.82$, $\beta = 1.37$, and $M = 0.625$.

A simple phenomenological form that captures experimental observations (Bassani and Batterman, 1987), again with $\underset{\sim}{e}_1$ in the direction of maximum principal stretch $\lambda_1 > 1$, is

$$\frac{\sigma_{y1}}{\sigma_{yi}} = \frac{1 + c_1(\lambda_1 - 1)^{\alpha_1}}{1 + c_2(\lambda_i - 1)^{\alpha_2}} \qquad ; \ i = 2,3 \tag{10}$$

where $(\lambda_i - 1)^\alpha \equiv \text{sign}(\lambda_i - 1)|\lambda_i - 1|^\alpha$. The constants c_1, c_2, α_1, and α_2 are arbitrary except for physical requirements such as h>0 in plane strain tension as noted above. Other reasonable forms for the ratio of the yield stresses are easily written down, but the form given in Eq. (10) seems to be rather flexible. Note that for $\lambda_1 > 1$ and $\lambda_i < 1$ (i=2,3) that $c_1 > 0$ and $c_2 > 0$ implies less stiffening in the 2 and 3 directions relative to that in the 1 direction.

Plane Strain Tension

Relative to the isotropic case where $\sigma_{y1} = \sigma_{y2} = \sigma_{y3}$, in plane strain tension ($\varepsilon_{22} = 0$) the anisotropic behavior described by Eqs. (1) - (10) leads to a softer $\sigma_1 - \varepsilon_1$ behavior. Figures 1a-d are plots of σ_1 and σ_2 vs. ε_1 in plane strain tension for various combinations of c_1, c_2, α_1, and α_2. One dashed curve in each figure corresponds to uniaxial tension which is the fundamental σ_e - ε_e curve that is taken to be the same in all cases. Plane strain behavior is given in Fig. 1 for the case where $c_1 = c_2 = 0$. As seen in these figures, in certain cases σ_2 decreases with increasing deformation at large enough ε_1. With positive c_1 and c_2, as c_1 increases so does σ_{y1}/σ_{y3} increase for a given ε_1, and as either or both c_1 and c_2 increase so does σ_{y1}/σ_{y2} increase for a given ε_1. When these ratios become large enough, which will in general occur at sufficiently large ε_1, σ_2 can become negative, and this is considered to be physically unrealistic. Therefore, for the strain levels considered in Figs. 1, i.e. $\varepsilon_1 \leqslant 1$, the magnitude of the anisotropic parameters considered in Eq. (10) have been limited by the requirement that $\sigma_2 > 0$.

Examples of the evolution of constants and the σ_e potential or yield surfaces are shown in Fig. 2 for selected values of the constants. These surfaces are obtained using Eq. (6) and neglecting elasticity, so that $\varepsilon_2 = 0$. The initial surfaces are Mises ellipses. The trajectory of the plane strain point $(\sigma_1, [h/(f+h)]\sigma_1)$ is also shown. In the isotropic case, the initial Mises surface expands in a self similar fashion, and the trajectory of the plane strain point is just a straight line, with the ratio $h/(f+h)$ (slope) remaining constant at a value of $1/2$.

The case of plane strain deformation where $c_1 = 1.0$, $c_2 = 0.0$, $\alpha_1 = \alpha_2 = 1.0$ gives qualitatively good agreement with experimental observations (Bassani and Batterman, 1987). Furthermore, experimental evidence presented by Brown et. al. (1968) indicates that the yield stresses in shear (with respect to the principal stretch directions) does not vary much, even at moderately high strains.

Plane Strain Neck Propagation

For steady neck propagation in an elastic material the propagation load P^* or nominal stress $n^* = P^*/A_o$, where A_o is the initial or undeformed cross-sectional area, can be easily calculated from the stress-strain relations and the conditions that exist is the fully necked and unnecked regions (Hutchinson and Neale, 1983, Fager and Bassani, 1986, and Batterman and Bassani, 1987). For completeness, the plane strain results of Bassani and Batterman (1987) are summarized here. Note that in plane strain the nominal stress $n_1 \equiv n = \sigma_1/\lambda_1 = \sigma_e/(A\lambda_1)$.

Figures 3a and 3b are plots of the plane strain neck propagation load (n^*) and draw ratio $(DR = \lambda_N/\lambda_U$, subscripts N and U denote the necked and unnecked regions respectively), both normalized by the respective isotropic values for various values of c_1, c_2, α_1, α_2. In general, the anisotropic variation of stiffnesses leads to a higher draw ratio and lower propagation load relative to the isotropic case (Bassani and Batterman, 1987).

PRESSURE SENSITIVITY

The yield criterion given in Eq. (1) when expanded to account for a Bauschinger effect (including the shearing terms) and the resulting pressure sensitivity is written as:

$$\sigma_e^2 = f(\sigma_2-\sigma_3)^2 + g(\sigma_3-\sigma_1) + h(\sigma_1-\sigma_2)^2$$

$$+ 2l\sigma_{23}^2 + 2m\sigma_{31}^2 + 2n\sigma_{12}^2$$

$$+ k_1\sigma_1 + k_2\sigma_2 + k_3\sigma_3 \tag{11}$$

The constants are determined in exactly the same manner as those in Eq. (2) and are expressed as:

$$2f = \sigma_{1t}^2 /(\sigma_{2t}\sigma_{2c}) + \sigma_{1t}^2/\sigma_{3t}\sigma_{3c}) - \sigma_{1t}/\sigma_{1c}$$

$$2g = \sigma_{1t}^2/(\sigma_{3t}\sigma_{3c}) - \sigma_{1t}^2/(\sigma_{2t}\sigma_{2c}) + \sigma_{1t}/\sigma_{1c}$$

$$2h = \sigma_{1t}/\sigma_{1c} - \sigma_{1t}^2/(\sigma_{3t}\sigma_{3c}) + \sigma_{1t}^2/(\sigma_{2t}\sigma_{2c})$$

$$(12)$$

$$2l = \sigma_{1t}^2/\tau_{y1}^2 \quad , \quad 2m = \sigma_{1t}^2/\tau_{y2}^2 \quad , \quad 2n = \sigma_{1t}^2/\tau_{y3}^2$$

$$k_1 = \sigma_{1t}(1 - \sigma_{1t}/\sigma_{1c}) \; , \quad k_2 = (\sigma_{1t}^2/\sigma_{2t}(1 - \sigma_{2t}/\sigma_{2c}),$$

$$k_3 = (\sigma_{1t}^2/\sigma_{3t})(1 - \sigma_{3t}/\sigma_{3c})$$

Where the subscripts t and c refer to tension and compression, respectively, and τ_{yi} are the yield stresses in shear, e.g. $\tau_{y1} = (\sigma_{23})_y$. As before, at yield in uniaxial tension, $\sigma_e = \sigma_{1t}$. The strains or strain rates are calculated, as before, from Eq. (3) with σ_e defined in Eq. (11). For brevity this will not be given here.

Evolution Of The Yield Stress Ratios

Now, with Eqs. (11) and (12), the ratio of the tensile yield stress in the 1 - direction (σ_{1t}) to σ_{1c} and both the tensile and the compressive yield stresses in the other two directions must be specified.

These relations are taken to be:

$$\sigma_{1c} = C_x\sigma_{1t} + \sigma_o$$

$$(13)$$

$$\sigma_{ic} = C_t\sigma_{it} \; ; \; i = 2,3$$

The k_i's in Eq. (11) can now be written as:

$$k = \sigma_{1t}[1 - \sigma_{1t}/(C_x\sigma_{1t} + \sigma_o)]$$

$$(14)$$

$$k_2 = (\sigma_{1t}^2/\sigma_{2t})(1 - 1/C_t) \; , \quad k_3 = (\sigma_{1t}^2/\sigma_{3t})(1 - 1/C_t)$$

The incompressibility condition $(k_1 + k_2 + k_3 = 0)$ leads to the following relation between C_x and C_t:

$$C_t = [(1 - \frac{\sigma_{1t}}{(C_x\sigma_{1t} + \sigma_o)})(\frac{\sigma_{1t}}{\sigma_{2t}} + \frac{\sigma_{1t}}{\sigma_{3t}})^{-1} + 1]^{-1}$$

$$(15)$$

As before, Eq. (10) is taken to define the ratios σ_{1t}/σ_{it}. Then combining Eqs. (10) and (13) leads to

$$\frac{\sigma_{1t}}{\sigma_{ic}} = \frac{1 + C_1(\lambda_1 - 1)^{\alpha_1}}{C_t(1 + C_2(\lambda_i - 1)^{\alpha_2})} \quad ; \; i=2,3$$

$$(16)$$

18

A comparison of this description of pressure sensitivity with experimental results is shown in Fig. 4 for $C_x=0$ and C_t given by Eq. (15).

FINITE ELEMENT RESULTS

The simple elastic analysis presented in the previous sections provide a means of evaluating the material behavior and identifying combinations of material constants which are reasonable for a complete elastic – plastic finite element analysis. A flow theory version of the proposed model has been implemented in the ABAQUS finite element code through the use of the UMAT (User MATerial) option which allows the user to incorporate his or her own constitutive behavior. Details of the finite element calculations are given by Batterman (1989) and a forthcoming paper.

Plane Strain Necking

The plane strain calculations were made using a rectangular mesh of isoparametric eight noded reduced integration elements. A reduction in width of 0.5% was introduced over the bottom quarter of the mesh in order to initiate the localization. Because large plastic strains dominate, for simplicity the deformation is overwhelmingly plastic, the elasticity is taken to be isotropic. The combination $c_1=1.0$, $c_2=0.0$, $c_x=0.0$, $\alpha_1=\alpha_2=1.0$ was chosen.

The finite element solutions also predicts that plastic anisotropy leads to an increased draw ratio (DR) and a decreased propagation load, as compared to the isotropic case. Figure 5 shows the fully necked (deformed) meshes for both material models at the same extension ratio (L/L_0) where only 1/4 of the necked sample is depicted. The overall load – deflection behavior and draw ratios for the three material models are also given. Neck localization is more abrupt in the anisotropic cases as is seen in the load-deflection and draw ratio (DR)-deflection curves.

Plane Strain Extrusion

In order to investigate the behavior of the constitutive models under predominantly compressive stressing, finite element studies of the plane strain extrusion problem were undertaken, again using ABAQUS with the same elements as in the case of necking with the IRS22 elements used to model the interface between the billet and the die. Frictionless contact between the die and the billet was considered in the analysis.

The slope of the die was taken to be shallow ($\alpha \approx 11.3^\circ$). The material was extruded through the die by applying constant displacement increments to the end of the specimen. Half of the deformed meshes during the transient and at steady state are shown in Fig. 6. Due to the displacement constraints imposed by the die, the deformed meshes resulting from each material description are virtually indistinguishable. In fact, there is also little difference in the load displacement curves (Fig. 6), which is in sharp contrast to the necking problem. The reason for this is twofold. First, due to the constraint imposed by the die, the deformation of the billet material is highly constrained. As a result, the anisotropy and pressure sensitivity are not as free to evolve as they are in the necking problem. Second, the reduction ratio (stretch) in this extrusion analysis is smaller than the natural draw ratios (stretches) in the necking problem. Therefore, since the anisotropy and pressure sensitivity evolve with stretching, the necking calculation displays more pronounced differences due to constitutive assumptions than in the extrusion calculation.

The evolution of the anisotropy in terms of the orientation of the maximum principal stretch directions is displayed in Fig. 7. Note that the maximum principal stretch direction "rolls over" as the material enters the die, and then aligns itself along the streamlines of the flow. Nevertheless, for the small reductions considered here the stretches in this "roll over region" are very close to one, so that the effect of the "rolling" anisotropy and pressure sensitivity are small. For larger reductions "rolling" may be more significant. In the necking problem, the orientation of the maximum principal stretch direction closely follows the streamlines, so that rotational effects are not significant.

SUMMARY

A constitutive framework has been developed that accounts for the deformation induced anisotropy and pressure sensitivity (Bauschinger effect) in polymers at large strains, and the effects of these on plane strain deformations has been demonstrated. For plane strain neck propagation, as the ratio of the yield stresses σ_{y1}/σ_{y2} and σ_{y1}/σ_{y3} become larger, which depends on the anisotropic parameters as well as the total stretch, the propagation load tends to decrease and the draw ratio tends to increase. The influence of the pronouced polymer Bauschinger effect or pressure sensitivity tends to moderate the effects of anisotropy relative to isotropy. The finite strain finite element analyses of neck propagation presented agree well with the simple elastic analysis.

Under the highly constrained deformations present in the extrusion problem the effects of anisotropy and pressure sensitivity are less pronounced. A more thorough investigation of extrusion that includes larger reductions needs to be made.

ACKNOWLEDGEMENTS

The assistance of Prof. N. Aravas in the finite element portion of this work and Dr. L. O. Fager with the neck propagation analysis and discussions with Prof. N. Brown are gratefully acknowledged. This supported of the National Science Foundation under grant MEA 83-52172 and from the Aluminum Corporation of America are also gratefully acknowledged.

REFERENCES

ABAQUS finite element code,,Hibbit, Karlsson and Sorensen, inc. of Providence R.I.

Backofen, W. A., 1972, <u>Deformation Processing</u>, Addison-Wesley, Reading, Massachusetts.

Bassani, J. L., and Batterman, S. D., 1987, "Deformation Induced Anisotropy And Neck Propagation In Polymers", AMD-Vol. 85, <u>Constitutive Modeling For Nontraditional Materials</u>, (Stokes and Krajcinovic, editors).

Batterman, S. D., 1989, "Evolution Of Anisotropy In The Mechanical Behavior Of Polymers", Ph.D. Thesis, University of Pennsylvania.

Brown, N., Duckett, R.A., and Ward, I.M., 1968, "The Yield Behavior of Oriented Polyethylene Terephthalate", <u>Philosophical Magazine</u>, Vol. 18, No. 153, pp. 483-502.

Caddell, R. M., and Woodliff, A. R., 1977, "Macroscopic Yielding of Oriented Polymers," Journal of Materials Science, Vol. 12, pp. 2028-2036.

Caddell, R. M., Raghava, R. S., Atkins A. G., 1973 "A Yield Criterion For Anisotropic And Pressure Dependent Solids Such As Oriented Polymers," Journal Of Materials Science, Vol. 8, pp. 1641-1646.

Fager, L. O., and Bassani, J. L., 1986, "Plane Strain Neck Propagation," International Journal of Solids and Structures, Vol. 22, No. 11, pp. 1243-1257.

Hill, R., 1951, The Mathematical Theory of Plasticity, Oxford University Press, New York.

Hill, R., 1979, "Theorectical Plasticity of Textured Aggregates," Math. Proc. Camb. Phil. Soc., Vol. 55, pp. 179-191.

Hutchinson, J. W., and Neale, K. W., 1983, "Neck Propagation," Journal of the Mechanics and Physics of Solids, Vol. 31, p. 405.

Malvern, Lawrence E., 1969, Introduction to the Mechanics of a Continuous Medium, Prentice-Hall, Inc. Englewood Cliffs, New Jersey.

Nagtegaal, J. C., Parks, D. M., Rice, J. R., 1974, "On Numerically Accurate Finite Element Solutions In The Fully Plastic Range," Computer Methods In Applied Mechanics And Engineering, Vol. 4, pp. 153-157.

Tugcu, P., and Neale, K. W., 1987, "Necking and Neck Propagation in Polymeric Materials Under Plane-Strain Tension," Int. J. Solids Structures, Vol. 23, p. 1063.

Yokouchi, M., Mori, J., and Kobayashi, Y., 1981, "Effects of Tensile Strain Rate on the Mechanical Properties of Constrained-Uniaxiallyand Simultaneous-Biaxially Drawn Poly(ethylene terephthalate) Sheets," Journal of Applied Polymer Science, Vol. 26, pp.3435-3446.

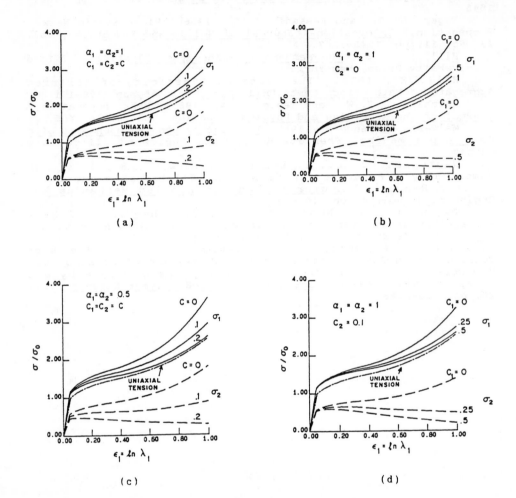

Fig. 1.   Plane strain ($\varepsilon_2=0$) tension stress-strain curves for various combinations of anisotropy parameters in Eq. (10);   solid curves: σ_1, single dash: σ_2, double dash: σ_1 in uniaxial tension ($\sigma_2 = \sigma_3 = 0$).

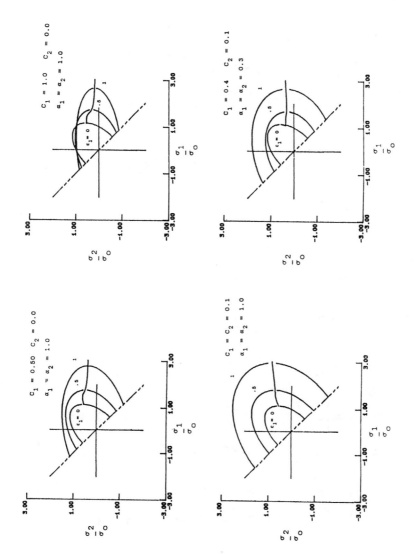

Fig. 2. Evolution of potential (yield) surfaces in plane strain tension at $\varepsilon_1 = \ln(\lambda_1) = 0.0$, 0.5, 1.0 for various combinations of anisotropy parameters in Eq. (10). The trajectory of the plane strain point throughout stress space is indicated.

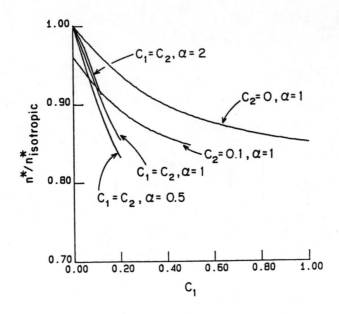

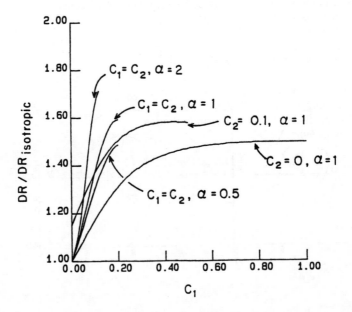

Fig. 3.   Propagation load (n^*) and draw ratio (DR) for plane strain neck propagation in a nonlinear elastic material with various combinations of anisotropy parameters.

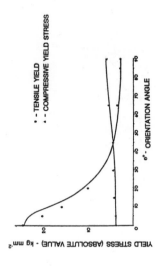

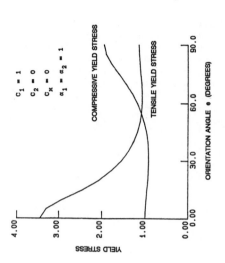

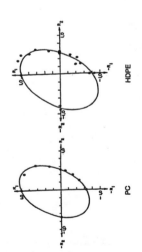

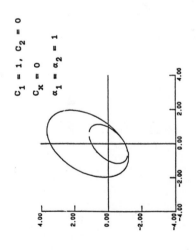

Fig. 4.   Comparison of the experimental results of Caddell et. al. with theoretical predictions.

25

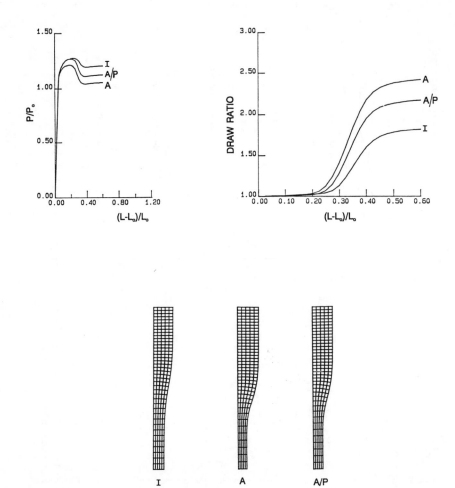

Fig. 5.   Load-deflection curve, draw ratio-extension curve, and
            fully necked mesh at steady state for: (I) isotropic, (A)
            anisotropic, and (A/P) anisotropic/pressure sensitive
            models.

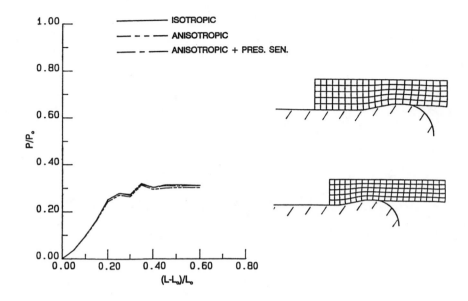

Fig. 6   Load-deflection curve and extruded mesh at two stages in the deformation for the three constitutive models.

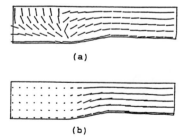

Fig. 7.   (a) Maximum principal stretch directions throughout the billet. (b) Orientation of the maximum principal stretch direction with lines scaled by $8(\lambda_1-1)$ (the factor 8 is arbitrary.)

AN EXPERIMENTAL AND ANALYTICAL INVESTIGATION OF THE LARGE STRAIN COMPRESSIVE AND TENSILE RESPONSE OF GLASSY POLYMERS

M. C. Boyce and E. M. Arruda
Department of Mechanical Engineering
Massachusetts Institute of Technology
Cambridge, Massachusetts

Abstract

In this investigation, the plastic flow of polycarbonate (PC) was examined by obtaining true stress-strain data over a range of strain rates at room temperature through homogeneous, uniaxial, constant strain rate compression testing to strains as high as 125%. Uniaxial compressive loading conditions give rise to a biaxial molecular orientation process which results in the observed strain hardening in compression. Uniaxial tensile tests on PC were also conducted. The necked region of the tensile specimen is being cold drawn resulting in a uniaxial state of orientation. Therefore, the observed macroscopic strain hardening in uniaxial tension distinctly differs from that obtained in uniaxial compression, giving different stress–strain curves. The major differences experimentally obtained between the large strain response in compression and tension indicate a need for an orientation-based model of the strain hardening process. The experimental program also acts to uncouple the effects of strain softening and strain rate providing more accurate data for future modelling of the true strain softening process.

Here, in order to uncover modelling areas which require further investigation and research, the constitutive model of Boyce, Parks, and Argon (1988) has been used to model and analyze the experiments discussed above. In the model, the strain hardening response is directly related to the state of molecular network stretch in the polymer. The material constants for the model are obtained from the compression test data and are found to simulate the observed rate dependent yield and post yield strain softening and hardening very well. The model is then utilized in a finite element analysis of the tensile tests on PC which were also conducted in our lab. Numerical results compared favorably with the experimental data including: load vs. contraction curves, natural draw ratio, and the axial stress–strain response of the cold drawing region. Some discrepancies between the simulations and the experiments were obtained in the large strain response indicating a need for more accurate modelling of the connection between orientation and strain hardening.

1 Introduction

Many polymers may be subjected to very large deformations (exceeding 100 percent strain) even at room temperature. The finite straining process acts to preferentially orient and thereby strengthen the polymer. Although this characteristic response of polymers is well known and indeed often exploited in solid phase polymer processing, it is not yet well understood or well characterized. In order to better understand the change in strength with polymer deformation and orientation, we have undertaken an experimental and modelling investigation into the strain hardening response of polycarbonate, a commercial glassy polymer, when subjected to two different large strain loading conditions.

In this investigation, the plastic flow of polycarbonate to very large strains (125%) over a range of strain rates and under different loading conditions (tension and compression) is examined. True stress-strain data were obtained for PC through homogeneous, uniaxial, constant strain rate compression testing. Under uniaxial compressive loading conditions, the underlying polymeric network chain structure is undergoing a biaxial orientation process which gives rise to the corresponding observed strain hardening behavior. Uniaxial tensile tests on PC were also conducted. During the uniaxial tensile test, the necked region of the polymer is being cold drawn and the network chain structure is undergoing a uniaxial orientation process. This results in a strain hardening response which is distinctly different from that observed in compression – *i.e.* the stress increases at a different rate with respect to imposed strain increment in tension than in compression. Therefore, these two uniaxial loading conditions give rise to distinctly different stress-strain curves. This demonstrates the need for a model of the strain hardening process which is based on the state of orientation in the material as opposed to a piecewise linear or power-law fit to the equivalent stress-equivalent strain response, for example, in uniaxial tension. The compression and tension experiments are discussed in section two.

In section three, the constitutive model of Boyce, Parks, and Argon (1988) is briefly summarized and then used to model and analyze the experiments of section two. In the model, the strain hardening response is directly related to the state of network stretch in the polymer. The material constants for the model are obtained from the compression test data and are found to simulate the observed rate dependent yield and post yield strain softening and hardening very well. The model is then utilized in a finite element analysis of the tensile tests on PC which were also conducted in our lab. The numerical results are compared with the experimental results and found to compare favorably, modelling the axial stress-strain response of the cold-drawing region, the global load-contraction curve, and the natural draw ratio. Although no attempt is currently made here to extend or modify this model, the analyses compared with the experimental results serve to uncover the strengths and weaknesses of the model and suggest avenues which require more experimental and analytical investigation.

2 Experimental Investigation

The true stress-strain curve can be considered to be an identifying feature of the macroscopic response of a material to imposed mechanical loading. In order to accurately model material behavior, a reliable set of experimental data over an adequate range of loading conditions is required. The true stress-strain testing of polymers is a formidable task due to the difficulty of conducting such tests to very large strains while maintaining a homogeneous state of deformation. The lack of a consistent set of true stress-strain data for polymers at large strains limits the development of constitutive models covering this behavior.

The tensile testing of thin sheets of material is still the most popular form of polymer testing [Haward (1987)]. However, during tensile testing, a neck develops creating a very inhomogeneous deformation field and thus making it difficult to accurately monitor the strain and obtain conditions of constant strain rate. Various techniques have been developed and employed to compensate for the occurrence of necking. These principally consist of monitoring the local stress and strain fields in the necked region [Hope, *et al.* (1980), G'Sell and Jonas (1979)].

True stress-strain data to large strains have also been obtained in simple shear [G'Sell, *et al.* (1983)]. This test method has proved to be very successful with the principal difficulty being the eventual cracking at the specimen ends. Buckling is also a possibility in these tests.

In this investigation, we choose to conduct both compression and tension testing to study the large deformation response of polymers. It is relatively easier to achieve a controlled homogeneous deformation in compression than in tension or torsion. Possible modes of inhomogeneous deformation which may occur in compression testing are barrelling, shearing, and buckling of the specimen. Barrelling can generally be avoided by the appropriate selection of a lubricant between the loading platens and the specimen surface. Buckling is avoided by keeping the height to diameter ratio small (about one). The test procedure followed in this project is described below and is similar to that described by Brown, *et al.* (1988) on metals with the exceptions being that the tests are conducted on polymers and to larger strains. The tension tests which are discussed later are conducted in order to obtain data on a different mode of deformation and polymer orientation for comparative purposes.

2.1 Specimen Preparation

Cylindrical compression specimens were machined from 12.7 mm polycarbonate bar stock to dimensions of 6.35 mm in diameter by 6.35 mm in length. Care was taken to insure the cylinder ends were smooth and parallel; in some cases these surfaces were lightly polished on a wheel with 0.3 micron alumina paste to remove machining grooves. These compression specimens were tested in the as-machined or as-polished condition.

Additional compression specimens were designed to determine if any orientation pre-existed in the bar stock material and hence in the test specimens. Two cubic compression specimens measuring 7.92 mm on a side were carefully machined from the bar stock. These specimens were inspected to insure the sides were smooth and parallel and no further preparation was deemed necessary. The orientation of these specimens relative to the longitudinal direction of the bar stock was noted as these specimens were designed to be compressed in a direction perpendicular to this longitudinal axis and therefore perpendicular to the applied load in the tension and compression tests.

A single compression specimen machined identically to the cylindrical specimens described earlier was annealed for four hours at $165^\circ C$ prior to testing at constant strain rate. The compression response of this specimen was also to determine the material characterization of the specimens used in the tension and compression experiments.

Polycarbonate tensile specimens were machined from the same 12.7 mm diameter bar stock used in the compression specimen preparation. The specimens were 152.4 mm in length, the diameter of the specimens in the grip regions was 12.7 mm. A section of the tensile specimens 12.7 mm in length was tapered to a diameter of 6.35 mm using a 1 inch radius tool. With this configuration the initial Bridgeman correction factor was 0.97. Tension tests were performed on these specimens in the as-machined condition and the load direction was parallel to that in the compression tests. The compression and tensile specimen geometries have been sketched in Figure 2.1A. In Figure 2.1B all the specimens described here have been inscribed within the bar stock to show the orientation of the applied load for each test condition.

2.2 Material Characterization

The extent of anisotropy (initial orientation of the molecular chains) in the polycarbonate material used for the compression and tension tests was determined from the true stress versus true strain responses plotted in Figure 2.2. The solid curve represents the typical result of the constant strain rate compression tests at -0.1/second; the specimen used to generate this curve was one of the as-machined cylindrical specimens used in all compression tests reported in the paper for the various rates. The response of this material has been compared to that of the annealed cylindrical specimen, plotted as short dashed lines in Figure 2.2, as well as the response of the transversely loaded cubes. The data show remarkable repeatability for the various test configurations and conditions represented in this figure; the yield points and amounts of softening recorded for all four of these tests confirm that the initial state of each of these specimens was similar and isotropic, i.e., the mechanical response of the material was independent of testing direction. No pre-orientation had existed in the material prior to testing and the bar stock had been supplied in the well-annealed condition.

2.3 Compression Tests

2.3.1 Testing Procedure

Compression testing was carried out using an Instron Model 1350 servohydraulic testing machine. The actuator displacement was controlled by means of a generated voltage signal sent from a personal computer, the computer recorded the feedback voltages corresponding to displacement and load. The controlling voltage signal varied in such a way so as to move the actuator to displace the specimen at a constant strain rate. Strain rates ranging from 1.0/second to 0.0001/second were used for the compression tests. A minimum of three tests were run at each rate except the slowest rate.

It was expected that compression test specimens would be subject to inhomogeneous deformation by either shearing or barrelling, therefore the testing apparatus was designed to minimize the occurrence of such deformations. The compression specimen sat atop a lubricated spherical seat allowing ball-and-socket motion of the apparatus. The stiff base of this seat was connected to the testing machine piston. Although carefully machined to insure all component surfaces were flat and parallel, the ball-and-socket motion was needed to correct any initial misalignment within the testing apparatus and minimize the incidence of shear during the test. The problem of barrelling was confronted with thin sheets of teflon foil placed between the specimen and the compression fixtures. One to two sheets of teflon on each side of the specimen was sufficient to eliminate barrelling at strains of -150%. Some specimens had sheared during the compression tests. Observation of the specimens during testing revealed no visible shear at strains of -90%, therefore if shear was reported for an individual run it occurred late in the test, after -90% strain.

2.3.2 Results of Compression Tests

The load versus time and displacement versus time record of the tests was used to determine the true stress versus true strain behavior of polycarbonate in compression. In converting the raw data, displacements due to the compliance of the testing apparatus were subtracted from the results.

Results of the compression tests are plotted for applied constant strain rates ranging from -1/second to -0.0001/second in Figure 2.3. The overall repeatability of each experiment was confirmed by conducting several tests at each strain rate. As was mentioned earlier, no shear was noted at strains up to -90%; the effect of shear late in the tests on the true stress versus true strain response appeared as a premature approach towards locking. The rate dependence of the plastic flow response of polycarbonate is clearly illustrated in these room temperature tests. The strain rate is found to affect the initial flow stress as expected. The amount of strain softening (drop in flow stress during plastic straining) appears to be relatively independent of the strain rate. In general, the strain hardening response at large strains appears to be independent of the applied strain rate with locking (asymptotically increasing stress) occurring at a compressive strain of approximately -1.25. However, some rate dependence of the response at very large strains is observed for the test at a rate of -0.0001/second; the duration of this test was over four hours. The rate of the test affects the behavior at very large strains, the effect being a delay in the amount of strain endured before the locking stretch is reached.

2.4 Tension Tests

2.4.1 Testing Procedure

Constant displacement rate tension tests were performed on the polycarbonate specimens for comparison with the simulated tensile loading results from finite element analysis. A lateral extensometer was mounted to the tensile specimen to measure the local diameter change at the region of smallest cross-section. Various tension tests were run on an Instron Model 1125 screw machine at crosshead speeds ranging from 12.7 mm/minute to 1.27 mm/minute. The load versus time and diametral contraction versus time data were recorded for each constant displacement rate test.

2.4.2 Results of Tension Tests

Results of the tension tests are shown in Figure 2.4. At yielding the specimen necks down in the immediate locality of the minimum cross-section and, upon yielding and softening, a very rapid decrease in diameter is recorded by the lateral extensometer. Once the neck stabilizes locally, drawing continues as material is pulled from the wider regions adjacent to the extensometer. During this time the load cell records very little increase in stress. Continued drawing becomes increasingly more difficult as the undrawn region is large in diameter and hardening ensues. This phenomenon, locking, is recorded as a sharp upturn in the load versus displacement curve reflecting the increased difficulty in further deforming the material, and eventually ends in fracture of the specimen.

The load versus diametral contraction data for the tension test results at an applied nominal displacement rate of 2.54 mm/minute have been converted to true local stress versus true local strain data and plotted in Figure 2.5A. Superposed upon this data is the stress versus strain response from a compression test at -0.01/second is taken from Figure 2.3, chosen for comparison because the nominal strain rate for the tension data shown is on the order of 0.01/second. Figure 2.5B serves to elucidate this point. In this figure the local strain rate versus local strain response of the tension test in Figure 2.5A has been plotted. The local nominal strain rate during this tension test is on the order of 0.01/second. During yielding in tension the local strain rate as measured by the lateral extensometer increases rapidly by an order of magnitude. The strain rate continues to increase rapidly then decreases rapidly during softening and, as drawing occurs, the strain rate decreases at a slower rate finally returning to the nominal rate during hardening.

The differences between the tension and compression data at moderate strain levels arise as a result of the varying local strain rate during tension testing. The strain rate has begun to increase by the time yielding occurs in tension. Thus, once flow in tension has begun, it is responding to a different strain rate than is the compression specimen. The flow stress in tension plateaus for an extended amount of strain because of the continued increase in strain rate. The softening recorded in tension is also affected by the observed strain rate transient; the true amount of softening is not picked up in the tension test because the expected drop in stress is competing with an increasing strain rate. This clearly demonstrates the need to conduct constant strain rate tests (as we have done in compression) in order to accurately obtain the true material response where the strain softening phenomenon is uncoupled from any effects due to changing strain rate.

The large strain response in tension also differs from that observed in compression. Locking occurs in tension at approximately 70% strain while in compression the locking strain is 125%. Locking in tension is a result of a uniaxial orientation process, whereas in compression a biaxially oriented state is achieved.

The tension data in Figure 2.4 have been converted to true local stress versus true local strain and plotted in Figure 2.5. The behavior during drawing and hardening are quite similar for all data with locking occurring at 70% strain in all cases.

3 Mathematical Modelling of Polymer Response

3.1 Constitutive Model

The three-dimensional constitutive model discussed below was developed from the concept that the total resistance to yield in glassy polymers is composed of two physically distinct sources: (1) an isotropic resistance to molecular chain segment rotation; and (2) an anisotropic resistance to molecular chain alignment. These two resistances lead to an internal variable formulation of the constitutive response, where the state description is given by a scalar and a second order tensor.

3.1.1 Isotropic Resistance

In the Boyce, Parks, and Argon (1988) model, the isotropic resistance to deformation is taken to result from intermolecular barriers to chain segment rotation. Argon (1973) has developed an expression for the plastic shear strain rate, $\dot{\gamma}^p$, which results once the free energy barrier to chain segment rotation in an equivalent elastic medium has been overcome:

$$\dot{\gamma}^p = \dot{\gamma}_o exp[-\frac{As_0}{\Theta}(1-(\frac{\tau}{s_0})^{\frac{5}{6}})];$$ (1)

where $\dot{\gamma}_o$ is the pre-exponential factor; A is proportional to the activation volume divided by Boltzmann's constant; Θ is the absolute temperature; $s_0 = \frac{.077}{1-\nu}\mu$ is the athermal shear strength; μ is the elastic shear modulus; ν is Poisson's ratio; and τ is the applied shear stress. This expression has been extended by Boyce, *et al.* (1988) to include the effects of pressure and strain softening. In the modification, the athermal shear strength s_0 is replaced by $s + \alpha p$, where p is the pressure, α is the pressure dependence coefficient, and s is now taken to evolve with plastic straining. The initial condition on s is the annealed value s_0. The shear resistance is phenomenologically modelled to decrease with plastic straining until reaching a "preferred" structure represented by s_{ss}, via:

$$\dot{s} = h(1-\frac{s}{s_{ss}(\Theta,\dot{\gamma}^p)})\dot{\gamma}^p;$$ (2)

where h is the rate of resistance drop with respect to the plastic strain, and s_{ss} generally depends on Θ and $\dot{\gamma}^p$. The evolution of s with plastic strain models the macroscopic response of true strain softening. Softening is considered to be the result of local dilation in the polymer [Haward, *et al.* (1980)]. Any fluctuations in local volume would impact the intermolecular barriers to deformation. This lead to the phenomenological evolution equation for s given above. Thus, s is taken to be an internal state variable which monitors the isotropic resistance of deformation.

3.1.2 Anisotropic Network Resistance

Once the material is stressed to the point of overcoming intermolecular barriers to chain motion, the molecular chains will tend to align along the direction of principal plastic stretch. This action decreases the configurational entropy of the system which, in turn, creates an internal network stress state. Following mechanics terminology, we call this internal stress the back stress tensor, **B**. Due to the rubbery network-like response of glassy polymers to plastic deformation (see [Haward (1973), Ward (1984), Parks, *et al.* (1984), Boyce, *et al.* (1988)], the back stress has been modelled using a statistical mechanics of rubber elasticity model. We have assumed that the initial network structure of the polymer is isotropic (as is the case in our experiments). Subsequent network stretching is taken to coincide with plastic stretching, and **B** is taken to be coaxial with the plastic stretch. Using the Wang and Guth (1952) model of rubber elasticity, **B**, expressed in terms of principal components, is given by:

$$\bar{B}_i = nk\Theta\frac{\lambda_L}{3}[\lambda_i^N \mathcal{L}^{-1}(\frac{\lambda_i^N}{\lambda_L}) - \frac{1}{3}\Sigma_j\lambda_j^N\mathcal{L}^{-1}(\frac{\lambda_j^N}{\lambda_L})];$$ (3)

where $\bar{B}_i$ is a principal component of the back stress tensor; λ_j^N are the principal network (and plastic) stretch components; $\lambda_L = \sqrt{N}$ is the tensile locking network stretch (or natural draw ratio), with N being the number of rigid links between physical molecular chain entanglements; n is the chain density; and $\mathcal{L}$ is the Langevin function defined by $\mathcal{L}(\beta_i) = coth\beta_i - \frac{1}{\beta_i} = \frac{\lambda_i^N}{\lambda_L}$. The model is functionally specified such that B_i becomes infinitely large as $\frac{\lambda_i^N}{\lambda_L}$ approaches 1. We point out here that the values for the back stress components determined from the function given in equation (3) are dominated by any tensile stretch components.

3.1.3 Three Dimensional Representation

We first consider the deformation of an initially isotropic body, B_0. The body is loaded to a state, B_t, and its deformed configuration may be described by the deformation gradient at this time, **F**. The deformation gradient may be multiplicatively decomposed into elastic and plastic components, $\mathbf{F} = \mathbf{F}^e\mathbf{F}^p$ [Lee (1969)]. Here, with no loss of generality [Boyce, Weber, and Parks (1989)], we take $\mathbf{F}^e$ to be symmetric, *i.e.* $\mathbf{F}^p$ represents the relaxed configuration obtained by elastically unloading without rotation to a stress free state. The rate quantities corresponding to this formulation begin with the velocity gradient:

$$\mathbf{L} = \dot{\mathbf{F}}\mathbf{F}^{-1} = \mathbf{D} + \mathbf{W} = \dot{\mathbf{F}}^e\mathbf{F}^{e-1} + \mathbf{F}^e\dot{\mathbf{F}}^p\mathbf{F}^{p-1}\mathbf{F}^{e-1}, \tag{4}$$

where $\mathbf{D}$ is the rate of deformation, and $\mathbf{W}$ is the spin. The velocity gradient of the relaxed configuration is given by $\mathbf{L}^p = \dot{\mathbf{F}}^p\mathbf{F}^{p-1} = \mathbf{D}^p + \mathbf{W}^p$. As discussed in Onat (1987), the spin of the relaxed configuration, $\mathbf{W}^p$, is algebraically defined as a result of the symmetry imposed on the elastic deformation gradient. The rate of shape change, $\mathbf{D}^p$, must be constitutively prescribed. The magnitude of $\mathbf{D}^p$ is given by the plastic shear strain rate, $\dot{\gamma}^p$, and the tensor direction of $\mathbf{D}^p$ is specified by $\mathbf{N} = \frac{1}{\sqrt{2}\tau}\mathbf{T}^{*\prime}$. The normalized deviatoric portion of the driving stress state, $\mathbf{T}^*$, is given by the tensor:

$$\mathbf{T}^* = \mathbf{T} - \frac{1}{J}\mathbf{F}^e\mathbf{B}\mathbf{F}^e, \tag{5}$$

where: $\mathbf{T}$ is the Cauchy stress tensor, given by $\mathbf{T} = \mathcal{L}^e[ln\mathbf{F}^e]$ (Anand, 1979), where $\mathcal{L}^e$ is the isotropic elastic modulus tensor; J is the volume change given by $det\mathbf{F}^e$. We refer to $\mathbf{T}^*$ as the driving stress state because it is only this portion of the stress which continues to activate plastic flow. The back stress portion is an internal stress which acts as a resistance. The back stress tensor is taken to be an isotropic function of the left plastic stretch tensor. The principal components of $\mathbf{B}$ are given by equation (3), where the λ_i^N are now replaced by the principal components of the left plastic stretch tensor. The effective equivalent shear stress, τ, is given by:

$$\tau = [\frac{1}{2}\mathbf{T}^{*\prime} \cdot \mathbf{T}^{*\prime}]^{\frac{1}{2}}. \tag{6}$$

The plastic shear strain rate, $\dot{\gamma}^p$, was constitutively prescribed in equation (1). However, τ is now taken to be the effective equivalent shear stress given above.

3.2 Determination of Material Constants for Polycarbonate

The constitutive model contains several material constants which must be obtained experimentally. These constants describe the elastic response of the material as well as the various dependencies of the yield and post yield behavior of the glassy polymer. A general procedure for obtaining the material properties from experimental data is given in Boyce, Parks, and Argon [1988]. Below we give a brief outline of this procedure and the properties obtained for polycarbonate from the experiments discussed earlier.

The elastic modulus was obtained from the tensile test data and found to be $E = 2300MPa$. The Poisson's ratio ν was taken to be 0.30. The initial athermal shear yield strength $s_0 = \frac{.077}{1-\nu}\mu$ is calculated from the elastic constants to be $s_0 = 97MPa$. From the compressive stress-strain data of Figure 2.3, the amount of strain softening is found to be relatively independent of strain rate. This lead to a constant steady state value for the athermal yield strength of seventy-five percent the initial value $s_{ss} = 0.75s_0 = 73MPa$. The strain softening was observed to occur over a plastic strain of approximately 20%, independent of the strain rate of the test, leading to a value for the softening slope of $h = 500MPa$.

The pressure dependence of yield is modelled with the pressure coefficient α described earlier. A value for α is obtained from the experimental data of Spitzig and Richmond (1979) on the initial yield of polycarbonate in tension and compression. This data gives a value of $\alpha = 0.08$.

The strain rate and temperature dependence of the initial yield are modelled with the material constants $\dot{\gamma}_0$ and A of equation (1). To obtain these constants, equation (1) is rearranged to be an equation for a line:

$$ln\dot{\gamma}^p = B + C(\frac{\tau}{s_0 + \alpha p})^{\frac{5}{6}}; \tag{7}$$

$$C = \frac{A}{\Theta}(s + \alpha p); \tag{8}$$

$$B = ln\dot{\gamma}_0 - C. \tag{9}$$

A least squares reduction of the data of Figure 2.3 for the compressive yield stress over a range of strain rates from $-0.0001/sec$ to $-1.0/sec$ result in a value for $\dot{\gamma}_0 = 2(10)^{15}/sec$ and a value for $A = 241K/MPa$.

The material properties required to describe the strain hardening response are the rubbery modulus $C_R = nk\Theta$ and the tensile locking stretch $\lambda_L = \sqrt{N}$ of equation (3). The tensile locking stretch is obtained from the limiting strain found from the compressive stress-strain curves. Noting that a biaxial state of orientation is obtained in uniaxial compression leads to the conclusion that the limiting stretch in compression, λ_C, must be inversely related to the square of the limiting tensile stretch:

$$\lambda_C = 1/\lambda_L^2. \tag{10}$$

Therefore, $\lambda_L = 1/\sqrt{\lambda_C}$, where $\lambda_C = exp(\epsilon_C)$, and ϵ_C is the limiting compressive strain. This is found from the data of Figures 2.3 to be relatively independent of strain rate and equal to -1.25 giving $\lambda_L = 1.87$. Some rate dependence of this property appears to manifest at the very low rate of $-0.0001/sec$ as is apparent in the locking (or dramatic increase in the strain hardening response) beginning at a somewhat larger compressive strain. A value of $C_R = 17MPa$ was fit to model the rate of hardening with respect to plastic straining which was observed in the tests.

The material properties used in the model are summarized below:

$$
\begin{array}{ll}
E & 2300MPa \\
\nu & 0.30 \\
s_0 & 97MPa \\
s_{ss} & 73MPa \\
C_R & 17MPa \\
h & 500MPa \\
\alpha & 0.08 \\
\dot{\gamma}_0 & 2(10)^{15}/sec \\
A & 241K/MPa \\
C_R & 17MPa \\
\lambda_L & 1.87
\end{array}
\tag{11}
$$

The compressive stress-strain curves over a range of strain rates simulated with the constituve model and the above properties are shown and compared to the experimental data in Figure 3.1. The strain rate dependence of the initial flow stress, the amount of strain softening and the strain increment in which softening occurs, as well as the strain hardening response are found to simulate the actual material response very accurately.

This constitutive model with the above properties has been incorporated into a nonlinear finite element code in order to make possible the solution to boundary value problems involving inhomogeneous deformation responses. Below, we examine the cold drawing of a tapered tensile specimen.

3.3 Numerical Simulation of the Uniaxial Tensile Testing of Polycarbonate

3.3.1 Description and Modelling of the Boundary Value Problem

The uniaxial tensile testing of a tapered cylindrical bar of polycarbonate was discussed in section one with a diagram of the tensile specimen shown in Figure 2.1A. The specimen is tapered in order to control the position along the specimen at which necking originates.

The finite element analysis of the tensile test takes advantage of the axisymmetry of the geometry, material and boundary conditions where only one-quarter of the specimen is modelled. The finite element model of the tensile test is shown in Figure 3.2. The mesh consists of 8-node, reduced integration, axisymmetric elements (ABAQUS type CAX8R). The element density is greatest in the tapered section of the bar where the plastic deformation will be concentrated. All nodes along the axis of symmetry are constrained to have no radial motion. The nodes located at the top of the specimen are constrained in the axial direction. These constraints are depicted by "rollers" in Figure 3.2. The nodes at the specimen center (the bottom of the mesh) are prescribed to move in the axial direction at a constant displacement rate of 2.54 mm/minute corresponding to that applied in one of the actual experiments. The material constitutive law used in the analysis is that discussed in section 3.1. The material properties used are those obtained from the compression data and discussed in section 3.2.

3.3.2 Results

The results for the numerical simulation of the uniaxial tensile testing of the tapered PC bar are shown in Figures 3.3 – 3.6. Figure 3.3 depicts the normalized load as a function of the diametral contraction of the point of minimum cross section which had an initial diameter of 6.35mm. The load is normalized with the maximum load calculated which was 2258N. This compares with an experimental load maximum of 2350N, giving a 4% difference between model and experiment. The same general trend in deformation response is obtained as was observed in the experiments. The load is computed to peak and then drop off sharply corresponding to a very localized deformation occurring in the region of minimum cross-section. The computed load drop was 76.5%, while the experimentally observed load drop was 70%. In the experiments, a much sharper localization process was observed then obtained in the numerical investigation which could account for the larger observed drop in load. As the deformation travels away (up and down the specimen) from the region of minimum cross-section, the load is computed to rise. A sharp rise is computed to begin at a contraction of 1.17mm (Figure 3.3), while a sharp rise is experimentally observed to begin at a contraction of 1.5mm (Figure 2.4). This will be further discussed below in terms of the local stress–strain response.

In Figure 3.4, the deformed mesh is depicted at five stages of deformation during the test. These five stages correspond to those marked on the load–contraction curve of Figure 3.3. The deformation process is further illustrated by the contours of plastic strain rate depicted in Figure 3.5 at the corresponding stages. The contours of plastic strain rate indicate where the specimen is currently in an active process of plastic flow. At stage one, the deformation response is localized at the region of minimum cross-section ($r=3.175$mm) corresponding to the peak in the load–contraction curve. This is apparent in Figure 3.5 by observing the high levels of plastic strain rate achieved at this location. A greater portion of the specimen is actively deforming by stage two as is clear in both the plot of the deformed mesh and the plastic strain rate contour. In the deformation plot, one can see that an axial section of the bar with an original radius between 3.175mm and 3.20mm is now contracting. This is also demonstrated by the contours of plastic strain rate which are spread over this region rather than concentrated at one axial location, *i.e.*, the deformation is no longer sharply localized. In stages 3, 4, and 5, the deformation is now traveling up into the portion of the specimen where the diameter begins to increase more rapidly as a function of axial position, resulting in the increasing load obtained in the load-contraction curve. The deformation is depicted in Figure 3.4, where one can observe the contraction and elongation of the specimen. The actively deforming zone can also be observed by examining the position of the peak plastic strain rate which moves up (and down) the specimen, always occurring in the "shoulder" region of the specimen.

Figure 3.6 depicts the model and experimental results for the local true stress-strain at the minimum cross-section of the tensile specimen and the response from the earlier compression test. As discussed in section two, significant differences are observed between the compressive response and the tensile response at both moderate and high strain levels. This is found in the experiments and is also predicted by the numerical simulation. The moderate strain level differences, primarily being the lower amount of strain softening observed in tension than compression, is due to the changing rate of straining with deformation in the tension test. At higher strains, the locking (dramatic strain hardening) phenomenon occurs at significantly lower strain levels than in compression. In the experiments, locking occurs at 70% tensile strain (giving a natural draw ratio of 2.0) and 125% compressive strain. The model predicts locking at 62% tensile strain (giving a natural draw ratio of 1.86) and 125% compressive strain. The differences between the tensile and compressive response at large strains is a result of the different molecular orientation processes occurring in the tension and compression loading conditions. The model underpredicts the natural draw ratio in tension by 7.5%. Therefore, although the model of the relationship between the strain hardening process and the state of network (plastic) stretch provides a clear representation of the different loading conditions, a more detailed examination of the orientation process is warranted in order to improve the accuracy of this model.

The lower modelled draw ratio results in the discrepancy between the model and experimental load-contraction curves noted earlier. A larger draw ratio would result in a greater load drop, where the minimum cross-section would draw-in to a greater extent prior to the localization process spreading up the specimen. This would cause a greater load drop. A larger natural draw ratio would also cause the load increase of the load-contraction curve to occur at a greater contraction. This would align the model results with the experimental results of the tensile test. However, a larger natural draw ratio would create a discrepancy between the model and experimental results of the compressive response. In order to obtain a more accurate representation of both loading conditions with the same material properties, the modelling of the orientation–strain hardening process must be further investigated.

4 Concluding Remarks

Uniaxial tensile and compressive tests were conducted on the glassy polymer polycarbonate over a range of strain rates. The strain rate dependence of yield and the post yield strain softening and strain hardening were observed in both sets of tests. The differences between the experimental compressive and tensile stress-strain curves at moderate strain levels clearly demonstrate the need to conduct constant strain rate tests in order to uncouple the phenomenon of true strain softening from any effects of possible transient strain rates. The constant strain rate compression tests serve to accurately and independently characterize the effects of strain rate and strain softening on the true stress-strain response of the polymer. The differences between the experimental compressive and tensile stress-strain curves at large strain levels where extreme strain hardening (locking) occurs at 70% tensile strain and 125% compressive strain demonstrates the need for an orientation-based hardening model (such as Parks, *et al.* (1984), Bassani and Batterman (1987), Boyce, *et al.* (1988)) if one is interested in accurately analyzing three-dimensional deformation processes.

The constitutive model of Boyce, Parks, and Argon (1988) was found to accurately detail the compressive response of the material including the specific features of strain rate dependence, strain softening, and strain hardening. The model was also found to reasonably predict the general and specific differences in response between the compressive and tensile loading conditions at all strain levels including the rate dependency, the softening and the hardening. The largest error was a 7.5% difference between the experimental and the predicted values of the natural draw ratio leading to the conclusion that, although the existing model predicts the major aspects of the material response well, a more detailed investigation relating the molecular orientation process to the strain hardening process is required.

Acknowledgements

This research has been funded by the National Science Foundation (MSM-8818233), the ALCOA Foundation, and the MIT Bradley Foundation.

References

1. Boyce, M.C., Parks, D.M., Argon, A.S., 1988, "Large Inelastic Deformation of Glassy Polymers, Part I: Rate Dependent Constitutive Model", *Mech. Materials 7*, 17.

2. Haward, R.N., 1987, "The Application of a Simplified Model for the Stress-Strain Curves of Polymers", *Polymer 28*, 1485.

3. Hope, P.S., Ward, I.M., Gibson, A.G., 1980, "The hydrostatic extrusion of PMMA", *J. Mat. Sci. 15*, 2207.

3. G'Sell, C., Jonas, J.J., 1979, "Determination of the Plastic Behavior of Solid Polymers at Constant True Strain Rate", *J. Mat. Sci. 14*, 583.

4. G'Sell, C., Boni, S., Shrivastava, S., 1983, "Application of the Plane Simple Shear Tests for Determination of the Plastic Behavior of Solid Polymers at Large Strains", *J. Mat. Sci. 18*, 903.

5. Brown, S.B., Kim, K.H., Anand, L., 1988, "An Internal Variable Constitutive Model for Hot Working of Metals", *Int. J. Plasticity*, .

6. Argon, A.S., 1973, "A Theory for the Low-Temperature Plastic Deformation of Glassy Polymers", *Phil. Mag. 28*, 39.

7. Haward, R.N., 1980, "The Effect of Chain Structure on the Annealing and Deformation Behavior of Polymers", *Coll. and Poly. Sci. 258*, 42.

8. Haward, R.N., **The Physics of Glassy Polymers**, App. Sci. Pub., Essex.

9. Ward, I.M., 1984, "The Role of Molecular Networks and Thermally Activated Processes in the Deformation Behavior of Polymers", *Poly. Eng. Sci. 24*, 724.

10. Wang, Y.Y., Guth, E.J., 1952, "Statistical Theory of Networks of Non-Gaussian Flexible Chains", *J. Chem. Phys. 20*, 1144.

11. Parks, D.M., Argon, A.S., Bagepalli, B.S., 1984, "Large Elastic-Plastic Deformation of Glassy Polymers", MIT Program in Polymer Science and Technology Report, MIT.

12. Lee, E.H., 1969, "Elastic-Plastic Deformation at Finite Strains", *ASME J. Appl. Mech. 56*, 1.

13. Boyce, M.C., Weber, G.G., Parks, D.M., 1989, "On the Kinematics of Finite Strain Plasticity", to appear in *J. Mech. Phys. of Solids*.

14. Onat, E.T., 1987, "Representation of Elastic-Plastic Behavior in the Presence of Finite Deformations and Anisotropy", *Int. J. Plasticity*.

15. Anand, L., 1979, "On H.Hencky's Approximate Strain Energy Function for Moderate Deformations", *ASME J. Appl. Mech. 46*, 78.

16. Spitzig, W.A., Richmond, O., 1979, "Effect of Hydrostatic Pressure on the Deformation Behavior of Polyethylene and Polycarbonate in Tension and Compression", *Poly. Eng. Sci. 19*, 1129.

17. Bassani, J.L., Batterman, S.D., 1987, "Deformation Induced Anisotropy and Neck Propagation in Polymers", in **Constitutive Modelling for Nontraditional Materials, ASME AMD-VOL85**, ed. V.Stokes and D.Krajcinovic, ASME, NY.

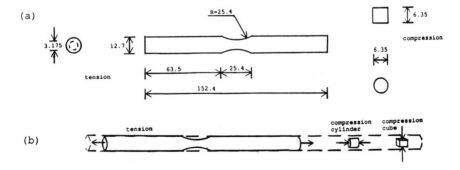

Figure 2.1. (a) Sketch of the tension and compression specimens, all dimensions in mm. (b) Schematic of all specimens inscribed within the bar stock material showing orientation of applied loads.

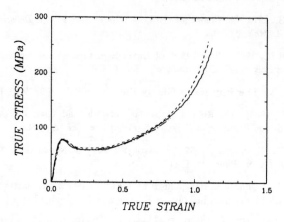

Figure 2.2. True stress vs. true strain data at a constant strain rate of -0.1/second for (a) original bar stock, (b) bar stock subjected to full anneal, (c) transversely loaded bar stock. Results indicate that the original state of the material was isotropic and fully annealed.

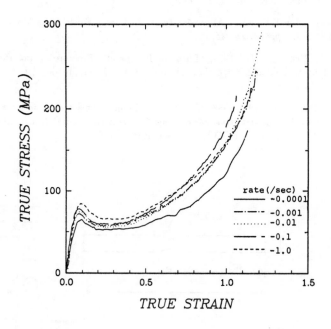

Figure 2.3. Experimental results of true stress vs. true strain for the strain rates of -0.0001/sec, -0.001/sec, -0.01/sec, -0.1/sec, -1.0/sec.

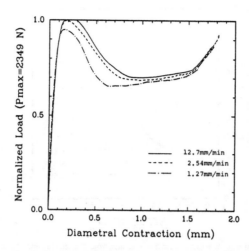

Figure 2.4. Normalized load vs. diametral contraction response for uniaxial tensile tests conducted at the applied displacement rates of 1.27 mm/min, 2.54 mm/min, and 12.7 mm/min. The loads are normalized by 2350N, and the initial diameter of measured contracting region was 6.35mm.

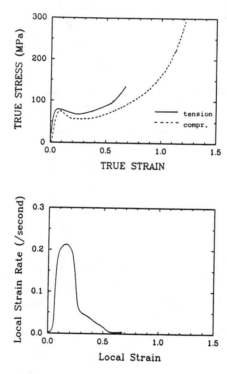

Figure 2.5. (a) True stress vs. true strain response for PC from (i) tension response of uniaxial tensile test at 2.54 mm/min displacement rate and (ii) compressive response from constant strain rate of -0.01/sec. (b) Local strain rate vs. strain for the tensile data of (a).

41

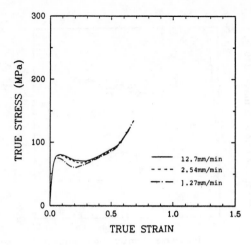

Figure 2.6 Local true stress vs. true strain response from the tensile tests of Figure 2.7 at the displacement rates of 1.27, 2.54, and 12.7 mm/min.

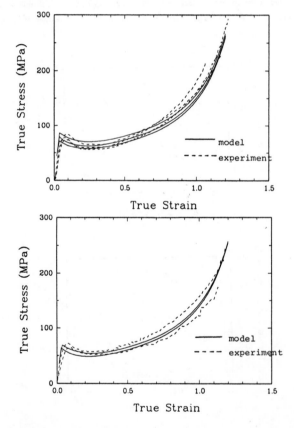

Figure 3.1. Model vs. experimental results from the compressive true stress vs. true strain response of PC over a range of strain rates: (a) -1.0/second, -0.1/second, -0.01/second, and (b) -0.001/second and -0.0001/second.

42

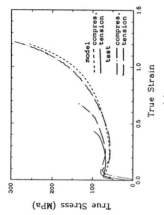

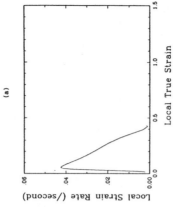

Figure 3.6. (a) Model and experimental results for the true stress-strain response for PC from the local tension response of the test at a displacement rate of 2.54mm/min and compressive response at a strain rate of -0.01/sec. (b) Model results for local strain rate vs. strain for model tensile data of (a).

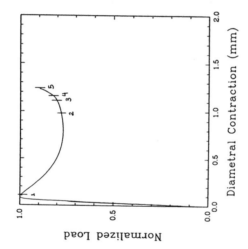

Diametral Contraction (mm)

Figure 3.2. Finite element model of the uniaxial tensile test. All elements are 8-node reduced integration axisymmetric elements.

Figure 3.3. Model results for the load vs. diametral contraction for a tensile displacement rate of 2.54 mm/minute. The load is normalized by 2258N and the original diameter was 6.35mm.

43

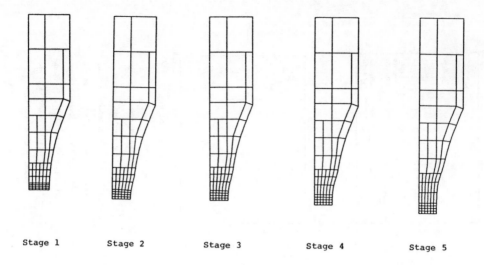

Stage 1   Stage 2   Stage 3   Stage 4   Stage 5

Figure 3.4. Plots of deforming mesh (one-quarter of the specimen) at five stages of the deformation process with stage, diametral contraction pairs of 1 – .1mm, 2 – .97mm, 3 – 1.1mm, 4 – 1.16mm, 5 – 1.23mm.

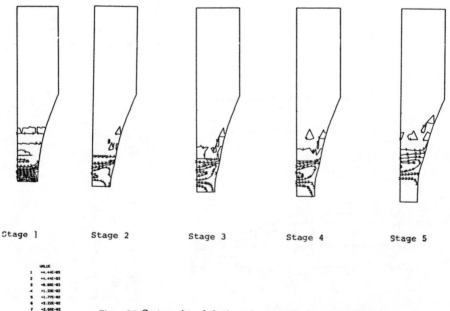

Stage 1   Stage 2   Stage 3   Stage 4   Stage 5

Figure 3.5. Contour plots of plastic strain rate at the five stages of deformation.

ON THE THEORY OF COLD DRAWING AND NECK FORMATION IN ELASTIC FILMS

B. D. Coleman and D. C. Newman
Department of Mechanics and Materials Science
Rutgers University
Piscataway, New Jersey

ABSTRACT

A discussion is given of the nonhomogeneous stretch of long, wide strips of film of incompressible, isotropic, elastic materials. It is shown how each of the coefficient functions in a recently proposed unidimensional theory of neck formation and cold drawing for bars of small but not zero thickness is determined, in the case of a strip of film, by the strain-dependent shear modulus of the three-dimensional material of which the bar is composed.

SHEAR DEFORMATIONS AND THE AXIAL STRETCHING OF FILMS

We are here concerned with bodies of incompressible, isotropic, simple, elastic materials. For such materials the stored energy density ψ (the Helmholtz free energy per unit of volume) is given by a function $\widehat{\psi}$ of the squares of the principle stretches u_i, i.e.,

$$\psi = \widehat{\psi}(b_1, b_2, b_3), \qquad b_i = u_i{}^2, \tag{1}$$

and this function is symmetric in the sense that

$$\widehat{\psi}(b_i, b_j, b_k) = \widehat{\psi}(b_1, b_2, b_3) \tag{2}$$

for every permutation (i, j, k) of $(1, 2, 3)$. For the derivative of $\widehat{\psi}$ with respect to its ith variable, we write $\partial_i \widehat{\psi}$:

$$\partial_i \widehat{\psi}(b_1, b_2, b_3) = \partial(b_1, b_2, b_3)/\partial b_i. \tag{3}$$

In applications $\widehat{\psi}$ is given as a smooth function through formulæ that can be interpreted off of the manifold of triples (b_1, b_2, b_3) of positive numbers b_i with $b_1 b_2 b_3 = 1$. Thus, in practice, there is no difficulty in computing the partial derivatives, $\partial_i \widehat{\psi}$, but values of these derivatives are meaningful only in those combinations that occur in expressions for $d\psi/d\zeta = d\widehat{\psi}(b_1(\zeta), b_2(\zeta), b_3(\zeta))/d\zeta$ along parameterized curves $\zeta \mapsto (b_1(\zeta), b_2(\zeta), b_3(\zeta))$ on which $b_1(\zeta) b_2(\zeta) b_3(\zeta) = 1$. Of special interest here is the case in which $b_1(\zeta) = \lambda(\zeta)^2$, $b_2(\zeta) = \lambda(\zeta)^{-2}$, $b_3(\zeta) = 1$, and we have

$$\frac{d\psi}{d\zeta} = 2\lambda \left[\partial_1 \widehat{\psi}(\lambda^2, \lambda^{-2}, 1) - \lambda^{-4} \partial_2 \widehat{\psi}(\lambda^2, \lambda^{-2}, 1) \right] \frac{d\lambda}{d\zeta}. \qquad (4)$$

Clearly, the combination $\partial_1 \widehat{\psi}(\lambda^2, \lambda^{-2}, 1) - \lambda^{-2} \partial_2 \widehat{\psi}(\lambda^2, \lambda^{-2}, 1)$ is meaningful; it occurs frequently in our analysis.

Consider now a three-dimensional body that in its undistorted reference configuration has the form of a long rectangular plate of small thickness D and large width W. Such a body may be called a *film*, or, in full, a *long, wide strip of film*. We employ a Cartesian coordinate system placed so that, in the reference configuration, the Y-axis is normal to the face of the film, the (X, Z)-plane is at mid-height in the film, and the Z-axis, called the *axis of the film*, runs parallel to the film's length. It follows that, in the reference configuration, each intersection of the film with a plane perpendicular to its axis, *i.e.*, each normal transverse cross section, is a rectangle of points with coordinates $|X| \leq W/2$, $|Y| \leq D/2$, $Z = $ constant. The deformation is assumed such that

$$x = X, \qquad y = \nu(Z)Y, \qquad z = \tilde{z}(Z), \qquad (5)$$

where x, y, z are the present coordinates of the material point with coordinates X, Y, Z in the reference configuration. The first equation here, $x = X$, expresses the idea that changes of width are negligible; the third equation, $z = \tilde{z}(Z)$, expresses the assumption that each normal transverse cross-sectional material plane remains planar and normal upon deformation, and, in view of the first equation, the second equation, $y = \nu(Z)Y$, asserts that in each such cross-sectional material plane the deformation is homogeneous. The measure of *stretch in the axial direction* is $\lambda = \lambda(Z) = d\tilde{z}/dZ$. As the condition that such a deformation be isochoric is

$$\lambda(Z)\nu(Z) = 1, \qquad (6)$$

the function $\tilde{z}$ determines the deformation.

When (5) holds, the principle stretches obey[1]

$$2u_1{}^2 = \left(\lambda^2 + (\nu_Z Y)^2 + \nu^2\right) + \left[\left(\lambda^2 + (\nu_Z Y)^2 + \nu^2\right)^2 - 4\lambda^2\nu^2\right]^{1/2},$$

$$2u_2{}^2 = \left(\lambda^2 + (\nu_Z Y)^2 + \nu^2\right) - \left[\left(\lambda^2 + (\nu_Z Y)^2 + \nu^2\right)^2 - 4\lambda^2\nu^2\right]^{1/2}, \qquad (7)$$

$$u_3{}^2 = 1,$$

and hence,

$$u_1{}^2 = \lambda^2 + (\lambda^4 - 1)^{-1}(\lambda_Z Y)^2 + O((\lambda_Z Y)^4),$$

$$u_2{}^2 = \lambda^{-2} - \lambda^{-4}(\lambda^4 - 1)^{-1}(\lambda_Z Y)^2 + O((\lambda_Z Y)^4), \qquad (8)$$

$$u_3{}^2 = 1.$$

Equation (1) then yields

$$\psi = \widehat{\psi}(\lambda^2, \lambda^{-2}, 1) + (\lambda^4 - 1)^{-1}\left[\partial_1\widehat{\psi}(\lambda^2, \lambda^{-2}, 1) - \lambda^{-4}\partial_2\widehat{\psi}(\lambda^2, \lambda^{-2}, 1)\right](\lambda_Z Y)^2$$
$$+ O((\lambda_Z Y)^4). \qquad (9)$$

By averaging ψ over a transverse cross section of the film, we find that

$$\Psi(Z) = \frac{1}{DW} \int_{-D/2}^{D/2} \int_{-W/2}^{W/2} \widehat{\psi}(b_1, b_2, b_3)\, dX\, dY$$

$$= \Psi_0(\lambda) - \frac{1}{2}\bar{\gamma}(\lambda)\lambda_Z^2 + O((\lambda_Z D)^4) \qquad (10a)$$

$$= \Psi_0(\lambda) - \frac{1}{2}\gamma(\lambda)\lambda_z^2 + O((\lambda_z D)^4), \qquad (10b)$$

where

$$\Psi_0(\lambda) = \widehat{\psi}(\lambda^2, \lambda^{-2}, 1), \qquad (11)$$

and

$$\bar{\gamma}(\lambda) = \frac{-D^2}{6(\lambda^4 - 1)}\left[\partial_1\widehat{\psi}(\lambda^2, \lambda^{-2}, 1) - \lambda^{-4}\partial_2\widehat{\psi}(\lambda^2, \lambda^{-2}, 1)\right], \qquad (12a)$$

$$\gamma(\lambda) = \lambda^2\bar{\gamma}(\lambda). \qquad (12b)$$

Under the assumption that the material surfaces $Y = \pm D/2$ sustain negligible loads, now familiar thermodynamical arguments[2] yield the following asymptotic expression for T, the *axial tension* in the film expressed in "engineering units", i.e., in force per unit of unstrained area,[3]

$$T = T(z) = \tau(\lambda) + \beta(\lambda)\left(\lambda_z\right)^2 + \gamma(\lambda)\lambda_{zz} + O\left(D^4\right); \qquad (13)$$

---

[1] $\nu_Z = d\nu/dZ$, $\lambda_Z = d\lambda/dZ$, etc. In the *spatial description*, λ is a function of z rather than Z, and we write $\lambda_z = d\lambda(z)/dz$, $\lambda_{zz} = d^2\lambda(z)/dz^2$, etc.

[2] See, e.g., Refs. [2,3,5,6].

[3] T is the total tensile force acting across a normal transverse cross section divided by DW, the area of the cross section in the reference configuration.

here

$$\tau(\lambda) = \frac{d}{d\lambda}\Psi_0(\lambda) = 2\lambda\left[\partial_1\widehat{\psi}(\lambda^2,\lambda^{-2},1) - \lambda^{-4}\partial_2\widehat{\psi}(\lambda^2,\lambda^{-2},1)\right], \qquad (14)$$

$\gamma(\lambda)$ is given by (12), and

$$\beta(\lambda) = \frac{1}{2}\frac{d}{d\lambda}\gamma(\lambda). \qquad (15)$$

It follows from (14) and (12) that

$$\gamma(\lambda) = -\frac{D^2\tau(\lambda)\lambda}{12(\lambda^4 - 1)}. \qquad (16)$$

Thus, we have the following asymptotic formula for the axial tension T in a stretched, long, wide strip of film:

$$T = T(z) = \tau(\lambda) + \beta(\lambda)\left(\lambda_z\right)^2 + \gamma(\lambda)\lambda_{zz}. \qquad (17)$$

This formula was proposed by Coleman [2,3] as generally appropriate to elastic bars of small, but not zero, thickness when subjected to non-homogeneous axial stretch. It is interesting that when equation (17), which we may take as the defining constitutive equation of a unidimensional body, arises (via the argument just outlined) as an asymptotic approximation for a three-dimensional body composed of an elastic material with a stored energy function, *the function τ determines the function γ, which, in turn, determines β, through the relations* (16) *and* (15).

When $\lambda_z = 0$, *i.e.*, when the deformation of the film is homogeneous, we have $T = \tau(\lambda)$, and the deformation is one of *pure shear* with the Z-axis giving the neutral direction and with $u_1 = \lambda$, $u_2 = \lambda^{-1}$, and $u_3 = 1$. We now derive a relation between the function τ and a modulus familiar in the theory of *simple shear*.

For a homogeneous simple shear, there is a Cartesian coordinate system for which

$$x = X, \qquad y = Y + \kappa X, \qquad z = Z \qquad (18)$$

with κ the *amount of shear*. When the material is elastic and isotropic, the shear stress T^{xy} is given by

$$T^{xy} = \kappa\hat{\mu}(\kappa^2) \qquad (19)$$

with $\hat{\mu}(\kappa^2)$ the *generalized shear modulus*. The principle stretches in the (X,Y)-plane, u_1, u_2, and the principle stresses in that plane, σ_1, σ_2, obey[4]

$$u_1{}^2 = u_2{}^{-2} = u_2{}^2 + 2\kappa\sqrt{1 + \frac{1}{4}\kappa^2} = 1 + \frac{1}{2}\kappa^2 + \kappa\sqrt{1 + \frac{1}{4}\kappa^2}, \qquad (20)$$

$$\sigma_1 - \sigma_2 = 2T^{xy}\sqrt{1 + \frac{1}{4}\kappa^2}. \qquad (21)$$

---

[4]See, *e.g.*, §54 of Ref. [7].

For a general incompressible, elastic, isotropic material, the relation $u_2 = u_1{}^{-1}$ implies

$$\sigma_1 - \sigma_2 = u_1 \tau(u_1) \tag{22}$$

with τ as in (14). It follows that the shear stress $\kappa\hat{\mu}(\kappa^2)$ in simple shear and the function τ are related as follows:

$$\kappa\hat{\mu}(\kappa^2) = \frac{u_1 \tau(u_1)}{2\sqrt{1 + \frac{1}{4}\kappa^2}}, \qquad u_1 = \left[1 + \frac{1}{2}\kappa^2 + \kappa\sqrt{1 + \frac{1}{4}\kappa^2}\right]^{1/2}. \tag{23}$$

Thus, in view of the relations (15) and (16), we can assert that, for a long, wide strip of film composed of an incompressible, isotropic, elastic, material with a stored energy function, all three of the material functions τ, β, and γ in Coleman's relation (17) can be determined from a graph of the shear modulus $\hat{\mu}(\kappa^2)$ *versus* κ.

AN EXAMPLE OF A MATERIAL SUSCEPTIBLE TO COLD DRAWING

In earlier work [4,5] we considered a material for which $\hat{\psi}$ has the special form

$$\psi = \hat{\psi}(b_1, b_2, b_3) = \sum_{i=1}^{3} \left[\frac{1}{2}\mu_1 b_i + \frac{1}{4}\mu_2 b_i^2 - 2\mu_3(b_i + 3)e^{-\frac{1}{4}(b_i - 1)}\right] \tag{24}$$

with μ_1, μ_2, μ_3 positive material parameters. For such a material, equation (14) yields

$$\tau(\lambda) = \mu_1(\lambda - \lambda^{-3}) + \mu_2(\lambda^3 - \lambda^{-5}) + \mu_3\lambda(\lambda^2 - 1)e^{-\frac{1}{4}(\lambda^2 - 1)}$$
$$+ \mu_3\lambda^{-3}(1 - \lambda^{-2})e^{\frac{1}{4}(1 - \lambda^{-2})}; \tag{25}$$

the functions $\hat{\mu}$, γ, and β may be obtained from this expression by use of equations (23), (16), and (15). In Figures 1 and 2 we show graphs of $\tau(\lambda)$, $\kappa\hat{\mu}(\kappa^2)$, $\gamma(\lambda)$, and $\beta(\lambda)$ with the parameters μ_i assigned the values

$$\mu_1 = 0.05, \qquad \mu_2 = 0.002, \qquad \mu_3 = 0.35. \tag{26}$$

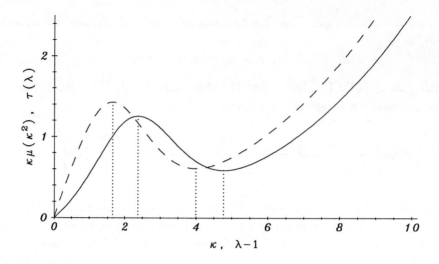

Figure 1. $\tau(\lambda)$ *versus* $\lambda-1$ $(---)$ and $T^{xy} = \kappa\hat{\mu}(\kappa^2)$ *versus* κ $(\underline{})$ according to equations (25), (26), and (23). The turning points in $\tau(\lambda)$ and $\kappa\hat{\mu}(\kappa^2)$ are indicated with dotted lines; for $\tau(\lambda)$ they occur at $\lambda_1 = 2.644$ and $\lambda_2 = 4.994$; for $\kappa\hat{\mu}(\kappa^2)$ they occur at $\kappa_1 = 2.377$ and $\kappa_2 = 4.767$. The graph of $\tau(\lambda)$ *versus* $\lambda - 1$ tells us that a film of a material obeying (24) is susceptible to cold drawing [2]–[5]. The graph of $\kappa\hat{\mu}(\kappa^2)$ *versus* κ suggests that a thick rectangular block of the same material should exhibit shear-band formation when subjected to mean shears of sufficient magnitude.

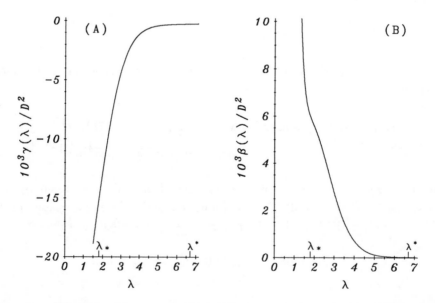

Figure 2. (A) $\gamma(\lambda)$ *versus* λ and (B) $\beta(\lambda)$ *versus* λ according to equations (25), (26), (16), and (15). The range of λ pertinent to theories of neck formation is $\lambda_* \leq \lambda \leq \lambda^*$ with λ_* and λ^* determined by (28). It should be noted that the functions γ and β are far from constant on that interval.

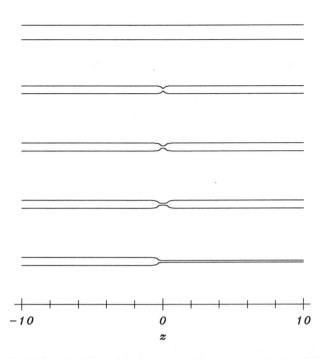

Figure 3. Profiles of a film showing neck formation, obtained by solving the equation of equilibrium (27) for a film of a material obeying (24) and (26) and hence with the functions τ, β, and γ as seen in Figures 1 and 2. From top to bottom: the undistorted film; neck-like solutions of (27) for $\delta = 1.0 \times 10^{-1}$, $\delta = 1.0 \times 10^{-5}$, $\delta = 1.0 \times 10^{-10}$; and a fully developed draw ($\delta = 0$).

In Figure 3 we show profiles of a film obtained from explicit solutions[5] of the equation of equilibrium $T(z) = T^{\circ}$ = constant, *i.e.*, the equation

$$T^{\circ} = \tau(\lambda) + \beta(\lambda)(\lambda_z)^2 + \gamma(\lambda)\lambda_{zz}, \tag{27}$$

for a material obeying (24) and (26). The distance scale is chosen so that $D = 1$, and the topmost figure is the unstretched film with $\lambda \equiv 1$.

Let λ_*, λ^*, and T^* be the (unique) roots of the equations

$$\int_{\lambda_*}^{\lambda^*} \left[\tau(\lambda) - T^* \right] \lambda = 0, \qquad \tau(\lambda_*) = \tau(\lambda^*) = T^*, \tag{28}$$

---

[5]The method of solving equation (27) for λ as a function of z and calculating the resulting profile is discussed in detail in references [2]–[5].

and let[6]

$$\delta = \frac{T^\circ - T^*}{\tau(\lambda_1) - T^*}. \tag{29}$$

It is clear from Figure 3 that as T° decreases from $\tau(\lambda_1)$ to T^*, *i.e.*, as δ decreases from 1 to 0, the corresponding neck at first both lengthens and deepens and then lengthens with its depth fixed, in accord with experience for materials that can be cold drawn [1]. The configuration for the case $\delta = 0$ ($T^\circ = T^*$) is called a *fully developed draw* and corresponds to a neck of infinite length; in that configuration,

$$\lim_{z \to -\infty} \lambda(z) = \lambda_*, \qquad \lim_{z \to +\infty} \lambda(z) = \lambda^*. \tag{30}$$

In the polymer industry, λ^* (or, occasionally, λ^*/λ_*) is referred to as the "draw ratio". Here, $\lambda_* = 1.79$ and $\lambda^* = 6.68$.

ACKNOWLEDGMENTS

This research was supported by the National Science Foundation under Grant DMS 88-15924 and by the Donors of the Petroleum Research Fund, administered by the American Chemical Society.

REFERENCES

[1] Carothers, W.H, and J.W. Hill, "Studies of Polymetrization and Ring Formation. XV. Artificial Fibers from Synthetic Linear Condensation Superpolymers," *J. Am. Chem. Soc.*, Vol. 54, 1932, pp. 1579–1587.

[2] Coleman, B.D., "Necking and Drawing in Polymeric Fibers Under Tension," *Arch. Rational Mech. Anal.*, Vol. 83, 1983, pp. 115–137.

[3] Coleman, B.D., "On the Cold Drawing of Polymers," *Comp. & Maths. with Appls.*, Vol. 11, 1985, pp. 35–65.

[4] Coleman, B.D. & D.C. Newman, "Constitutive Relations for Elastic Materials Susceptible to Drawing," in *Proceedings of the Symposium on Constitutive Modeling for Nontraditional Materials*, V.K. Stokes and D. Krajcinovic, Eds., Vol. 85, Applied Mechanics Division Series, American Society of Mechanical Engineers, New York, 1987, pp. 47–58.

---

[6] λ_1 is the value of λ at which $\tau(\lambda)$ has a local maximum. See Figure 1 and its caption.

[5] Coleman, B.D. & D.C. Newman, "On the Rheology of Cold Drawing. I. Elastic Materials," *J. Polymer Sci., Polymer Phys. Ed.,* Vol. 26, 1988, pp. 1801–1822.

[6] Coleman, B.D. & D.C. Newman, "On Waves in Slender Elastic Rods," *Arch. Rational Mech. Anal.,* in press.

[7] Truesdell, C. & W. Noll., The Non-Linear Field Theories of Mechanics, *Encyclopedia of Physics,* Vol. III/3, Springer, Berlin-Heidelberg-New York, 1965.

SOLID STATE COMPRESSION OF POLYPROPYLENE ABOVE ITS CRYSTALLIZATION TEMPERATURE

P. T. Wang and C. H. Chen-Tsai
Alcoa Laboratories
Alcoa Center, Pennsylvania

ABSTRACT

Solid state processing of thermoplastic materials has gained significant attention from both academic research and plastic industries. The physical and mechanical properties of processed material can be improved drastically by using conventional metal forming techniques with appropriate processing conditions, which include the control of temperature, stress/load, strain and strain rate. In this paper, the compressive deformation behavior of polypropylene at temperatures between the crystallization and melting, and the physical/structural changes associated with the process are investigated.

A constitutive model is proposed in which the initial flow stress is a function of the Zener-Hollomon parameter which accounts for temperature and strain rate effects. The subsequent flow stress levels depend on a strain hardening function which is related to molecular orientation. Two microstructure variables, molecular orientation and degree of crystallinity, are monitored throughout the deformation process.

INTRODUCTION

Use of polymers is widespread because of their low density, relatively low cost and ability to be modified to meet a variety of performance characteristics. Most polymers consist of a group of molecular chains which are comprised of repeating carbon-carbon units along their backbone. If the chains exist as entangled coils, the physical and mechanical properties are less desirable. In recent structural applications which require high elastic modulus and impact strength, the polymeric chains had to be fully extended. These properties can be achieved through solid state processing (1-3).

In general, the thermoplastics are fabricated below melting temperature for semicrystalline materials and slightly above glass transition temperature for amorphous materials. The fabrication methods, similar to those for metals, include extrusion, drawing, forging and rolling. Significant improvement in mechanical properties can be achieved through solid state processing as compared to the melting process because via solid state processing the molecular chains are oriented to the preferred direction as long as the structure is flexible (4,5) This flexibility in the solid state is influenced by its chain configuration, the volume fraction of crystallinity and the processing temperature.

The most effective temperature range for processing semicrystalline polymers into oriented structures lies between the alpha crystallization temperature of the polymer, T_{ac}, and the melting point, T_m (6). In this temperature range, the molecular chains have the most mobility but the material remains solid. The alpha crystallization temperature corresponds to a secondary transition temperature of the polymer, where crystal subunits are capable of being moved within the larger crystal unit (7). Above this temperature, lamellae in crystalline phase slip easily and the amorphous chain segments uncoil rapidly. As a result of this local movement, an extended chain crystal is formed. The alpha crystallization temperature for polypropylene is about 100 C.

In studying the solid state processability of a polymer, it is critical to understand the material constitutive behavior. This includes the evolution of microstructure during deformation (8-9), the relaxation and recovery response (10), and the effect of thermomechanical deformation history at large strains (11-13). Extensive work has been done in both the melt process above T_m and solid deformation below the glassy transition temperature, T_g. To date, very little research work has covered the solid state deformation for semicrystalline polymers such as polypropylene under compression at various strain rates and at temperatures between T_{ac} and T_m.

The aim of this work is to investigate the solid state processing of polypropylene in this temperature range where the goal is to produce an orthotropical sheet. Topics in this paper include the monotonic compression of polypropylene at various strain rates and temperatures, the development of phenomenological constitutive equations and the identification of microstructure variables and their evolution.

MATERIAL

Commercially available fractional melt flow polypropylene (PP), Himont 6823 and experimental grade of ultra high molecular weight PP (UHMWPP) from Aristech were used for the study. Axisymmetric compression tests of cylindrical specimens made from PP and UHMWPP were conducted with the intent of observing deformation behavior and developing a constitutive model. All the tests were performed at controlled constant true strain rates and isothermal conditions. The specimens were also compressed to different strains, 60%, 120% and 180%, for structural and physical property evaluations. Optical microscopy was applied to follow the structural changes that occurred during deformation. Differential Scanning Calorimetry (DSC) was used to probe the influence of deformation and thermal history on crystallinity and crystal structure of the specimens.

AXISYMMETRIC COMPRESSION

The deformation response of PP was measured by performing axisymmetric compression tests. Cylindrical specimens 25.4 mm in dia. and 12.7 mm in height were machined from sheets. The specimens were heated to the deformation temperature in either a silicone oil bath or a sand bath, which is maintained at a temperature a few degrees above the deformation temperature. This procedure produced a sufficient heat-up rate (about 15 mins. to reach 150 C from room temperature) and a relatively uniform temperature distribution in the specimen without causing overheating. Any overheating would have been evidenced by observation of local melting.

All the compression tests were performed under isothermal conditions and at constant true strain rates. The temperature for individual tests varied between 110 and 155 C and the strain rate varied between 0.01 and 10/sec.

After deformation the polymeric specimens had a tendency to lose their deformed shape and possibly suffered microstructural changes. To preserve the microstructure it was necessary to quench the specimen to room temperature rapidly. However, quenching the specimens after deformation can lead to significant shape distortion. To avoid this problem the specimens were quenched in situ under load.

A standard slab analysis showed that friction can contribute up to 5% of the flow stress. During testing, the friction was reduced to minimum by using a hotter die and applying a lubricant. Hence, the true stress was calculated assuming frictionless conditions.

PHENOMENOLOGICAL CONSTITUTIVE MODEL

The basic crystal structure of a semicrystalline polymer consists of chain folded lamellae radiating outward from a nucleation site to form a spherulite. In between crystallites, amorphous phases are present and provide part of the load-bearing strength for the bulk polymer. The viscoelastic load-bearing strength varies depending upon the degree of microstructure orientation in this amorphous region (8).

A typical stress-strain curve can be schematically broken down into three parts as shown in Figure 1. The first part is fairly linear and may be attributed to elastic deformation. This response may be associated with the behavior of tie molecules which provides intermolecular forces between amorphous and crystalline structures. This type of deformation occurs up to true strain values of 5 to 10%.

The second part at the stress-strain curve exhibits the least amount of work hardening. Depending on the evolution of structure, the stress-strain curve could be undergoing either little or no hardening (solid line) or softening (dashed line). In this regime, which includes strains up to 120%, the amorphous structure starts to uncoil and orientate itself, and the crystalline structure (lamellae) begins slipping.

The last part of the stress-strain curve is the so-called orientation hardening, where the stress increases drastically. At this point, molecular segments of both amorphous and lamella components attempt to orientate along the material flow direction. The spherulites become very indistinct in shape. In fact, they may disintegrate due to extensive stretching. Material deformed to this extent has enhanced mechanical properties such as tensile and impact strengths.

Based on this experimental work a flow-theory for the constitutive model is proposed. Material isotropy and the idea of effective stress and strain are adopted. In addition, the plastic strain is approximated by the total strain to simplify the analysis. Nevertheless, an elastic strain and its modulus can be incorporated if required.

In general, the incremental form of the constitutive equation can be expressed in the following form:

$$d\bar{\sigma} = g(\bar{\sigma},\bar{\epsilon}) \, d\bar{\epsilon} \tag{1}$$

where $\backslash O(\sigma,^-)$ is an effective stress associated with the deviatoric Cauchy stress state $\sigma\backslash S(',ij)$ such that $\bar{\sigma} = \sqrt{\frac{3}{2}\sigma'_{ij}\sigma'_{ij}}$, and $\bar{\epsilon}$ is an effective strain associated with the logarithm of principle stretch, λ_i, where $\bar{\epsilon} = \sqrt{\frac{2}{3}\ln\lambda_i \cdot \ln\lambda_i}$. Note that the summation convention is employed. It is assumed that the above function g can be decomposed into

$$g(\bar{\sigma},\bar{\epsilon}) = \bar{\sigma} \cdot h(\bar{\epsilon}) = \bar{\sigma} \cdot A/n \, \bar{\epsilon}^{n-1} \tag{2}$$

where linear dependencies on both stress and a function, h, depending on strain are proposed.

Substituting Equation (2) into Equation (1) and integrating gives:

$$\bar{\sigma} = \bar{\sigma}_o \cdot \exp(A \cdot \bar{\epsilon}^n) \tag{3}$$

Here $\bar{\sigma}_0$ is the initial flow stress before orientation hardening (i.e., before stage 3) and it is assumed to depend on Z, the Zener-Hollomon parameter (14), which is defined by

$$Z = \dot{\bar{\epsilon}} \, \exp(\Delta E/RT) \tag{4}$$

where $\dot{\bar{\epsilon}}$ is the effective strain rate, ΔE is the apparent activation energy for plastic flow, R is the gas constant and T is the absolute temperature.

Equations (3) and (4) form the basic constitutive equations describing the bulk polymer response subjected to finite deformation at temperatures between T_{ac} and T_m. There are two features in this proposed constitutive equation. First, the initial flow stress is strain rate and temperature dependent, which takes into account the initial yield response as a function of Z. Second, the orientation hardening occurring in the later stage of deformation is strain but not strain rate or temperature dependent.

The value of activation energy, ΔE, can be determined by a pair of deformation conditions which shows a coincidence of stress-strain curves at a constant Z. Equation (4) leads to

$$Z_c = \frac{}{\dot{\varepsilon}_1} \exp\left(\frac{\Delta E}{RT_1}\right) = \frac{}{\dot{\varepsilon}_2} \exp\left(\frac{\Delta E}{RT_2}\right) \tag{5}$$

Thus the activation energy of PP at the current test conditions can be determined. In Figure 2, a pair of deformation conditions with a coincidence of stress-strain curve is found at temperature and strain rate values of 120 C-0.01/s and 140 C-10/s. Although a discrepancy is observed in the early stage of deformation, the stress-strain responses are almost identical and overlap after a strain of 0.6. Since strains in the range of 100-180% are of interest, the data shown in Figure 2 are considered identical for present purposes. The activation energy thus obtained is about 466.5 kJ/K-mole.

Krausz and Eyring (15) investigated PP under constant tensile creep stress with temperatures ranging from -80 to +80 C. They suggested the activation energy is stress dependent with a value of 234.6 kJ/K-mole using a linear extrapolation to the zero stress level. To the contrary, the current data indicate that a nonlinear extrapolation at a lower stress level may be necessary since the structure changes with deformation at or above the crystallization temperature (100 C). The determined activation energy at current test conditions is much higher than that of Krausz and Eyring, and may be due to the increase of molecular orientation at finite strain and the difference of crystallinity.

The material constants A and n in Equation (3) can be determined in the following manner. Equation (3) suggests that a logarithmic stress vs. nth power of strain should be linear and A is the slope of this linear response. The intercept stress, found by setting strain equal to zero, is equivalent to the initial flow stress $\bar{\sigma}_0$. It is found that at n=3, the slope, A, is virtually unchanged for various temperatures and strain rates and has a value equal to 0.1697. At very small strains, however, there is a small peak due to the softening behavior mentioned earlier. Disregarding this peak, the intercept stress can be determined. The intercept stress results from the linear extrapolation of the stress-strain curve to the vertical axis where the strain is set to zero. The intercept stress, $\bar{\sigma}_0$, is plotted against the logarithm of Z in Figure 3 where the functional form is expressed as

$$\bar{\sigma}_0(Z) = -19.5049 + 1.0138 \times 10^{-5} (\ln Z)^3 \tag{6}$$

It shows that a fairly good fit can be obtained between the initial flow stress and the Zener-Hollomon parameter except at both extremes of the abscissa.

Using Equations (3) and (6), the stress-strain response at various temperatures and strain rates is calculated and compared with the experimental data in Figure 4. Reasonable agreement is obtained except at 155 C where deviations occur at the lower strain rates.

STRUCTURE AND PROPERTY CHARACTERIZATION

DSC technique was applied to follow the polymer crystalline phase changes along deformation. The possible changes which may occur in a solid state compressively deformed PP specimen are: (1) crystallinity change caused by specimen heat-up, strain induced crystallization and specimen cool-down, and (2) crystal transformation caused by heat and deformation history. The specimens which were inspected by DSC are listed in Table 1.

Table 1. Specimens inspected by DSC and their thermomechanical history.

Specimen #	POL #	T (C)	Strain Rate (s^{-1})	Strain (%)	Peak T (C)	ΔH (J/g)
1	UHMWPP	-	-	-	166.70	79.13
2	351	130	1.00	180	168.22	94.60
3	354	140	1.00	180	166.02	95.62
4	359	155	0.10	180	171.94	93.92
5	360	155	0.01	180	173.82	113.62
6	Reg PP	-	-	-	162.45	76.46
7	142	120	0.01	180	161.79	91.38
8	143	120	0.01	120	161.81	81.75
9	144	120	0.01	60	147.35	83.17
10	152	120	10.0	180	157.29	73.97
11	153	120	10.0	120	160.36	85.20
12	154	120	10.0	60	162.11	83.82
13	90	120	1.00	180	159.73	77.54
14	147	120	0.10	180	162.55	83.82
15	94	130	1.00	180	161.04	82.77
16	93	130	0.10	180	159.00	86.98

Figure 5 shows the DSC melting traces of both UHMWPP and regular PP specimens before subjection to thermomechanical deformation. The UHMWPP melts at slightly higher temperature range and has a greater crystallinity to start up with than the regular PP. Both specimens were discs made by injection molding. Peak melting temperatures and heats of fusion of the specimens can be found in Table 1. The strain rate effect is shown in Figure 6. The regular PP specimens were deformed at strain rates equal to 0.01, 0.10, 1 and 10/sec. at 120 C. The composite plot indicates that the slower the deformation the more opportunity for PP to crystallize due to thermal annealing and strain induced crystal growth. Thus, the specimen deformed at 0.01/sec. strain rate gives the highest crystallinity among the four. The shape of the melting trace of this specimen also suggests the crystal structure is the most perfect among the four. One would expect that the high deformation rate will destroy some of the crystal organization and leave no time for segments of long molecules to recrystallize. The existence of the second peak/shoulder at lower melting temperature indicates the presence of the more random crystal superstructure (spherulites).

Specimens were also prepared to different strains at a constant strain rate. Figure 7 shows the effect of strain at 0.01 s^{-1} strain rate. The heat of fusion (ΔH) data suggest that at low strain (60%) the crystallinity built up from specimen preheating is present. But at 120% strain, apparently the crystals are disrupted and thus there is a crystallinity drop. The crystallinity builds up again giving enough time for annealing during subsequent deformation to 180% strain. All the specimens were quenched to near room temperature after deformation strain had been achieved. Thus, very little crystallinity change is expected during the quenching process. However, the trend of crystallinity change was not found in specimens deformed at high strain rate (10.0 s^{-1}). In fact, the trend is almost inverse (see Figure 8). The 180% strain deformed specimen has a crystallinity lower than the original nonheated, nondeformed specimen. The disruption of the crystal structure is shown by the increased amount of the low melting crystals. So far, we have demonstrated the presence of two competing mechanisms: the thermally induced crystallization and the strain induced or disrupted crystallization. These competing processes can influence the strain hardening behavior of the material during deformation and thus alter the general trend of the stress-strain curves discussed in the Phenomenological Constitutive Model section.

For the UHMWPP specimens, similar influence of the two competing mechanisms is found. The lower the strain rate at constant strain and deformation temperature, the higher the crystallinity can be built through the thermal-mechanical process. The effect of deformation temperature can also be seen by comparing the melting behavior and the heat of fusion data (see Table 1).

Besides DSC experiments, optical microscopy inspections of ultramicrotomed specimens were also performed. The spherulite structure is found compressed in the compression axis, and expanded in the plane which is perpendicular to the same axis. At 180% strain, the spherulites are so disrupted that no clear boundary can be observed on the optical scale. On the other hand, the DSC data suggest crystal perfection in the slow strain rate cases. This perfection is expected to occur in the crystallite/lamellar level. A quantitative study by employing X-ray diffraction techniques of the microstructure evolution in such thermomechnical deformation process is underway (16). The two internal variables, crystallinity and molecular orientation, will be incorporated into the constitutive equation to describe the material response to axisymmetric compression.

CONCLUSIONS

A well-balanced investigation on solid state processing of polypropylene at temperatures above crystallization and below melting has been established. It includes basic material testing and constitutive modeling. The methodology is general enough that it can be applied to other process applications. By no means is the study complete for all of these ingredients. Nevertheless, the current results indicate that:

• Thermomechanical deformation of PP is extremely sensitive on temperature and strain rate. In general, the stress-strain response can be captured by an elastic yield at the early stage, a constant flow stress at the intermediate stage and an orientation hardening at the final stage.

• A constitutive equation with orientation hardening can describe the PP flow behavior for various temperature and strain rate ranges during monotonic loading conditions. It has the initial flow stress as a function of Zener-Hollomon parameter and the orientation hardening as a function of strain. Deformation history and anisotropy are not yet incorporated but will be subjects for future research.

• Two competing mechanisms are identified which occur during the thermomechanical deformation process. They are: (1) the thermally induced crystallization, and (2) the strain induced or disrupted crystallization. This finding provides some interpretation to the material stress-strain behavior also found in the study.

ACKNOWLEDGMENT

The authors wish to thank Drs. O. Richmond, L. A. Lalli and D. V. Humphries for many valuable discussions. Special appreciation is due to Mr. R. Rolles for his timely assistance on the completion of process validation. Thanks are also due to Dr. V. M. Sample for conducting the compression experiments, and to Messrs. W. L. Burton and J. Cornetta for material property evaluations.

REFERENCES

1.   Hope, P. S., Gibson, A. G., Parsons, B. and Ward, I. M., "Hydrostatic Extrusion of Linear Polyethylene in Tubular and Noncircular Sections," Polymer Eng. Sci., Vol. 20, No. 8, 1980 May, pp. 540-545.

2.   He, T. and Porter, R. S., "Uniaxial Draw of Poly(4-methyl-pentene-1) by Solid State Coextrusion," Polymer, Vol. 28, 1987 May, pp. 946-950.

3.   Wang, P. T., Chen-Tsai, C. H. and Sample, V. M., "Biaxial Orientation of Polypropylene Under Solid State Compression," 46th Annual Technical Conference of the Society of Plastic Engineers, 1989 May 01-04, New York, New York.

4.   Linder, B. E. and Samuels, R. J., "Prediction of Axial and Off-Axis Yield Behavior of Isostatic Polypropylene Film from Structural State Parameters," Polym. Eng. Sci., Vol. 25, No. 14, 1985 Mid-October, pp. 875-887.

5.   Thakkar, B. S., Broutman, L. J. and Kalpakayian, S., "Impact Strength of Polymer 2: The Effect of Cold Working and Residual Stress in Polycarbonates," Polym. Eng. Sci., Vol. 20, No. 11, 1980 July, pp. 756-762.

6.   Aharoni, S. M. and Sibilia, J. P., "Crystalline Transitions and the Solid-State Extrusion of Polymers," J. Appl. Polym. Sci., Vol. 23, pp. 133-140 (1979).

7.   Aharoni, S. M. and Sibilia, J. P., "On the Conformational Behavior and Solid-State Extrudability of Crystalline Polymers," Polym. Eng. Sci., 1979 May, Vol. 19, No. 6, pp. 450-455.

8. Samuels, R. J., "Polymer Structure: The Key to Process-Property Control," Polym. Eng. Sci., Vol. 25, No. 14, 1985 Mid-October, pp. 864-874.

9. Samuels, R. J., "The Influence of Structure on the Failure of Isostatic Polypropylene Films," Polym. Eng. Sci., 1979 Mid-February, Vol. 19, No. 2, pp. 66-76.

10. Rizzo, G. and Spadaro, G., "Deformation Recovery Behavior of a Solid Polymer After Tensile Yielding," Polym. Eng. Sci., Vol. 24, No. 18, 1984 December, pp. 1429-1432.

11. Spitzig, W. A. and Richmond, O., "Effect of Hydrostatic Pressure on the Deformation Behavior of Polyethylene and Polycarbonate in Tension and in Compression," Polym. Eng. Sci., Vol. 19, No. 16, 1979 December, pp. 1129-1139.

12. Titomanlio, G. and Rizzo, G., "Experimental Study of Stress-Relaxation Behavior of Polycarbonate After Yielding," Poly. Bull., 4, pp. 351-356 (1981).

13. Mittal, R. K. and Singh, I. P., "Large Deformation Behavior of Some Thermoplastics," Polym. Eng. Sci., Vol. 22, No. 6, 1982 April, pp. 358-364.

14. Zener, C. and Hollomon, J. H., "Effect of Strain Rate Upon the Plastic Flow of Steel," J. of Applied Physics, American Institute of Physics, Vol. 15, 1944, p. 22 .

15. Krausz, A. S. and Eyring, H., Deformation Kinetics, Wiley-Interscience Publication, Chapter 4.

16. Hirsch, J. R. and Wang, P. T., "Texture and Strength Evolution in Deformed Polypropylene," ASM International Conference, 1989 October, Indianapolis, Indiana.

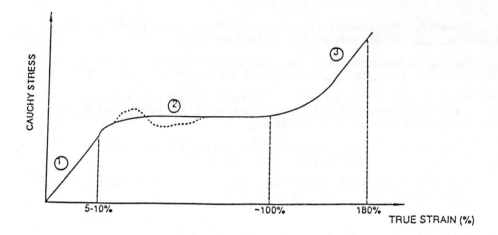

① Elastic

② Amorphous Part Uncoiled

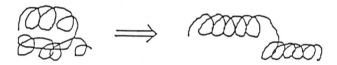

and Crystalline Part Slipped and Uncoiled

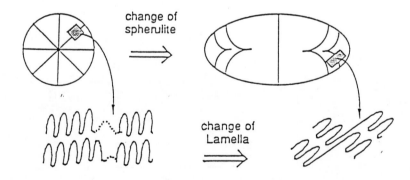

③ Orientation Hardening

Spherulite broken down, Chain conformation aligned

**Figure 1.   Stress-Strain-Structure Characteristics
for Semi-Crystalline Polymers**

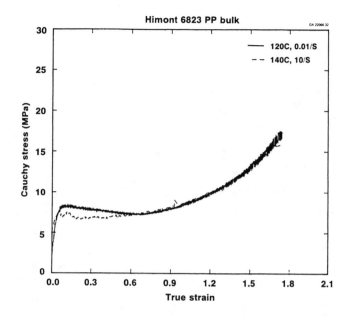

Activation Energy Determination
Figure 2

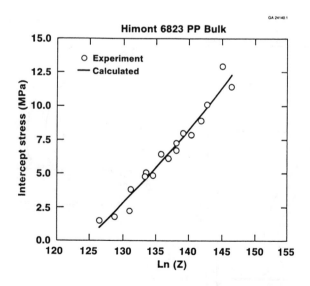

Zener-Hollomon Parameter vs. Intercept Stress
Figure 3

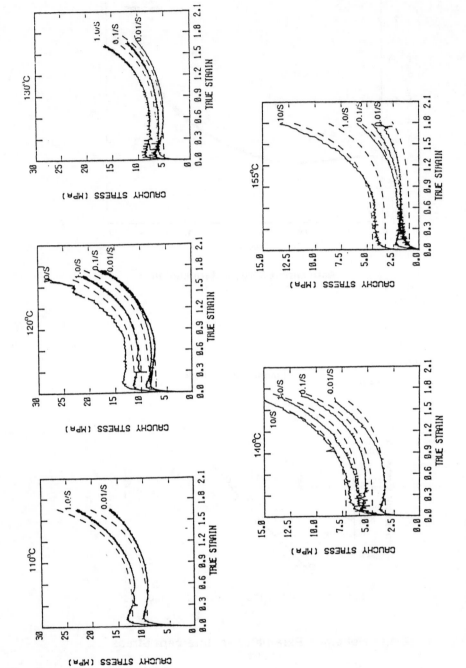

Figure 4. Comparison between experimental and calculated stress-strain responses of Himont 6823 PP at various temperatures and strain rates.

64

DSC Traces of UHMWPP and Reqular PP Control Samples
Figure 5

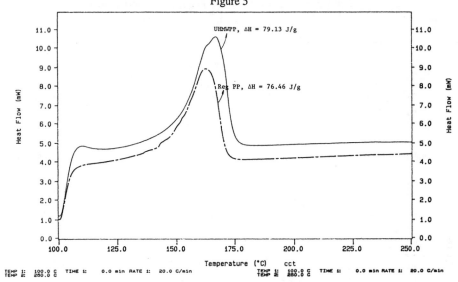

UHMWPP, ΔH = 79.13 J/g

Reg PP, ΔH = 76.46 J/g

TEMP 1: 100.0 C   TIME 1:   0.0 min RATE 1: 20.0 C/min   TEMP 1: 100.0 C   TIME 1:   0.0 min RATE 1: 20.0 C/min
TEMP 2: 250.0 C   TEMP 2: 250.0 C

Composite Plot of DSC Traces for Reqular PP Specimens
Deformed at Different Strain Rates at 120 °C
Figure 6

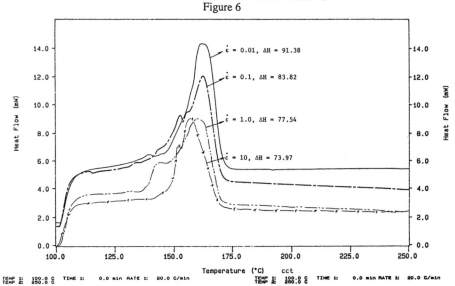

$\dot{\varepsilon}$ = 0.01, ΔH = 91.38

$\dot{\varepsilon}$ = 0.1, ΔH = 83.82

$\dot{\varepsilon}$ = 1.0, ΔH = 77.54

$\dot{\varepsilon}$ = 10, ΔH = 73.97

TEMP 1: 100.0 C   TIME 1:   0.0 min RATE 1: 20.0 C/min   TEMP 1: 100.0 C   TIME 1:   0.0 min RATE 1: 20.0 C/min
TEMP 2: 250.0 C   TEMP 2: 250.0 C

Composite Plot of DSC Traces for Reqular PP Specimens
Deformed to Different Strains at Strain rate Equal to 0.01/sec
at 120 °C
Figure 7

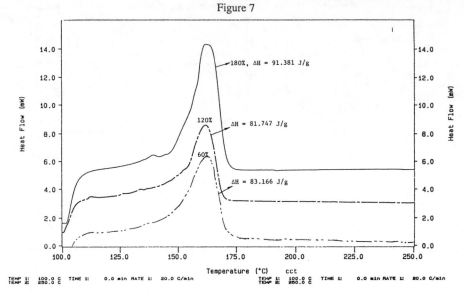

Composite Plot of DSC Traces for Reqular PP Specimen
Deformed to Different Strains at Strain Rate Equal to 10/sec
at 120 °C
Figure 8

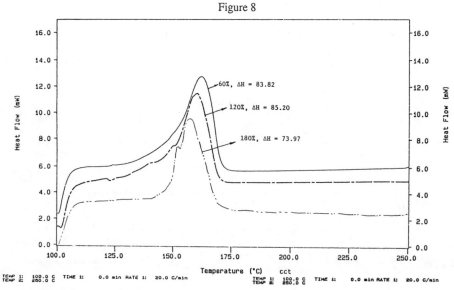

KINEMATICS OF DAMAGE ZONE ACCOMPANYING CURVED CRACK

W.- L. Huang, B. Kunin, and A. Chudnovsky
Department of Civil Engineering, Mechanics and Metallurgy
University of Illinois at Chicago
Chicago, Illinois

ABSTRACT

This paper presents a methodology of a quantitative characterization of the kinematics of evolution of a damage zone surrounding the tip of a slowly propagating crack. On the basis of the Crack Layer Theory, the evolution of the damage zone is modelled as a combination of a few elementary motions: translation, rotation, isotropic expansion and distortion. A procedure for evaluating the rates of the elementary motions on the basis of direct measurements is developed. The procedure is illustrated for curved crack layer growth in the vicinity of a hole in commercial polystyrene. The important role of the damage zone in determining the main crack trajectory and speed is clearly demonstrated.

1. INTRODUCTION

It is well established that in most engineering materials a slowly propagating crack is preceded by an evolving damage zone. It was also observed in [1] (as well as below) that the damage zone is a controlling factor for both the crack trajectory and the rate of crack growth. The Crack Layer (CL) Theory [2,3] was developed to model the damage zone evolution together with the main crack growth. An Active and Inert Zones are distinguished within CL. The Active Zone (AZ) is defined as that part of CL around the crack tip, where the damage growth takes place (see Fig. 1). The consecutive configurations of the AZ at 790, 800, 810, etc. cycles and the corresponding envelope of the CL are shown in Fig. 2. The curvature of CL as well as non-monotonic changes of AZ configuration, easily seen in Fig. 2, are caused by the presence of a hole. There is also an Inert Zone formed as a wake of the Active Zone (see Fig.1 and 2). This zone is distinguished within CL by the zero rate of damage growth.

The CL theory proposes to model AZ evolution by a few time dependent elementary motions: translation, rotation, expansion, and distortion. The rates of AZ translation, rotation, and deformation are considered as generalized fluxes, and the conjugate forces (driving forces) are identified. Rotation takes place only for the curved crack layer. Studies on AZ kinematics for the rectilinear CL in polystyrene, polycarbonate and polypropylene have been recently reported [4-7].

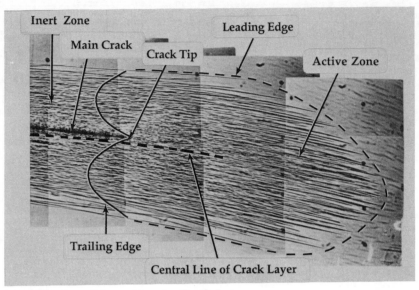

Fig. 1   Schematics of a Crack Layer.

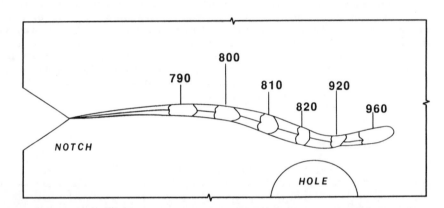

Fig. 2   Crack Layer evolution in a vicinity of a hole. The number of cycles counted from the beginning of the test is indicated above of the corresponding AZ position. The middle line represents the main crack trajectory.

The objective of the present work is to formulate kinetic equations of CL evolution, i.e. relationships between the rates of AZ translation, rotation, and deformation and the corresponding driving forces. The research consists of two coupled projects: (1) observation and characterization of the CL kinematics; and (2) evaluation of CL driving forces.

In this paper, we report the results related to the first project, namely, a characterization of the fatigue CL growth in the vicinity of a hole. This setup allows one to observe in one experiment a damage formation in a vicinity of the crack tip under variable stresses.

2. MATERIAL AND EXPERIMENTAL SETUP

Commercial polystyrene obtained from Transilwrap Corporation (Cleveland, Ohio) was used in this study. Rectangular 0.25mm sheets were cut and machined to single edge notched specimens with 80mm gauge length and 20mm width. A 60 degree V-shaped notch was milled into the edge to 1mm depth at the midspan of the gauge length.

The tension-tension fatigue experiments were conducted at a 1.1 Ton (2.5 Kip) capacity servohydraulic Instron Testing System at room temperature. Sinusoidal waveform loading with frequency 0.65 Hz, maximum stress 17.0 Mpa (= 0.44 σ_f) and load ratio ($\sigma_{max}/\sigma_{min}$) 0.1 were employed.

The damage evolution and crack propagation were observed and recorded using Hamamatsu video system attached to a Questar long range travelling microscope.

After testing, specimens were polished to a thickness of 20-30μm. The density and orientation of crazes within AZ were analyzed by using a Zeiss microscope (with transmission light).

3. RESULTS

3.1 Kinematics of the Main Crack

The CL propagates in a rectilinear fashion until it enters the domain of the stress field excited due to the presence of a hole. At this stage CL begins to curve. This is illustrated in Fig. 3 by superposing in one plot three crack trajectories traced from different specimens. The trajectories begin at differently positioned notches (reference line at the distance of 5/3 R, 7/3 R, and 9/3 R from the hole center, respectively; the hole radius R being 1.5mm). Figures 4 - 6 show the speed of the main crack along each of the three trajectories shown in Fig. 3.

3.2 Crack-Damage Interaction

Intense crazing within AZ appears as a response to the stress concentration at the crack tip. On the other hand, AZ extends ahead of the main crack. Therefore, the crazes at the AZ leading edge "sense" the changes in the stress field earlier than the main crack. As a result, the AZ rotates prior to the main crack rotation. The last qualitative statement is illustrated in Fig. 7 where the curvatures of the main crack and of the CL are shown along the crack trajectory. One observes (in the curvature plot, Fig. 7) that the CL turns first and the main crack follows. Notice the apparent oscillation of the CL curvature relative to the main crack curvature, as if the AZ guides the main crack with a strong feedback

3.3 Characterization of AZ Morphology

We characterize the AZ morphology by the craze density $\rho(\underline{x})$ and the average AZ craze orientation ω. Specifically, the craze density at a point $\underline{x}$ is the total area of the craze middle planes within an elementary volume V (its center at $\underline{x}$) divided by V, i.e. $\rho(\underline{x}) = A(V)/V$, thus the dimension of ρ is m^2/m^3. The size of the elementary volume is chosen much smaller than the typical craze size, thus all crazes are parallel within V. Craze orientation at $\underline{x}$ is the angle $\omega(\underline{x})$ between the crazes in V and the direction tangent to the crack at its tip (x_1-direction of the local coordinate system).

69

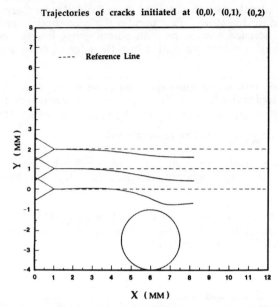

Fig. 3   A superposition of the main crack trajectories in different specimens grown from the notches positioned at various distances from the hole.

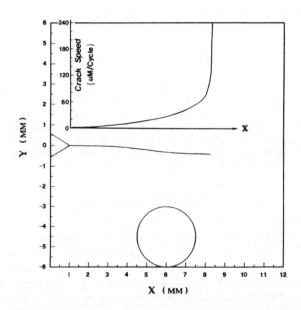

Fig. 4   The farthest of the hole crack trajectory from Fig. 3 is shown together with the plot of the crack speed along it. The crack speed is monotonically increasing with the crack length.

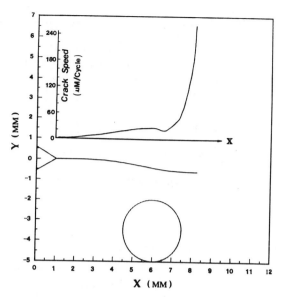

Fig. 5   The middle crack trajectory from Fig. 3 is shown together with the plot of the crack speed along it. The crack speed is reduced in the "shadow" (for tensile stress) above the hole.

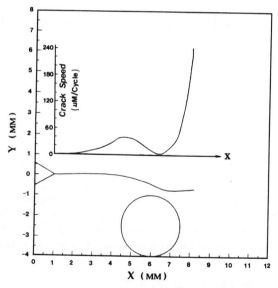

Fig. 6   The closest to the hole crack trajectory from Fig. 3 is shown together with the plot of the crack speed along it. The corresponding CL is presented in Fig. 2. Notice that the crack slows down and then is arrested ($\dot{l} \approx 0$)for a short time above the hole. Furthermore, one can see from Fig. 2 an order of magnitude drop in average crack speed at this position.

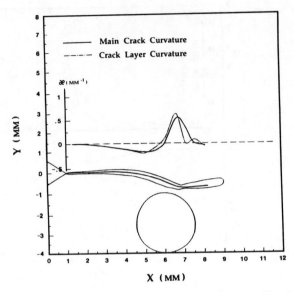

Fig. 7   The closest to the hole crack trajectory from Fig. 3 is shown together with the envelope of CL. The curvatures of the crack trajectory and the middle line of the CL are plotted above.

We introduce the following quantitative measure of the total crazing within AZ:

$$R = \int\limits_{V_{AZ}} \rho(\underline{x}) \, dV \quad ,$$ (1)

and the average craze orientation

$$\omega = \frac{1}{R} \int\limits_{V_{AZ}} \omega(\underline{x})\rho(\underline{x}) \, dV.$$ (2)

As an illustration, let us consider the morphology of the AZ formed at 790 cycles (Fig. 2). A micrograph of this zone is shown in Fig. 8.a. To characterize the craze distribution, we introduce integral parameters such as R and higher order "inertia moments" instead of craze density $\rho(\underline{x})$. The "gravity center" x_{ci} and the second order "central inertia moment" I_{ij} are conventionally defined:

$$x_{ci} = \frac{1}{R} \int\limits_{V_{AZ}} x_i \rho(\underline{x}) \, dV \quad , \qquad i=1,2$$ (3)

$$I_{ij} = \frac{1}{R} \int\limits_{V_{AZ}} (x_i - x_{ci})(x_j - x_{cj})\rho(\underline{x}) \, dV \quad , \qquad i,j=1,2$$ (4)

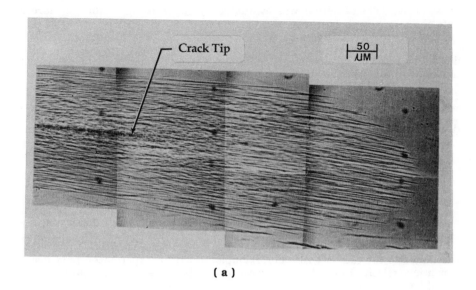

(a)

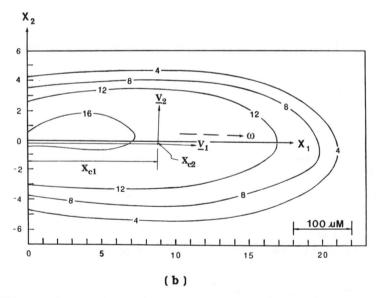

(b)

Fig. 8   (a) A micrograph of the crazed zone ahead of the crack tip immediately after 790 cycles. (b) An outline of the AZ from the above micrograph; shown are: contours of equal craze density, position of the AZ gravity center $\underline{x}_c$, eigenvectors $\underline{v}_1, \underline{v}_2$ of the AZ central inertia moment matrix $\underset{\sim}{I}$, and the average AZ craze direction (dashed line)

In Fig. 8.b, the contours of equal level of the craze density $\rho(\underline{x})$ are shown together with the location of $\underline{x}_c$, the eigendirections of $\underline{I}$ and the average craze direction ω. (The density ρ was evaluated by superposing a 10mm square grid onto the micrograph from Fig. 8.a printed at X400 magnification.) Comparing consecutive AZ positions, one can evaluate the rates $\dot{\underline{x}}_c$, $\dot{\underline{I}}$, and $\dot{\omega}$, and use these to derive the rates of AZ elementary motions. The methodology of the derivation is described in the next section.

3.4 Kinematics of AZ

The self-similarity assumption regarding AZ evolution [2,3,5] states that the damage distributions within the AZ at times t and $t + \Delta t$ are related by an affine transformation:

$$\rho_{t+\Delta t}(\underline{x}) = \rho_t(\underline{A}^{-1}_{t,\Delta t} \cdot (\underline{x} - \underline{x}_{c,t+\Delta t}) + \underline{x}_{c,t}) \tag{5}$$

Here we use local coordinates attached to the crack tip as shown in Fig. 9; $x_{c,t}$, $x_{c,t+\Delta t}$ are the coordinates of the AZ "gravity center" at the times t and $t + \Delta t$ respectively; $\underline{A}_{t,\Delta t}$ is a 2X2 matrix, which characterizes the AZ deformation from the time t to the time $t + \Delta t$. For example, if the AZ experienced pure isotropic expansion (relative to the AZ gravity center), and became 1.1 times larger, then we would have

$$A = \begin{bmatrix} 1.1 & 0 \\ 0 & 1.1 \end{bmatrix}.$$

A straightforward calculation yields the relation between the AZ inertia moment matrices at t and $t + \Delta t$:

$$\underline{I}_{t+\Delta t} = \underline{A}_{t,\Delta t} \underline{I}_t \underline{A}^T_{t,\Delta t} \ , \tag{6}$$

where $\underline{A}^T$ is the transpose of $\underline{A}$.

For a short time increment Δt, the AZ deformation is close to the identity transformation and can be written as

$$\underline{A}_{t,\Delta t} = \underline{1} + \dot{\underline{A}}_t \Delta t \ , \tag{7}$$

where $\underline{1} = \begin{bmatrix} 1 & 0 \\ 0 & 1 \end{bmatrix}$ is the identity matrix. From Eqs. (6) and (7), we find:

$$\dot{\underline{I}}_t = \frac{1}{\Delta t}(\underline{I}_{t+\Delta t} - \underline{I}_t) = \dot{\underline{A}}_t \underline{I}_t + \underline{I}_t \dot{\underline{A}}^T_t \ . \tag{8}$$

One evaluates the components of the matrices $\underline{I}_t$ and $\underline{I}_{t+\Delta t}$ experimentally. The matrix $\dot{\underline{A}}$ is to be determined. For simplicity, consider the AZ transformation $\underline{A}_{t,\Delta t}$ consisting only of rotation and dilations in the tangent and normal directions of the crack at its tip (i.e. in the local coordinate directions). This implies that the matrix $\overset{\cdot}{\underset{\sim}{A}}_t$ has the following form:

$$\overset{\cdot}{\underset{\sim}{A}}_t = \begin{bmatrix} \dot{a}_1(t) & \dot{\theta}(t) \\ -\dot{\theta}(t) & \dot{a}_2(t) \end{bmatrix} \tag{9}$$

Thus the rates of elementary AZ movements are: $\dot{e} = \frac{1}{2}(\dot{a}_1 + \dot{a}_2)$ for AZ isotropic expansion, $\dot{d} = \frac{1}{2}(\dot{a}_1 - \dot{a}_2)$ for AZ distortion as well as $\dot{x}_{c1}$, $\dot{x}_{c2}$ for AZ translations parallel and perpendicular to the main crack, respectively. In addition to these, the rate $\dot{\omega}$ characterizes the rotation of average craze orientation and $\dot{\theta}$. characterizes the rotation of eigendirections of AZ central inertia tensor $\underset{\sim}{I}$. Finally, the list of rates characterizing the evolution of the entire CL consists of the above $\dot{e}, \dot{d}, \dot{\underset{\sim}{x}}_c, \dot{\omega}, \dot{\theta}$ together with the rate of the main crack length growth, $\dot{\ell}$, and the rate of the main crack rotation, $\dot{\omega}_o$ (i.e. rotation of the tangent to the main crack at its tip).

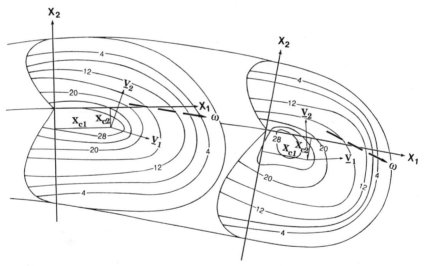

Fig. 9   Two consecutive AZ configurations formed at 813 and 820 cycles (see also Fig. 2). The contours of equal craze density (solid lines), the position of the local coordinate system (x_1, x_2) (its origin at the crack tip, x_1-direction is tangent to the crack), position of the AZ gravity center $\underset{\sim}{x}_c$, eigenvectors $\underset{\sim}{v}_1, \underset{\sim}{v}_2$ of the AZ central inertia moment $\underset{\sim}{I}$, and the average AZ craze direction ω (dashed arrow) are shown.

3.5 Example

Figure 9 shows two AZ positions separated by 7 cycles or $\approx$11sec ($\Delta t = \Delta N/\nu = 7/0.65$), and the corresponding craze density distributions, craze orientations, etc. For each of the two positions, the total amount of damage R, the components of the gravity center $\underset{\sim}{x}_c$, the AZ central inertia tensor $\underset{\sim}{I}$, and etc. are shown in Table 1.

Table 1

Quantities	Position 1	Position 2
No. of cycles, N	813	820
Crack length, l [mm]	4.66	5.08
Crack orientation (relative to the reference line), ω_o [rad]	0.21	0.33
Total amount of damage in AZ, R [mm^2]	18.5	15.7
AZ gravity center (in local coordinates), $\underline{x}_c$ [μm]	(139,-48)	(113,-58)
Components of AZ central inertia tensor, $\underset{\sim}{I}$ [10^{-3} mm^2]	$\begin{bmatrix} 9.3 & -1.3 \\ -1.3 & 6.2 \end{bmatrix}$	$\begin{bmatrix} 5.9 & -0.2 \\ -0.2 & 7.1 \end{bmatrix}$
Average AZ craze orientation (in local coordinate) ω [rad]	0.17	0.29

Table 1 represents the various parameters characterizing CL of its propagation for two specific stages: one immediately after 813 cycles, another immediately after 820 cycles. The "global coordinate system" is attached to the specimen, as shown in Figs. 4 - 6. The "local coordinate system" is attached to the crack tip, as shown in Fig. 9.

The average (over ≈11 seconds) rates of elementary CL motions evaluated based on Table 1 are presented in Table 2.

Table 2

Rates	Average Values
Crack growth	$< \dot{l} > = (l^{(2)} - l^{(1)})/\Delta t = 39$ μm/sec
Crack tangent rotation	$< \dot{\omega}_o > = (\omega_o^{(2)} - \omega_o^{(1)})/\Delta t = 0.011$ rad/sec
AZ translation	$< \dot{x}_{c1} > = (x_{c1}^{(2)} - x_{c1}^{(1)})/\Delta t = -2.4$ μm/sec $< \dot{x}_{c2} > = (x_{c2}^{(2)} - x_{c2}^{(1)})/\Delta t = -0.94$ μm/sec
Eigendirection of the AZ central inertia tensor	$< \dot{\theta} > = -0.030$ rad/sec
AZ expansion	$< \dot{e} > = -0.004$ sec^{-1}
AZ distortion	$< \dot{d} > = -0.017$ sec^{-1}
Average craze orientation (in local coordinate)	$< \dot{\omega} > = (\omega^{(2)} - \omega^{(1)})/\Delta t = 0.011$ rad/sec

4. SUMMARY AND CONCLUSIONS

1. Evolution of CL in the vicinity of a hole (i.e. in heterogeneous stress field) is analyzed. The methodology of CL characterization is demonstrated by evaluating the rates of Active Zone translation, rotation, etc.

2. Crack-damage interaction is a controlling factor for the Crack Layer evolution. This is illustrated by comparative analyses of crack trajectories,

crack growth rates along the trajectories, and the curvatures of the CL and the main crack.

3. The stress state produces a noticeable effect on CL growth. It is clearly illustrated by the crack arrest (Fig. 6) taking place in the close vicinity of the hole, "shielded" from applied tension.

4. There are energy release rates associated with each elementary motion of the AZ as well as with the motion of the crack tip. For example, the energy release rates associated with translations of the AZ (i.e. increments in $\underline{x}_c$) can be expressed in terms of the Eshelby's energy-momentum tensor P_{ij} (see [8] for details) :

$$J_i^{AZ} = \int_{\partial V_{AZ}} P_{ij}\, n_j d\Gamma \ , \qquad i, j = 1, 2$$

where ∂V_{AZ} is the boundary of the AZ. It is worthy to note that J_2, associated with the AZ translation normal to the crack trajectory, becomes as significant as J_1, well known in conventional fracture mechanics.

5. The methodology presented above is an important step toward the formulation of the kinetic equations of CL growth, i.e. correlation between the fluxes ($\dot{\underline{x}}_c$, $\dot{\theta}$, etc.) and the corresponding force.

ACKNOWLEDGMENTS

Financial support from the Air Force Office of Scientific Research (Contract No. AFSOR-89-0105) and Dow Chemical Company, (Texas, Polycarbonate Research) is gratefully acknowledged.

REFERENCES

1. A.Chudnovsky, K. Choui, and A. Moet, "Curvilinear Crack Layer Propagation," Journal of Materials Science Letters , 6, 1987, pp. 1033-1038.

2. A. Chudnovsky, "The Crack Layer Theory," NASA Contractor Report , 174634, 1984.

3. A. Chudnovsky, "Crack Layer Theory," Proceedings of the 10th U.S. National Conference on Applied Mechanics, J. P. Lamb, Ed., ASME, 1986, pp. 97.

4. J. Botsis and X.Q. Zhang, "On the Kinematics of Crack-Damage Evolution During Fatigue Fracture," International Journal of Fracture, to appear.

5. J. Botsis and B. Kunin, "On Self-Similarity of Crack Layer", International Journal of Fracture, Vol. 35, 1987, pp. R51-R56.

6. N. Haddaoui, A. Chudnovsky, and A. Moet, "Ductile Fatigue Crack Propagation in Polycarbonate," Polymer, Vol. 27, 1986, pp.1377-84.

7. A. Chudnovsky, A. Moet, R. J. Bankart, and M. T. Takemori, "Effect of Damage Dissemination on Crack Propagation in Polypropylene," Journal of Apply Physics, Vol. 54, No. 10, 1983, pp. 5562-67.

8. A. Chudnovsky and Shaofu Wu, "Effect of Crack-Microcracks Interaction on Energy Release Rates," International Journal of Fracture, to appear.

CRAZE DISTRIBUTION WITHIN A CRACK LAYER IN POLYSTYRENE

X. Q. Zhang, B. L. Gregory, and J. Botsis
Department of Civil Engineering, Mechanics, and Metallurgy
University of Illinois at Chicago
Chicago, Illinois

ABSTRACT

Results of analysis on craze distribution along the trailing edge and within the active zone of a crack layer in fatigued polystyrene thin specimens are reported. It is shown that: (i) the distributions of crazes along the trailing edge of the active zone are related by a scaling parameter; (ii) the craze density along the trailing edge as well as within the active zone is constant; (iii) the weighted center of gravity of the active zone is found to be ahead of the crack tip. The distance of the center from the crack tip increases with crack length; (iv) the transformation of points with equal craze density within the active zone is found to be linear during the initial phases of the fracture process. The results agree with the kinematics of damage proposed by the crack layer model.

INTRODUCTION & BACKGROUND

Experimental investigations reported in the literature during the last decade have shown that a zone of intense damage accompanies the process of crack propagation under fatigue and creep. This zone is usually called a process zone [1,2], active zone [3], etc. The size of an active zone is usually small as compared to the crack length and the specimen width [4 - 8]. The energy spent for its formation however, may be substantially greater than the energy spent for crack growth. This effect is usually manifested in the deceleration of a large crack and the history dependence of conventional fracture toughness parameters. Consequently, the nature of damage and its extent, by and at large, determine the useful time of a structure and the toughness of the material.

Experimental observations in different materials has shown that the evolution of a damage zone displays characteristics which are independent of the particular damage elements [4-8]. These observations suggest that the evolution of a crack and the surrounding damage may be described in phenomenological terms. Thus instead of attempting to model a fracture process in terms of the kinetics of a damage field, it is proposed to employ integral damage parameters to characterize a crack and its associated damage zone as a macroscopic entity, a crack layer (CL), and introduce the corresponding driving forces using general principles of irreversible thermodynamics [3,9].

A schematic of a CL is shown in Fig. 1. CL consists of an active zone and an inert zone. The active zone is defined as the part of the CL where the time rate of a damage parameter is nonzero; $\dot{\rho} > 0$. The inert zone is the part of the CL complementary to the active zone with $\dot{\rho}=0$ and $\rho>0$. The two zones are separated by the trailing edge.

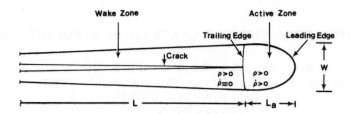

Fig. 1. A schematic of a crack layer

The CL model characterizes a fracture process in terms of the constitutive response of an active zone. Accordingly, the kinematics of an active zone are essential. At present, the kinematics are deduced from a hypothesis of self similarity of damage evolution. This hypothesis states that points with equal damage density, transform according to a time dependent affine transformation of the space variables. Thus assuming that crack propagation coincides with translation of the active zone, a fracture process is described in terms of translation, rotation and homogeneous deformation of the active zone.

On the basis of the of the aforementioned hypothesis, points with equal level of damage at two consecutive configurations of the active zone, are related with a linear time dependent transformation of the space variables [9,10] given by:

$$\rho_t(D_t(\underline{x}_t-\Delta l_t))=\rho_{to}(\underline{x}_{to}) \tag{1}$$

Here ρ_t is the value of a damage parameter at a point $\underline{x}_t$ at time t; D_t is a 2x2 matrix for a plane problem. Its coefficients depend on time, material structure and loading history; Δl_t is the crack length increment. Matrix D_t can be uniquely decomposed into a product of three time dependent matrices: a scalar matrix (expansion of an active zone), a symmetric positive definite matrix with determinant equal to one (distortion of an active zone), and an orthogonal matrix (rotation of the active zone). Accordingly, the evolution of a damage field $\rho_t(\underline{x})$ can be reduced to the evolution of six scalar parameters for a plane problem; four components of the matrix D_t and two components of the vector Δl_t.

An important implication of the above decomposition is that it allows for the expression of the energy release rates associated with the translation, rotation, isotropic expansion and distortion of an active zone in the form of J_i, L, M, and N_{ij} integrals [3,9]. Analysis of process zone kinematics and identification of the corresponding energy release rates have also been reported in [11]. Reference [11] treats a process zone as a homogeneous solid capable of undergoing rigid body motion and homogeneous deformation. The results are similar to the CL approach described in [12,13]. However, there are no solid experimental results to justify the particular form of the self similarity hypothesis of damage evolution.

In this paper we report preliminary experimental results and analysis of crazing distribution in polystyrene(PS) single edge notched specimens subjected to fatigue fracture. The experimental results consist of crazing measurements along the trailing edge and within the active zone while the analysis involves a quantitative comparison of histograms of crazes along the trailing edge and within the active zone. The implications of the results on the kinematics of damage proposed by the CL model are discussed.

EXPERIMENTAL PROCEDURES

Commercial PS obtained from Transilwrap Corporation (Cleveland, Ohio) is the material employed in these studies. The specimen dimensions are: gauge length=80mm, width=20mm, and thickness=0.25mm. A $60°$ degree V-shaped notch 1mm in depth is milled into the edge of each test piece at the mid-span of the gauge length. Details of the specimen preparation can be found in [14].

Rectilinear fatigue crack propagation experiments are conducted an a dual actuator servohydraulic Instron Mechanical Testing System at ambient temperature in a laboratory atmosphere. The input loading wave form is a sinusoidal function of frequency 1.10Hz, fixed to yield a maximum stress of 15.2MPa and a load ratio of 0.28.

Craze accumulation and crack growth are observed by means of a Questar long range microscope. The fracture process is recorded in real time using a Hamamatsu video system (recording speed 30 frames/s) which is attached to the Questar microscope. Damage analysis is undertaken to characterize the density and distribution of crazes along the trailing edge and within the damage zone. Sections parallel to the specimens are prepared for observation by employing standard metallographic grinding and polishing techniques [15].

RESULTS AND DISCUSSION

Figure 2 illustrates the evolution of active zone width w and length l_a with crack length l (Fig. 2a), and the variation of their ratio w/l_a as a function of dimensionless crack length l/l_c (Fig. 2b), (l_c being the critical value of l). Note that both w and l_a increase monotonically with l. Moreover, the changes of w with l decrease and approach zero near the critical CL propagation (Fig. 2a).

The increase in w, l_a and the changes in w/l_a with the crack length suggest, at first, that during rectilinear CL propagation fracture occurs by the translation, isotropic expansion, and distortion of the active zone. However, craze distribution within the active zone is non-homogeneous. Thus identification of the kinematic parameters should result from the study of the evolution of the craze distribution.

It has been shown [14] that during rectilinear CL propagation in PS, crazing within an active zone is uniformly distributed in the thickness direction, and increases monotonically toward the fracture surfaces. In addition, it is found that crazing is symmetrically distributed around the crack path. This is attributed to the symmetries of specimen, applied load and the homogeneity of the material. Hence analysis of the craze distribution can be performed on the symmetric half of a CL. Moreover, no changes in orientation of crazes with respect to the crack path is observed. That is no rotation of the active zone occurs. Thus a scalar parameter suffices to characterize damage. Accordingly we represent crazing density $\rho[mm^2/mm^3]$, as the total area of craze middle planes per unit test volume.

Craze measurements are carried out on sections parallel to the plane of the specimen. These sections are prepared by standard metallographic and polishing techniques. Optical micrographs of these sections are covered by a mesh of rectangles whose size is approximately 12x12mm. In every rectangle, the number of intersections of crazes with a vertical test line is counted [16]. Craze density is obtained by use of the formula: $\rho=nbt/abt$, where ρ represents the amount of area of mid-planes of crazes per unit test volume; n is the number of intersections of crazes with the vertical test line at the respective rectangle; a,b are the height and width of a rectangle, respectively, and t is the thickness of the specimen.

Characterization of damage distribution is carried out along the trailing edge and within the active zone at different configurations. Note that an accurate definition of the trailing edge requires a detailed analysis for the stresses and strains of a crack surrounded by an array of crazes. Attempts toward this goal have been recently reported [17,18]. Here, we approximate the trailing edge by a straight line through the crack tip and normal to the crack growth direction.

Figure 3 represents the histograms of craze distribution in vertical cross sections at seven locations along the crack path. Craze distributions are examined by comparing their central moments. As a result of the present symmetry, the odd moments of the distributions are negligible. Thus from the practical standpoint it suffices to compare their second and fourth moments only [19].

Figure 4, shows the ratio of the second moment μ_2, to the square root of the fourth moment $\sqrt{\mu_4}$ (Fig 4a), of the distributions in Fig. 3, with the crack length. An average crazing density $<\rho>$ is defined as the quotient of the total number of crazing along the trailing edge of the active zone to the respective width of the active zone. The evolution of $<\rho>$ is plotted in Fig. 4b as a function of the crack length. Within experimental error, both $\mu_2/\sqrt{\mu_4}$ and $<\rho>$ are constant for the entire range of CL growth. Their constancy suggests that the distribution of crazes along the trailing edge are related by a scalar parameter.

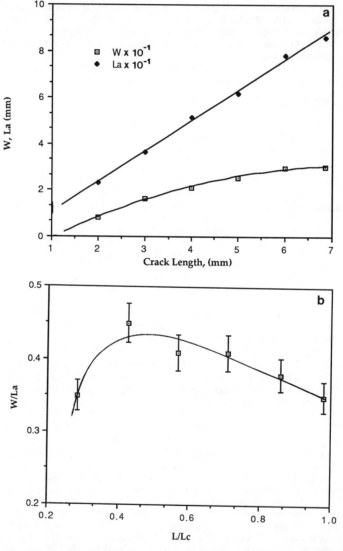

Fig. 2. (a) Evolution of active zone width w and length l_a with crack length. (b) w/l_a plotted against dimensionless crack length.

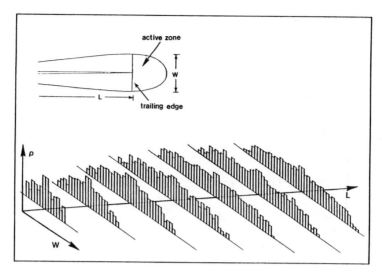

Fig. 3. Histograms of craze density in vertical cross sections at six locations along the crack path.

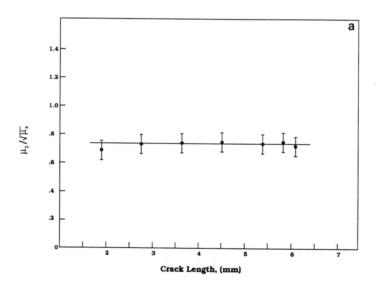

Fig. 4a. Evolution of the ratio $\mu_2/\sqrt{\mu_4}$ with crack length.

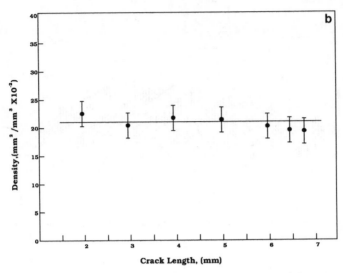

Fig. 4b. Average craze density $\langle\rho\rangle$, along the trailing edge of the active zone plotted against the crack length.

Measurements of craze density within the active zone at three consecutive configurations yield the contours of equal crazing density shown in Fig. 5. The polygonal contours indicate the experimental measurements and the continuous lines their elliptical approximations with reference to the coordinate system attached at the center of gravity of the zone.

Analysis of the experimental data shows that the center of the active zone for the three configurations examined in this study is 0.103, 0.165, and 0.206mm, respectively. These results suggest that crack growth does not coincide with the movement of the active zone. Thus an additional kinematic parameter should be introduced to describe a fracture process in PS. For a zone of area A, an average craze density $\langle\rho_o\rangle$, is defined as, $\langle\rho_o\rangle=\Sigma/A$, where Σ is the total number of crazes within the zone.

For the three configurations of the active zone displayed in Fig. 5, the evolution of $\langle\rho_o\rangle$ as a function of crack length is shown in Fig. 6. Within experimental error, the data in Fig. 6 indicate that $\langle\rho_o\rangle$ is constant.

Although, the constancy of $\langle\rho\rangle$, $\langle\rho_o\rangle$, and the relation by a scaling parameter of the distributions along the trailing edge can be regarded as necessary conditions for the transformation of points with equal craze density to be linear(relation 1), they are not sufficient.

A complete experimental characterization of the kinematics of a damage zone requires comparison of damage distribution within the zone at different configurations. In the particular case under study, one can examine the evolution of the ratios, $\lambda_{ij}=a_{jk}/a_{ik}$ and $\mu_{ji}=b_{jk}/b_{ik}$, where a_{ik}, b_{ik} are the major and minor axes of an ellipse, in the coordinate system attached at the center of gravity of the zone. The indices i,j, and k, refer to the configurations and contour number, respectively.

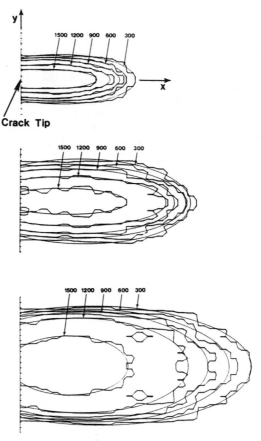

Fig. 5. Contours of equal damage density (polygonal lines) $\rho[mm^2/mm^3]$, and their elliptical approximations (continuous lines).

Values of λ_{ji} and μ_{ji} corresponding to the contours of equal damage density from the first configuration to the second one, and from the second to the third configuration, are shown in Fig. 7a and 7b, respectively.

The data in Fig. 7a indicate that both λ_{21} and μ_{21} are constant for the contours with density, $\rho=300$, 600, 900, and 1,200 mm^2/mm^3. Deviations are observed in the ratios of contours with the highest density ($\rho=1,500mm^2/mm^3$). It is likely however, that the experimental error associated with the measurements of crazing density in the close proximity of the crack tip is relatively large. This is because the high craze density there limits optical microscopy resolution and consequently renders accurate measurements difficult.

The constancy of λ_{21} and μ_{21} (Fig. 7a) indicates that from the first to the second configuration the evolution of damage can be represented by (1). Matrix D_t is a symmetric matrix since no rotation of the active zone takes place. Furthermore, the system of coordinates employed in the analysis constitutes the principal system. Thus D_t is diagonal with elements $d_{11}=1/\lambda_{21}$ and $d_{22}=1/\mu_{21}$. Moreover, λ_{21} and μ_{21} are practically equal (Fig 7a). This implies that between the first and second configurations the active zone evolves by translation and isotropic expansion, the distortion being insignificant. Accordingly, the energy release rates are given by the J_1 and M integrals [3,9].

Inasmuch as the transformation of points with equal damage is given by (1), the coefficients λ_{21} and μ_{21} can also be obtained from the inertia moments $I_{\bar{x}}$ and $I_{\bar{y}}$

85

along the axes parallel to x, y (Fig. 5) intersecting at the center of gravity of the zone. Indeed, the analysis shows that, $I_{\bar{x}}^{(2)}/I_{\bar{x}}^{(1)}=\lambda_{21}^2$ and $I_{\bar{y}}^{(2)}/I_{\bar{y}}^{(1)}=\mu_{21}^2$ (the superscripts refer to configuration). Hence, the coefficients of the transformation can be obtained from the contour analysis or from the ratios of moments of inertia of the zone.

Between the second and third configurations the data suggest that the transformation of the contours with equal density cannot be approximated by a linear function. This result however, should be considered only approximate for the crack increment between these two configurations is rather large as compared to the size of the active zone ($\Delta l \sim 1.2$, $l_a \sim 1$mm). Therefore additional configurations between the second and the third one should be examined.

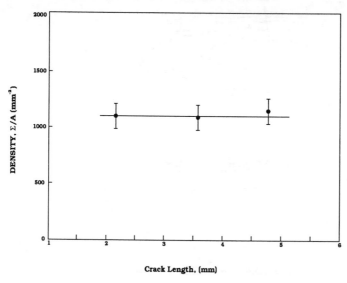

Fig. 6. Evolution of $<\rho_o>$ as a function of the crack length.

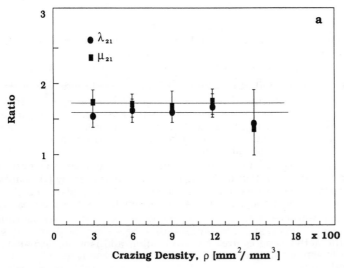

Fig. 7a. Variation of the ratios λ_{21} and μ_{21} as a function of craze density (see text for details).

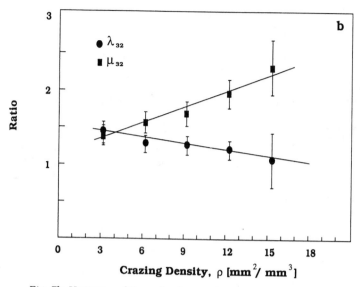

Fig. 7b. Variation of the ratios λ_{32} and μ_{32} as a function of craze density (see text for details).

CONCLUDING REMARKS

Analysis of experimental results on craze distribution along the trailing edge and within the active zone of a CL in PS show that; (i) the distributions of crazes along the trailing edge of the active zone are related by a scaling parameter; (ii) the craze density along the trailing edge as well as within the active zone is constant; (iii) the center of the active zone is found to be ahead of the crack tip. The data suggests that crack growth and active zone movement do not coincide; (iv) the transformation of points with equal craze density within the active zone is found to be linear during the initial phases of the fracture process. The results are in favor of the hypothesis of damage evolution proposed by the crack layer model [3,9].

Damage accumulation however, is a loading and material dependent process. Therefore, in order to make general statements regarding its evolution, experimental data under a wide spectrum of loading conditions and for a variety of material structures should be available.

ACKNOWLEDGEMENT

The authors wish to acknowledge the partial financial supports from the Department of Civil Engineering, Mechanics and Metallurgy and the Research Board of the University of Illinois.

REFERENCES

1. Burech, F. E., in <u>Fracture 1977</u>, Vol.3, Pergamon Press, London, 1977, p. 929.
2. Claussen, N., <u>Journal of the American Ceramic Society</u>,1976, Vol.56, p. 49.
3. Chudnovsky, A., 'Crack Layer Theory', 1984, NASA Contractor Report 17463.
4. Botsis, J. ,A. Chudnovsky and A. Moet, <u>XXIX ANTEC-SPE,</u> Chicago IL, 1983, p. 444.

5. Chudnovsky, A., Moet, A., Bankert, R. J., and Takemori, T. M., Journal of Applied Physics, 1983, Vol. 54, p. 5567.
6. Chudnovsky, A., and Bessendorf, M., 'Crack Layer Morphology and Toughness Characterization in Steels', 1983, NASA Contractor Report.
7. Haddaoui, N., Chudnovsky, A., and Moet, A., Polymer, 1985, Vol. 27, p. 1377.
8. Nguyen, P., and Moet, A., Journal of Vinyl Technology, 1985, Vol. 7, p. 140.
9. Chudnovsky, A., Proceedings of the 10th U.S. National. Conference on Applied Mechanics, 1986, J. P. Lamb Ed. ASME, p. 97.
10. Botsis, J. ,and Kunin, B., International Journal of Fracture, 1987, Vol. 35, p. R51.
11. Aoki, S., Kishimoto, K.,and Sakata, M.,Journal of Applied Mechanics, 1981, Vol. 48, p. 825.
12. Chudnovsky, A., Dunaevsky, V. ,and Khandogi, V. A., Archives of Mechanics, 1978, Vol. 165, p. 30.
13. Khandogi, V. A. , and Chudnovsky,A., in 'Dynamics and Strength of Aircraft Structures' 1978, K. M. Kurshin, Ed., (in Russian), Novosirbisk, p. 148
14. Botsis, J., Chudnovsky, A., and Moet, A., International Journal of Fracture, 1987, Vol. 33, p. 263.
15. Holik, A. S., Kambour, R., P., Fink, D., G.,, and Hobbs, S. Y., in Microstructural Sciences, 1979, Vol. 7, LeMay, Fallon and McCalls Eds, Elsevier North Holland, p. 357.
16. Underwood, E. E., in Quantitative Stereology, 1970, Addison-Wesley, p. 23.
17. Chudnovsky, A., and Ben Ouezdou, M., International Journal of Fracture, 1988, Vol. 37, p. 3.
18. Chudnovsky, A., and Wu S., 'Crack Layer Translational Energy Release Rates', International Journal of Fracture, 1989, to appear.
19. Pease, E., M., J., and G. P. Wadsworth, G., P., in Calculus with Analytic Geometry', 1968, The Ronald Press Company, New York.

FRACTURE OF SHORT FIBRE COMPOSITES WITH ENHANCED FIBRE LENGTH DISTRIBUTIONS

M. J. Carling
Polymer Products Department
E. I. DuPont de Nemours Incorporated
Wilmington, Delaware

J. G. Williams
Department of Mechanical Engineering
Imperial College
London, United Kingdom

ABSTRACT

Short glass fibre reinforced Nylon composites were prepared from feedstock made by either an extrusion or a pultrusion process. The material made by the pultruded process had a significantly improved fibre length distribution compared to the extruded feedstock material. The pultruded material used thicker fibres however. We examined the effects of competition between the beneficial increase in fibre length and the detrimental increase in fibre radius on the fracture properties of the material. The toughness of the material was measured over a wide range of loading rates from impact, through Kc and Gc type tests to slow crack growth tests. The moisture content of the materials was varied from dry to saturated (equivalent to 100% rel. humidity). It was found that there was little difference in fracture toughness when the materials were dry. However when moist the pultruded material showed significant improvements over the extruded material. This behavior is explained in terms of the fibre length distributions. When dry there is little difference in the proportion of fibres longer than the critical length. As the critical length increases when wet though, the proportion of fibres above the critical length does not fall as rapidly in the pultruded material as in the extruded material (a result of the different shape of the fibre length distribution curves) and so the pultruded material exhibited much improved wet properties.

INTRODUCTION

Much of the work done to improve the properties of short fibre composites has concentrated on increasing the lengths of the fibres in the finished article. Fibre composites with enhanced fibre length distributions are now commercially available and we shall examine these materials in comparison to conventional short fibre composites to determine what effect changes in fibre length distribution have . We shall first review some of the relevant literature on the effect of fibre length on composite properties.

The way in which the stress field in a composite is modified by the presence of a stiffer fibre was first described by Cox (1952) and by Dow (1963) in the shear lag analysis. Cox showed that the shear stress along the fibre matrix interface, which transfers the load to the fibre, is at a maximum at the ends of the fibres. The shear stress falls to essentially zero a distance along the fibre when the matrix and fibre strains become equal. Subsequent calculations of the stress fields around short fibres have been mainly concerned with quantifying the simplifying assumptions made in the Cox model - the basic result remains however. Smith and Spencer (1970) repeated the calculations without assuming equal transverse moduli and showed that the shear stress concentration at the fibre end was more severe than that predicted by Cox. More recently Termonia (1984), using computational methods, was able to take into account the effect of stress concentration at the sharp corners of the fibres and showed that up to 25% of the tensile stress in the fibre could come from stress transfered at the ends of the fibres.

Experimental verification of these results was first provided by Tyson and Davies (1965) using Dural bars embedded in epoxy. Galiotis et al. (1984) and Robinson et al. (1987) were able to directly measure the stresses in polydiacetylene fibres embedded in epoxy using an elegant Raman spectroscopy technique. All of these workers found that although the pattern of stress distribution was that predicted by the analytical solutions, the magnitude of the stress concentration at the fibre ends was higher than expected and that non linear visco-elastic and plastic deformation needed to be considered.

From this type of stress distribution a critical fibre length can be postulated. This is a length such that sufficient force can be transferred across the fibre / matrix interface to fracture the fibre. For a given fibre diameter, if the fibre is less than the critical length there is not sufficient surface area for the shear stress to transmit the required load and the fibre will only debond from the matrix. The critical length is proportional to the diameter of the fibre and so for fibres of different diameters it is better to talk of a critical aspect ratio (length / diameter) than a critical length.

The stiffness of a short fibre composite should increase with increasing fibre length until the stiffness equals that of a continuous fibre composite (at very long fibre lengths). This improvement in stiffness should be predicted by the Halpin Tsai equations (1969). In practice increases in fibre length have given either no improvements (Ramsteiner, 1981; Dingle, 1974) or disappointing improvements (Folkes and Kells, 1985) in the stiffness.

An expression for the effect of fibre length on the strength of short fibre composites was developed by Bowyer and Bader(1972). Their model predicts that the strength of short fibre composites should increase with fibre length in a similar way as the modulus . By using a very wide range of fibre lengths Hancock and Cuthbertson (1970) showed that this behavior indeed occurred. In more commercial type materials few improvements have been found (Curtis et al., 1985), this was attributed to the lack of any improvement to the ease with which cracks initiated at the fibre corners.

The first work on finding the optimum fibre length for toughness was done by Cottrell (1964) who showed that the optimum length was one just less than the critical length of the fibre. Below the critical length the toughness should increase with the square of the fibre length. This work was extended by Kelly (1970) to fibre lengths greater than the critical length. Above the critical length the toughness was proportional to the reciprocal of the fibre length. Experimental work by Alfred and Schuster confirmed this (1973).

The one remaining variable to be considered is the radius of the fibre. The critical length of a fibre increases as the radius of the fibre increases. A short fibre composite with thicker fibres will have more fibres below the critical length than one with thinner fibres and so should have worse properties. This was confirmed by Ramsteiner and Theyssohn (1985), for strength and unnotched impact measurements although curiously notched impact values were unaffected by changes in fibre diameter.

MATERIALS

Two classes of short glass fibre reinforced Nylons were examined; conventional extruded feedstock materials and pultruded feedstock materials. The extruded feedstock materials are ICI Maranyl A190 composites. Their average fibre length is about 0.25mm and few fibres are longer than 0.5mm. The pultruded feedstock material, ICI Verton, has an enhanced fibre length distribution. Its average fibre length is about 0.45mm but the distribution of fibre lengths is much broader and fibres with lengths of up to 4mm survive the injection moulding. A comparison of the two fibre length distributions is shown in Figure 1. There is another significant difference between the materials. The conventional extruded feedstock material has a fibre diameter of 10μm. The pultruded material has a fibre diameter of 17μm, presumably to survive the rigors of the pultrusion process.

The conventional feedstock material is designated EF30 and has 30% glass fibres. The pultruded feedstock material with 30% glass fibres is designated PF30 and the material with 50% glass fibres is designated PF50. The materials were in the form of 1/8" plaques.

EXPERIMENTAL

The toughness of these materials was tested over a wide range of speeds. Kc and Gc tests were performed on razor notched bend specimens (4" by 1/2" by 1/8") tested on a screw driven testing machine. The toughness in impact was measured on an instrumented pendulum. Slow crack growth tests were done in three point bending rigs with dead weight loading. The crack growth rates were calculated from the changes in displacement with time.

In order to vary the critical length of the fibres in the materials, the specimens were treated to give three moisture levels in the matrix. The swelling of the matrix when combined with any chemical degradation of the fibre matrix interface would reduce the level of adhesion between the fibre and the matrix. This effect could be seen in the fracture surfaces of the composites: the fibre surfaces in the moist materials were much cleaner than in the dry materials. We do not claim that the only effect of moisture on short fibre composites is a reduction in interfacial adhesion, but rather that differences in the relative response of each composite must be due to their different fibre length distributions. The increase in moisture content increases the critical length, and so gives a qualitative indication of the effect of critical length on the toughness of these composites. The conditions were :-

i) Dry - Specimens were heated in vacuum at 90 deg C until there was no further weight loss

ii) 50% moisture - Specimens were heated in a mixture of 100 parts water to 125 parts Potassium Acetate until the weight stabilized . The matrix would then contain about 2.3% water

iii) 100% moisture - The specimen was heated in water until saturated. The matrix would then contain about 8.6% water.

RESULTS

The effect of humidity on the critical stress intensity factor (Kc) of these composites is shown in Fig.2. These are results for the crack running in the L direction, i.e. parallel to the initial direction of flow into the mould. It can be seen that there is little difference between the short fibre composite and the enhanced fibre length composite in the dry or 50% humidity case. It is only when the matrix is saturated and the critical lengths are longer that one sees a distinct advantage in the longer fibre composite. When the fracture toughness (Gc) results are compared (Fig. 3) the longer fibre composite is actually worse than the conventional short fibre composite at the lower moisture content and it is only with the saturated matrix that the advantage of the longer fibres are again seen. A similar pattern of behavior is seen when the material is tested in the transverse direction although the differences between the different cases are less marked.

The results from the impact tests shown in Table 1 for two loading directions. Again there is very little difference between the two materials when they are dry. The addition of moisture when loading the specimens at impact speeds, however, serves to increase the toughness of both of the materials because of plasticisation of the matrix. We must separate the two factors that contribute to the increase in toughness; the plasticisation of the matrix and the fibre length distribution changes. Both materials contain the same nylon, the same amount of water and the same fibre content. We can reasonably expect that the improvement in toughness due to the plasticisation of the matrix should be similar in each case. The increase in toughness of the conventional material is not nearly as great as that recorded by the enhanced material however. Looking at Table 1 we see that the conventional EF30 material increases its toughness by an average of 18%. The longer fibre PF30 improves by 55% and 30% depending on the direction. It is highly unlikely that this improvement is due solely to the same mechanism as the 18% improvement in the EF30 composite. Rather the longer fibres themselves begin to play a more significant role in the swollen matrix, as they did in the slower Kc tests above.

The effect of moisture on the slow crack growth behavior of nylon composites can be quite catastrophic. The plasticised matrix appears to allow the redistribution of stress after local failure of a fibre/matrix bond or of a fibre much more rapidly than a dry nylon matrix does and so the deleterious effect of moisture on toughness is much more significant at long loading times in the slow crack growth tests than in the relatively rapid Kc type tests (Carling, 1988). The enhanced fibre length material again shows no improvement in slow growth behavior when dry or with a 50% moisture content. Again though, it is a substantial improvement on the conventional material when wet and this can be seen in Figure 4.

DISCUSSION

In each loading case the results clearly show that the enhanced fibre length material is the better material only when the matrix is saturated. There is little difference between the two materials when dry. Indeed the enhanced fibre length material is slightly worse when the energy to fracture, Gc, is measured. Why should the material with a clearly superior fibre length distribution only show advantages is specific conditions? The answer lies in the increase in thickness of the fibres in the enhanced fibre length material and in the relative shapes of the fibre length distribution curves.

We must first determine the critical length of the glass fibres in the nylon melt. As we have no means of doing this we must turn to the literature to determine this value and presume that the fibre matrix interface has been optimized in each material. Bader and Bowyer (1972) report shear strengths of 45 MPa at a nylon /glass interface. Plueddeman (1986) and Fraser (1975) report shear strengths of 35 MPa and Folkes (1982) reports critical lengths of 250 µm. Putting the last two values into the equations for critical pull-out length, we calculate a fibre strength of 1.75 GPa, a value at the bottom end of the range usually given for the strength of glass but reasonable for a fibre that has been both extruded and injection moulded.

Using the value of 35MPa for the shear strength we calculate a critical fibre length for the enhanced fibre length material of 425 µm (for fibres of 17 µm diameter) and a critical fibre length for the conventional material of 250 µm. We can now go back to the fibre length distribution curves and calculate the proportion of fibres with lengths above the critical length. We can further show how this proportion would change if the critical length were to change as a result of a lowering in the shear strength of the interface. This data can be seen in Figure 5.

It can be seen that, when dry, there is not a great difference in the proportion of fibres with lengths greater than the critical length even though there are much greater differences in the absolute fibre lengths. It is not the absolute length of the fibres that is important in short fibre composites but rather the aspect ratio, (the ratio of fibre length to fibre diameter) that is crucial. However, as the fibre matrix shear strength decreases, because of moisture, swelling or thermal expansion then the conventional short fibre composites break down more quickly as they have fewer longer fibres reinforcing them. The enhanced material has a broader fibre length distribution and so continues to retain a reasonable number of fibres above the critical length even after significant reductions in the fibre matrix interface strength. This is the reason that the wet properties of this material is better than the wet properties of the conventional material.

There are other factors that limit the effectiveness of the thicker fibres in reinforcing the material. The first is that a material with the same weight fraction of thicker fibres, has simply less fibres. In our case with 10 and 17 µm fibres the enhanced material has only 34.6% of the fibre length in it that the conventional material does. This limits the opportunities for crack arrest, the amount of pullout possible and other energy saving mechanisms.

There are problems in the enhanced material with fibre bundling. Because of the increased fibre lengths the flow into the mould is more nearly plug flow and fibres join into bundles. When the material fractures, whole bundles pull out rather than individual fibres and so the energy that would have been absorbed in individual fibre pull out is lost and the material fails in a more brittle manner than it should. Micrographs of the fibre surfaces suggest that the adhesion to the matrix in the enhanced material may not be as good as in the conventional material and this may also bias the results against the enhanced material.

CONCLUSIONS

Increasing both the fibre length and the fibre diameter in a glass fibre / nylon composite provides little improvement in toughness when the matrix is dry.

The toughness of the composite is improved by these changes when the matrix is wet and the fibre / matrix adhesion is lower. This is because more fibres remain with lengths equal to or greater than the critical length because of the broader fibre length distribution in the pultruded feedstock material than do in the extruded feedstock material.

The pultruded feedstock material appears to have problems with fibres bundling together in the center of mouldings resulting in the bundles rather than individual fibres pulling out. In addition the fibre matrix adhesion in the pultruded feedstock material may not be as good as that seen in the extruded feedstock material.

ACKNOWLEDGEMENTS

I would like to thank R Bailey and R Moore of ICI Wilton for providing the materials and E.I. duPont de Nemours for funding this work.

REFERENCES

R.E. Alfred and D.M. Schuster (1973)	J. Mater. Sci.	Vol.8 p.245
W.H. Bowyer and M.G. Bader (1972)	J. Mater. Sci.	Vol.7 p.377
M.J. Carling (1988)	Ph.D. Thesis, Univ. of London,	
A.H. Cottrell (1964)	Proc. Royal Soc.	A282 p.2
H.L. Cox (1952)	Brit. J. Appl. Phys.	Vol.3 p. 72
P.T. Curtis, M.G. Bader and J.E. Bailey (1970)	J. Mater. Sci.	Vol.5 p.762
T.Dingle (1974)	Proc. 4th Int. Conf. on Carbon Fibres, Plastics Institute, , Paper 11	
N. Dow (1963)	GEC Report R63 SD61	
M.J. Folkes (1982)	Short Fibre Reinforced Thermoplastics, J. Wiley and Sons,	
M.J.Folkes and D.Kells (1985)	Plast. and Rubber Proc and Appl.	Vol.5 p.125
W.A. Fraser, F.N. Anker and A.T. diBennedetto (1975) SPI Conf. Reinf. Plast.		Vol.30 22-A
C. Galiotis, R.J. Young, P.H.J. Yeung and D.N. Batcheldor (1984) J. Mater. Sci.		Vol.19 p.3640
J.C. Halpin (1969)	J. Comp. Mat.	Vol.3 732
P. Hancock and R.C. Cuthbertson (1970)	J. Mater. Sci.	Vol.5 p.762
A. Kelly (1970)	Proc. Royal Soc.	A319 p.33
E.P. Pleuddeman (1986) Mech. Properties of Reinforced Thermoplastics Eds. D.W. Clegg and A.A. Collyer, Elsevier		
F. Ramsteiner (1981)	Composites	Vol.12 p.344
F. Ramsteiner and R Theyssohn (1985)	Comp. Sci. and Tech.	Vol.24p.231
I.M. Robinson, R.J. Young, C Galiotis and D.N. Batcheldor (1987) J.Mater. Sci.		Vol.22p.3642
G.E. Smith and A.J.M. Spencer (1970)	J.Mech. Phys. Solids	Vol.18 p.81
Y. Termonia (1984)	J. Mater. Sci.	Vol.22 p.504
W.R. Tyson and R.J. Davies (1965)	Brit. J. Appl. Phys.	Vol.16 p.199

TABLE 1
IMPACT TOUGHNESS

Material	Dry	Wet
	Kc (MPa√m)	
EF30L	2.10	2.45
EF30T	3.11	3.72
PF30L	2.16	3.36
PF30T	3.16	4.10

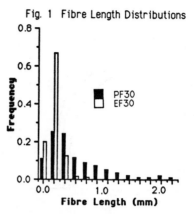

Fig. 1 Fibre Length Distributions

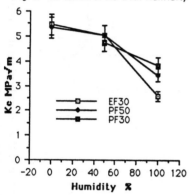

Fig. 2 Variation of Kc with Humidity

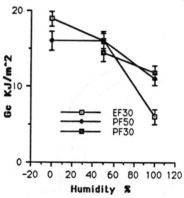

Fig.3 Variation of Gc with Humidity

Fig. 4 Slow Growth Curves

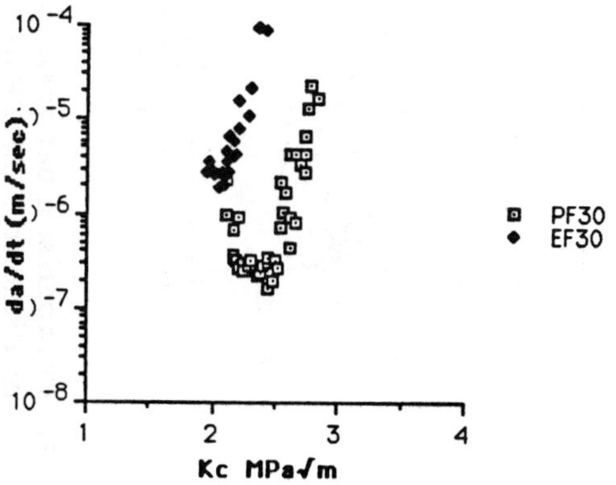

Fig. 5 % Fibres above Critical Length

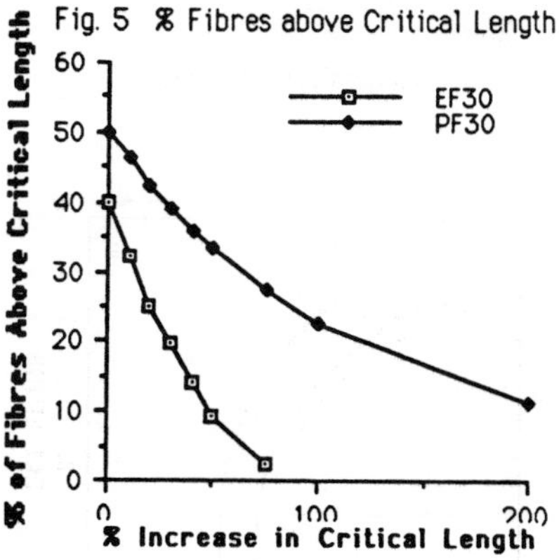

95

IMPACT DAMAGE MECHANISMS AND MECHANICS OF LAMINATED COMPOSITES

F.- K. Chang, H. Y. Choi, and S.- T. Jeng
Department of Aeronautics and Astronautics
Stanford University
Stanford, California

Abstract

An investigation was performed to study impact damage of laminated composites caused by a line-nose impactor. The major objective of the study was to fundamentally understand the failure mechanisms in composites caused by impact, and to identify the essential parameters causing the damage in composites. The primary concern of the failure modes were matrix crackings and delaminations. An unique and special impact test facility was designed and built for the investigation. The major characteristic of the facility was the use of a rectangular barrel so that a line-nose impactor could be adopted to produce uniformly distributed, transient dynamic loadings across the specimen's width and substantially simplify impact damage patterns. T300/976 Graphite/ Epoxy prepregs were selected to fabricate specimens. All specimens were X-rayed and C-scanned before and after impact to examine damage caused by impact. An analytical model was also developed for simulating the impact response of the specimens and to determine the important parameters dominating the impact damage in composites. An excellent agreement was found between the data and the predictions. Based on the experiments and the numerical calculations, it can be concluded that 1). matrix cracks were the initial impact damage mode, (2). delamination was initiated by the matrix cracks during impact, 3). laminates with inherent cracks, resulting from manufacturing, are much more susceptible to impact than laminates without, 4). residual thermal stresses are crucial for impact damage in composites, and 5). the interlaminar shear stresses and in-plane tensile stresses are the dominating stresses causing the initial matrix cracking.

INTRODUCTION

Low velocity impact could cause internal damage in laminated composites, such as matrix crackings and delaminations. These types of damage are hard to detect without the use of X-ray or C-scan machines, and could potentially cause significant reduction of mechanical properties

97

of the materials (1–11). Therefore, considerable work has been performed in the literature to study damage in fiber-reinforced laminated composites caused by low-velocity impact (12–20). Numerous experiments were conducted by researchers using either drop weight tests or air guns with point-nose impactors. Several analytical models were also proposed to estimate impact damage.

However, the impact damage results produced by a point-nose impactor are three-dimensional and very complex, involving the dynamic interaction between matrix crack propagation and delamination growth. As a consequence, although a considerable amount of test data has been produced, the current understanding of impact damage is still very limited. The knowledge of impact damage on damage mechanisms and mechanics is very premature. Hence, the analyses proposed on the basis of the previous experiments are very preliminary.

Therefore, the objective of this investigation is to design a new impact tester to simplify impact damage so that impact damage mechanisms can be fundamentally understood and the essential parameters governing the impact damage event in composites can be identified. During the investigation, a new impact tester was designed and built. Tests were performed to generate impact damage in composites. An analytical model was also developed to simulate the impact test and to evaluate the impact damage.

A NEW IMPACT TESTER

A new impact testing facility was designed and built during this investigation. The major apparatus of the facility consists of a pressure tank, a precision-made barrel, a high precision timer, optical fiber photoelectric sensors, and supporting fixtures as shown in Figure 1. The essential characteristic of the design which is different from any others is the use of a rectangular barrel and the impactors. Because of the use of the rectangular barrel, the impactors were designed into two parts, a rectangular base and the noses which can be changed (see Figure 2). Different types of noses can be mounted to the base to produce different types of damage patterns.

The impactor is driven by compressed air from the air tank through the rectangular barrel. The velocity of the impactor can be controlled by selecting the proper weights of the base and the nose of the impactor and adjusting the air pressure from the air tank. The impactor will hit on the target and rebound back into the barrel without leaving the barrel. Therefore, this facility can be used to evaluate the impact damage in composites as a function of the weight and velocity of the impactor and nose shape of the impactor. The range of the velocity of the impactor is between 2 and 30 m/sec, depending upon the weight of the impactor.

In this investigation, a line-nose impactor was chosen for the study as shown in Figure 2. The use of the line-nose impactor will result in an uniformly distributed transient dynamic loading across the specimen, which is clamped on two parallel free edges. It was expected that such uniform loading would produce a consistent and uniform damage pattern through the specimen width; hence, substantially simplifying impact damage mechanisms from a three-dimensional to a two-dimensional event.

EXPERIMENTS AND RESULTS

Extensive impact tests were performed to study impact damage in laminated composites subjected to a line-nose impact. Different ply orientations and various thicknesses of the specimens were selected for the tests. During impact, different weights and velocities of the impactor were also used as test parameters.

T300/976 Graphite/Epoxy prepregs were selected to fabricate specimen panels. An autoclave was used to cure the panels. Panels were then sliced into specimens by a diamond-coated saw. All the specimens were X-rayed before testing to evaluate internal pre-existing damage caused by curing or cutting. Initially, a standard cure cycle was selected to cure the panels. However, for some ply orientations, such as $[0_6/\pm45_4/90_5]_s$, the cured panels under the standard cure cycle contained significant internal matrix cracks as shown in Figure 3. As a consequence, an altered cure cycle with a slower heat-up rate and longer curing time was chosen for most of the panels to minimize the matrix cracks due to the residual stresses. No matrix cracks were found in these panels.

Overall, more than one hundred tests were performed during this investigation. Significant results have been produced. Due to a limited space available, details of the test results are presented elsewhere (21), hence, only some typical results are presented in this paper.

Matrix Cracks and Delaminations

Figure 4 shows a typical schematic of the impact damage pattern in $[0_n/90_m]_s$ laminated composites impacted by a line-nose impactor. A few matrix cracks were generated in 90 degree plies and could be clearly seen from the sides of the specimen in two possible positions; one was near the center but away from the impacted area and the other was located near the ends of the clamped areas. An enlarged view of the impact damage near the center region is drawn in Figure 4. It strongly indicated that matrix cracking was the initial failure mode of impact damage in laminated composites. A photograph of a life-size specimen of $[0_6/90_2]_s$ after impact is shown in Figure 5. Apparently, delaminations were initiated from these matrix cracks (refered to as "critical" matrix cracks) and propagated along the interfaces between 0 degree plies and 90 degree plies. Figure 6 shows the X-radiograph of a $[0_6/90_2]_s$ specimen before and after impact. No pre-matrix cracks were found in the specimen before impact. However, extensive micro-matrix cracks were detected in the X-radiograph made after impact. These cracks could not be seen by the naked eyes and were most likely generated during delamination growth. the cracks were confined to the extent of the size of delaminations. Similar results were also found for specimens with other ply orientations.

Thermal Residual Stresses

Due to the thermal expansion coefficient mismatch, manufacturing produces significant thermal residual stresses in laminated composites (22). The amount of residual stresses depends strongly on the degree of the thermal coefficient mismatch, the ply orientation, and the cure cycle. For some ply orientations, residual stresses could well exceed the transverse tensile strength of each individual ply in the laminate and result in premature matrix cracks as shown in Figure 3 for a $[0_6/\pm45_4/90_5]_s$ specimen under a standard cure cycle.

Figure 3 also shows the X-radiograph of a $[0_6/\pm45_4/90_5]_s$ specimen with premature matrix cracks after impact. No additional matrix cracks were found after the impact, but delaminations were initiated from one of the existing matrix cracks and propagated into the interfaces with a similar pattern as shown in Figure 4. It was also found that for the specimens with pre-matrix cracks, the impact energy required to initiate damage was substantially lower than for those without pre-matrix cracks. However, for laminates without premature matrix cracks, the residual stresses still exist in the materials and could also have significantly affected impact damage as will be verified in the analysis in Section IV.

Impact Energy Threshold

By examining impacted specimens, it was found that the impact energy, the mass and the velocity of the impactor, significantly affect the impact damage. There apparently exists an impact energy threshold beyond which damage occurs. Figure 7 presents the measured delamination sizes in $[0_6/90_2]_s$ and $[0_3/\pm45_4/90_3]_s$ specimens as a function of impact energy. For $[0_6/90_2]_s$ laminates, the energy threshold is about 9 Joules beyond which significant delaminations were produced in the laminate. No damage (neither matrix cracks nor delaminations) was found in these specimens tested below that energy. The impact energy threshold for $[0_3/\pm45_4/90_3]_s$ laminates is about 17 Joules. Similar phenomena was also found for the specimens with other ply orientations.

ANALYSIS AND NUMERICAL SIMULATIONS

During the investigation, an analytical model was also developed for modeling the impact damage in laminated composites. A two-dimensional transient dynamic finite element analysis was developed which can be used to calculate transient stresses, strains, and deformations inside the laminate during impact. The analysis was based on plane strain condition. Owing to the importance of the residual stresses to impact damage, the residual stresses resulting from manufacturing were calculated and incorporated into the analysis. The material properties used for the calculations are listed in Table 1.

The equilibrium equation at instant time $= t$ in a variational form can be expressed as ($\underline{12}$)

$$0 = \int_A \rho u_{i,tt}\delta u_i\,da + \int_A \sigma_j^M\,\delta c_j^M\,da - \int P\overline{T}_i\delta u_i\,da \quad \begin{cases} i = 1,2 \\ j = 1,3,5 \end{cases} \tag{1}$$

where u_i and $u_{i,tt}$ are the displacements and the accelerations ($u_{i,tt} = \partial^2 u_i/\partial t^2$), respectively. $\overline{T}_i$ are the contact force distribution during impact. σ_j^M and c_j^M are the mechanical stresses and strains in contracted notations, i.e.,

$$\begin{aligned} \{\sigma_1^M, \sigma_3^M, \sigma_5^M\} &= \{\sigma_{11}^M, \sigma_{33}^M, \sigma_{13}^M\} \\ \{\epsilon_1^M, \epsilon_3^M, \epsilon_5^M\} &= \{\epsilon_{11}^M, \epsilon_{33}^M, \gamma_{13}^M\} \end{aligned} \tag{2}$$

The mechanical stresses are related to the mechanical strains through the following equation

$$\sigma_j^M = [C_{jk}]\epsilon_k^M \tag{3}$$

where $[C_{jk}]$ is the stiffness matrix of composites based on the plane strain assumption. The detail expression of each component of the matrix can be found in ($\underline{23}$).

The total stresses σ_j in composites during impact is the summation of the mechanical stresses σ_j^M and the thermal residual stresses σ_j^R produced during manufacturing; i.e.,

$$\sigma_j = \sigma_j^M + \sigma_j^R \tag{4}$$

The thermal residual stresses σ_j^R can be calculated separately from the following equation ($\underline{24}$)

$$\sigma_j^R = [C_{jk}](c_k^o - \alpha_k\Delta T) \tag{5}$$

where c_k^o are the thermal residual strains caused by manufacturing and α_k and ΔT are the thermal expansion coefficients and the temperature, respectively.

In order to solve Eq. (1), the contact load distribution $\overline{T}_i$, between the impact and the composite, must be know first. In this investigation, Hertzian contact law was adopted to simulate the contact load distribution. Because the use of a cylindrical line-nose impactor (see Figure 2), the Hertzian contact law has a considerably different expression from the one most commonly used for a spherical point-nose impactor (12–20). Accordingly, the contact load distribution $\overline{T}_i$ $(= f)$ is related to the indentation depth α $(= ds - dp =$ the change in the distance between the center of the impactor nose and the contact point of the plate) by the expression (25)

$$\alpha = f(\delta_2 + \delta_p)\left\{1 - ln[fr(\delta_s + \delta_p)]\right\}; \qquad \text{when } d_s \geq d_p$$
$$f = 0; \qquad\qquad\qquad\qquad\qquad\qquad \text{when } d_s < d_p$$

(6)

where d_s is the impactor displacement, d_p is the plate displacement measured at the center of the plate surface, which is opposite to the impact surface, and δ_s and δ_p are the constants defined as

$$\delta_s = \frac{1 - \nu_s^2}{\pi E_s}$$

(7)

$$\delta_p = \frac{1}{\pi E_{yy}}$$

and r, ν_s and E_s are the local radius, the Poisson ratio, and the Young modulus of the impactor, respectively. E_{yy} $(= E_{zz})$ is the modulus of elasticity of the impacting composite ply in the direction transverse to the fibers.

In order to analyze the impact damage mechanisms, failure criteria were adopted in the model for predicting initial damage, especially the matrix cracking. Physically, there are three major stress components contributing to initial matrix cracking under the given loading condition considered in this investigation, as shown schematically in Figure 8. These are the interlaminar shear stress σ_{13}, in-plane tensile stress σ_{11}, and out-of-plane normal stress σ_{33}. In this investigation, three-dimensional Hashin matrix tensile failure criterion was selected for predicting initial failure (26). The criterion can be expressed as follows:

$$\frac{1}{Y_T^2}(\sigma_{yy} + \sigma_{zz})^2 + \frac{1}{S^2}(\sigma_{yz}^2 - \sigma_{yy}\sigma_{zz})$$
$$+ \frac{1}{S^2}(\sigma_{xy}^2 + \sigma_{xz}^2) \geq e_M$$

(8)

$$e_M \geq 1 \quad \text{failure}$$
$$e_M < 1 \quad \text{no failure}$$

where Y_T is the in situ ply transverse tensile strength (27,28) and S is the in situ ply shear strength (27–30). The subscripts x and y indicate the directions parallel and normal to the fiber direction, respectively. The subscript z denotes the direction normal to the ply $(x - y)$ surface.

Table 2 shows, for a given mass, the predicted velocities, with and without inclusion of thermal residual stresses, required to cause initial matrix cracking for T300/ 976 Graphite/epoxy composites as compared with the test data. The predictions with consideration of thermal residual stresses agreed with data very well. However, for the calculations without thermal residual stresses, the predictions over estimated the results by as much as 200 percent. Clearly, the introduction of the thermal residual stresses in the analysis is crucial for prediction of impact damage in laminated composites.

Figure 9 shows the predicted strength ratio (e_M) of $[0_6/90_2]_s$ and $[0_3/\pm 45_4/90_3]_s$ specimens based on the Hashin criterion as a function of position. It indicated that the peak strength ratio occurs at two possible positions; one is close to the impacted area and the other is near the ends of the specimens. These predicted locations are also consistant with the experimental findings (see Figure 10). Similar results were also found for other specimens. It is worth pointing out that the predicted matrix cracking near the center did not occur directly under the impacted area but a distance away from the area. By carefully examining the stress distributions near the center impacted area as shown in Figure 11, the interlaminar shear stress and the in-plane tensile stress were comparably higher than the out-of-plane normal stress. The out-of-plane normal stress σ_{33} decreased rapidly once it was away from the impacted area (see Figure 11). Apparently, the interlaminar shear stress and the in-plane tensile stress are the dominating stresses which cause matrix cracking during impact. Accordingly, the contribution of the out-of-plane normal stress to the initiation of impact damage is negligible.

To understand how the initial "critical" matrix cracks can initiate delaminations, material properties within the elements where matrix crack failure has been predicted were reduced according-ing to the property degradation models developed previously (27,31). The stresses and strains were recalculated in the finite element analysis at the same instant time again. Figures 12–14 show the stress distributions along the specimen length before and after material reduction. Peak out-of-plane normal tensile stresses along the interfaces were found comparably higher than the others immediately adjacent to the damaged elements. This clearly indicated that delamination would be initiated by the highly concentrated normal tensile stress due to mode I fracture. Two peak stresses along the upper and lower interfaces of 0 degree and 90 degree plies were found as shown in the Figure 12, which could initiate upper and lower interface delaminations propagating in two opposite directions. The results coincided with the physical findings of the delamination propagation as shown in Figures 4 and 5.

CONCLUDING REMARKS

An investigation was performed to study impact damage in laminated composites. Both experiments and analysis were involved in the study. Based on the test data and the numerical calculations, the following remarks can be made:

1) matrix crack was the initial failure mode,

2) delamination was initiated by the initial "critical" matrix cracks,

3) residual thermal stresses could substantially reduce impact resistance of composites,

4) there apparently exists an impact energy threshold above which impact damage occurs,

5) interlaminar shear stresses and in- plane tensile stresses are the dominating factors causing initial matrix cracks;

6) delamination growth was dominated by suddenly increased out-of-plane normal stress (Mode I fracture) as a result of matrix cracking.

ACKNOWLEDGEMENTS

The support of Army Research Office Contract No. DAAL 03-87-K-0115 and the National Science Foundation grant MSM 87-02892 is gratefully appreciated. The authors would also like to thank Mr. R. J. Downs for helping design and build the impact test facility.

REFERENCES

1. Talreja, R., "Transverse Cracking and Stiffness Reduction in Composite Laminates," J. of Composite Materials, Vol. 19, (1985), pp. 355.

2. Sun, C. T. and Jen, K. C., "On the Effect of Matrix Cracks on Lamiante Strength," J. of Reiforced Plastics and Composites, Vol. 16, (1987), pp. 208–222.

3. Peters, P. W. M., "The Strength Distribution of 90° Plies in 0/90/0 Graphite-Epoxy Laminates," J. of Composite Materials, Vol. 18, (1984), pp. 545–556.

4. Garg, A. G., "Delamination—A Damage Mode in Composite Structures," Engineering Fracture Mechanics, Vol. 29, (1988), pp. 557–584.

5. Bowles, D. E., "Effect of Microcracks on the Thermal Expansion of Composite Laminates," J. of Composite Materials, Vol. 17, (1984), pp. 173–187.

6. Aronsson, C. and Bäcklund, J., "Tensile Fracture of Laminates with Cracks," J. of Composite Materials, Vol. 20, (1986), pp. 287–307.

7. Dvorak, G. J., Laws, N. and Hejazi, M., "Analysis of Progressive Matrix Cracking in Composite Laminates, I. Thermoelastic Properties of a Ply with Cracks," J. of Composite Materials, Vol. 19, (1985), pp. 216–234.

8. Manders, P. W., Chou, T., Jones, F. R. and Rock, J. W., "Statistical Analysis of Multiple Fracture in 0°/90°/0° Glass Fiber/Epoxy Resin Laminates," J. Material Science, Vol. 18, (1983), pp. 2876–2889.

9. Garrett, K. W. and Bailey, J. E., "Multiple Transverse Fracture in 90° Cross-Ply Laminates of a Glass Fibre-Reinforced Polyester," J. of Material Science, (1977), pp. 157–168.

10. Parvizi, A. and Bailey, J. E., "On Multiple Transverse Cracking in Glass Fibre Epoxy Cross-Ply Laminates," J. of Material Science, Vol. 13, (1978), pp. 2131–2136.

11. Parvizi, A., Garrett, K. W. and Bailey, J. E., "Constrained Cracking in Glass Fibre-Reinforced Epoxy Cross-Ply Laminates," J. of Material Science, Vol. 13, (1978), pp. 195–2015.

12. Wu, H. T. and Chang, F. K., "Transient Dynamic Analysis of Laminated Composite Plates Subjected to Transverse Impact," J. of Computers and Structures, (To appear in March 1989 issue).

13. Wu, H. T. and Springer, G. S., "Measurements of Matrix Cracking and Delamination Caused by Impact on Composite Plates," J. of Composite Materials, Vol. 22, No. 6, (1988), pp. 518–532.

14. Wu, H. T. and Springer, G. S., "Impact Induced Stresses, Strains and Delaminations in Composite Plates," J. of Composite Materials, Vol. 22, No. 6, (1988), pp. 533–560.

15. Joshi, S. P. and Sun, C. T., "Impact-Induced Fracture in Quasi-Isotropic Laminate," J. of Composite Technology and Research, Vol. 19, No. 2, (1986), pp. 40–46.

16. Glosse, J. H. and Mori, P. B. Y., "Impact Damage Characterization of Graphite/Epoxy Laminates," Proceedings of the Third Technical Conference of the American Society for Compostes, Seattle, WA, (1988), pp. 334–353.

17. Joshi, S. P., "Impact-Induced Damage Initiation Analysis: An Experimental Study," Proceedings of the Third Technical Conference of the American Society for Compostes, Seattle, WA, (1988), pp. 325–333.

18. Ross, C. A. and Malvern, L. E., Sierakowski, R. L. and Taketa, N., "Finite-Element Analysis of Interlaminar Shear Stress Due to Local Impact," Recent Advnaces in Comp. in the United States and Japan, ASTM STP 864, (J. P. Vinson and M. Taya, Eds.) American Society for Testing and Materials, Philadelphia, PA, (1985), pp. 335–367, (1985).

19. Aggour, H. and Sun, C. T., "Finite Element Analysis of a Laminated Composite Plate Subjected to Circularly Distributed Central Impact Loading," J. Computers and Structures, Vol. 28, No. 6, (1988), pp. 729–736.

20. Sun, C. T. and Rechak,S., "Effect of Adhesive Layers on Impact Damage in Composite Laminates," Composite Materials: Testing and Design (Eighth Conf.), ASTM STP 972 (J. D. Whitcomb, Ed.), American Society for Testing and Materials, Philadelphia, PA, (1988), pp. 97–123.

21. Chang, F. K. and Choi, H. Y., "Impact Damage Mechanics of Laminated Composites by a Line-Nose Impactor," J. of Composite Materials, (Submitted).

22. Flaggs, D. L. and Kural, M. H., "Experimental Determination of the In Situ Transverse Lamina Strength in Graphite/Epoxy Laminate," J. of Composite Materials, Vol. 16, (1982), pp. 103–116.

23. Chang, F. K. and Springer, G. S., "The Strengths of Fiber Reinforced Composite Bends," J. of Composite Materials, Vol. 20, No. 1, (1986), 30–45.

24. Tsai, S. W. and Hahn, H. T., Introduction to Composite Materials, Technomic Publishing Co., (1980).

25. Goldsmith, W., Impact Theory and Physical Behavior of Colliding Solids, Edward Arnold Ltd., London, U.K., (1960).

26. Hashin, Z., "Failure Criteria for Unidirectional Fiber Composites," J. Applied Mechanics, Vol. 47, (1980), pp. 329–334.

27. Chang, F. K. and Lessard, L. B., "Damage Tolerance of Laminated Compostes Containing an Open Hole and Subjected to Compressive Loadings: Part I—Analysis," J. of Composite Materials, (Submitted).

28. Lessard, L. B. and Chang, F. K., "Damage Tolerance of Laminated Compostes Containing an Open Hole and Subjected to Compressive Loadings: Part II—Experiment," J. of Composite Materials, (Submitted).

29. Chang, F. K. and Chen, M., "the In Situ Ply Shear Strength Distributions in Graphite/Epoxy Laminated Composites," J. of Composite Materials, Vol. 21, (1987), pp. 708–732.

30. Chang, F. K., Tang, J. M. and Peterson, D. G., "The Effect of Testing Methods on the Shear Strength Distribution in Laminated Composites," J. of Reinforced Plastics and Composites, Vol. 6, (1987), pp. 304–318.

31. Chang, F. K. and Chang, K. Y., "A Progressive Damage Model for Laminated Composites Containing Stress Concentrations," J. of Composite Materials, (1987), Vol. 21, pp. 834–855.

Table 1. Material properties of T300/976 used in the calculations.

Moduli	Symbol (unit)	
In-plane longitudinal modulus	E_{xx} (Gpa)	156
In-plane transverse modulus	E_{yy} (Gpa)	9.09
In-plane shear modulus	G_{xy} (Gpa)	6.96
Out-of-plane shear modulus	G_{yz} (Gpa)	3.24
In-plane poisson's ratio	ν_{xy}	0.228
Out-of-plane poisson's ratio	ν_{yz}	0.400
Density	ρ (Kg/m^3)	1540
Thermal expansion coefficient	α_x((μm/m)/K)	1.07
	α_y((μm/m)/K)	25.2

Strength	Symbol (unit)	
Longitudinal tension	X_T (Mpa)	1520
Longitudinal compression	X_C (Mpa)	1590
Transverse tension	Y_T (Mpa)	49
Transverse compression	Y_C (Mpa)	252
Ply longitudinal shear *	S_C (Mpa)	105

Impactor	Symbol (unit)	
Modulus	E (Gpa)	207
Poisson's ratio	ν	0.3
Nose radius	r (mm)	1.5

* Shear strength measured from a cross-ply [0/90]$_s$ composites

Table 2. The velocities of an impactor required to initiate "critical" matrix cracks in T300/976 composites. Comparison between the data and the predictions.

Ply orientation	mass (Kg)	length (cm)	Predicted Velocity (m/s)		Test Velocity (m/s)
			Without residual stress	With residual stress	
[0$_7$/90$_1$]s	0.08	10	40	25	22±2.0
[0$_6$/90$_2$]s	0.08	10	45	20	16.5±1.0
[0$_3$/±45$_4$/90$_3$]s	0.12	10	50	21	17±0.2

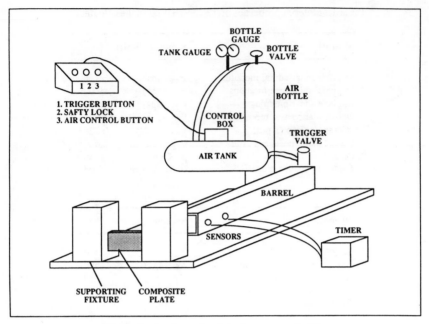

Figure 1. A schematic of the impact test facility.

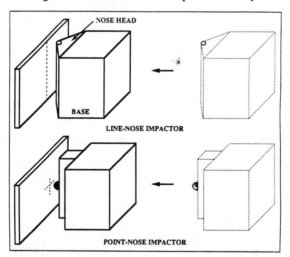

Figure 2. A Description of the impactor. (Above): line-nose impactor. (Below):
point-nose impactor.

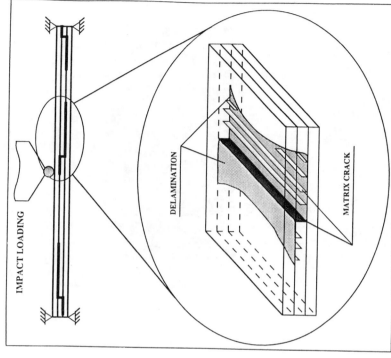

Figure 4. A schematic of the typical impact damage pattern in $[0_n/90_m]_s$ composites impacted by a line-nose impactor.

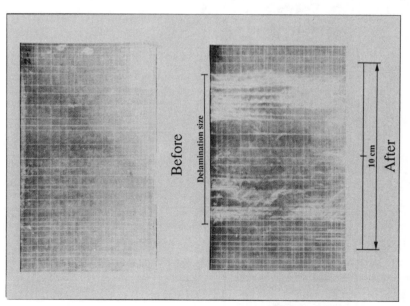

Figure 3. X-Radiographs of $[0_6/\pm45_4/90_5]_s$ specimen before and after impact.

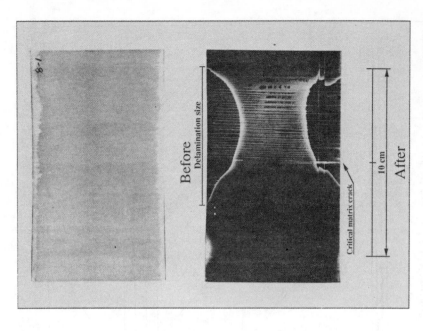

Figure 6. X-Radiographs of $[0_6/90_2]_s$ specimen before and after impact.

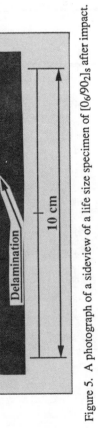

Figure 5. A photograph of a sideview of a life size specimen of $[0_6/90_2]_s$ after impact.

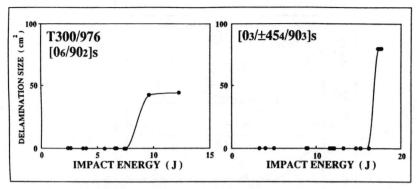

Figure 7. The measured delamination sizes in $[0_6/90_2]_s$ and $[0_3/\pm45_4/90_3]_s$ impacted specimens as a function of impact energy.

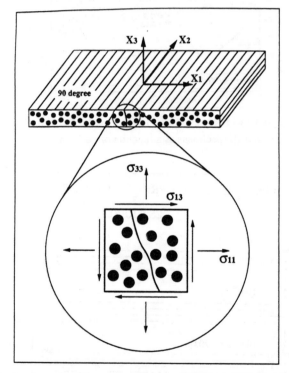

Figure 8. A schematic of the major stress components contributing the matrix cracking in 90 degree layers.

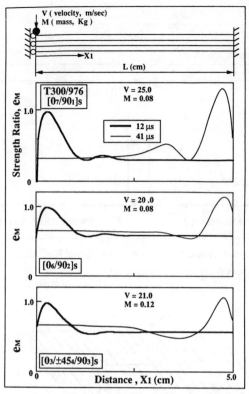

Figure 9. The predicted strength ratio (e_M: Hashin criterion) of $[0_6/90_2]_s$ and $[0_3/\pm45_4/90_3]_s$ specimens as a function of position.

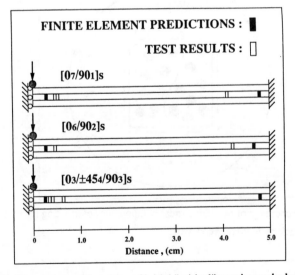

Figure 10. Comparison of locations of initial "critical" matrix cracks between the data and the predictions.

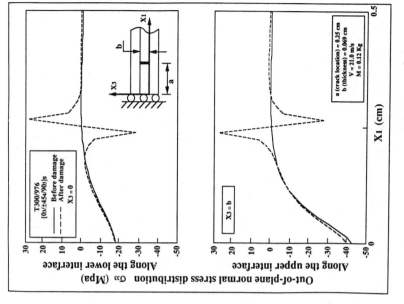

Figure 12. Comparison of instant out-of-plane normal stress distributions along the upper and lower interface of 90 degree plies before and after material degradation within the damaged area (No thermal stresses were included).

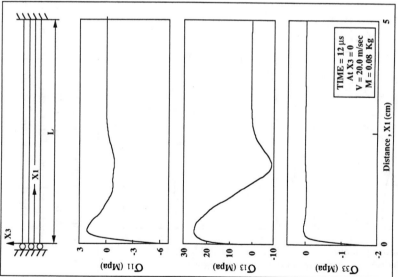

Figure 11. Instant stress distributions near the center impacted area of a [0₆/90₂]s specimane (No thermal residual stresses were included).

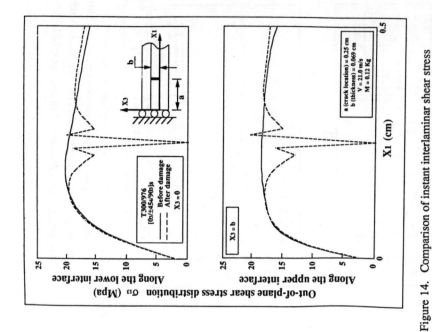

Figure 14. Comparison of instant interlaminar shear stress distributions along the upper and lower interface of 90 degree plies before and after material degradation within the damaged area (No thermal stresses were included).

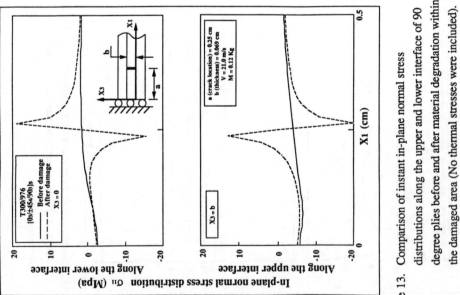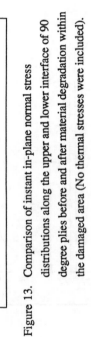

Figure 13. Comparison of instant in-plane normal stress distributions along the upper and lower interface of 90 degree plies before and after material degradation within the damaged area (No thermal stresses were included).

112

A GENERAL FAILURE THEORY FOR FIBRE-REINFORCED COMPOSITE LAMINAE

K. W. Neale
Faculté Sciences Appliquées
Université de Sherbrooke
Sherbrooke, Québec, Canada

P. Labossière
Ècole de Génie
Université de Moncton
Moncton, New Brunswick, Canada

ABSTRACT

The failure of thin, fibre-reinforced composite laminae under plane stress loading conditions is considered. A general parametric failure criterion, inspired by a general formulation recently proposed for the plastic yielding of sheet metals, is presented. The criterion satisfies the essential conditions that must be met in developing phenomenological strength criteria for composite materials. This entirely general formulation encompasses all previously proposed failure theories (e.g. tensor polynomial) for fibre-reinforced orthotropic laminae. The inherent advantages and drawbacks of the proposed parametric approach are discussed, and an example of application is presented in detail.

NOMENCLATURE

$g(\mu)$	Yield vector modulus in Budiansky's yield criterion
F_i, F_{ij}, F_{ijk}	Strength parameters in the tensor polynomial theory
P, P'	Lamina strength under biaxial tension and compression
S	Lamina shear strength
X, Y	Uniaxial tensile strengths
X', Y'	Uniaxial compressive strengths
α, β, ϕ	Orientation of the strength vector
$\Gamma(\alpha, \beta, \phi)$	Modulus of the strength vector
σ_1, σ_2	Normal stresses in the plane of the lamina
σ_b, σ_s	Yield strength under biaxial stress and under shear stress
σ_p, σ_q	Principal stresses in the isotropic plane
τ_{12}	Shear stress in the plane of the lamina

INTRODUCTION

In many applications, fibre-reinforced materials consist of thin laminae reinforced either unidirectionally or bi-directionally and assembled together to form a laminate. These components are generally subjected to bending and in-plane loadings, and the laminae are usually analyzed on the assumption of plane stress conditions. To characterize the strength properties of a particular lamina, strength or failure theories are employed. Among the many failure criteria that have been proposed for unidirectional and bi-directional composite laminae, the maximum stress and maximum strain theories are still the most widely used in practice (1). This is mainly because they are the easiest to apply. Also, simple hypotheses concerning failure modes can be associated with the failure parameters in these theories. Among researchers, however, the tensor polynomial theory of Tsai and Wu (2) has received the widest attention. This is due to the mathematical generality of the tensorial formulation, which can be manipulated according to well-established rules. Moreover, Wu (3) has shown it to encompass every failure criterion previously proposed for fibre-reinforced laminae.

Although the tensor polynomial equation can be expanded to any degree, the quadratic form remains the most popular choice so far. In plane stress, this produces an ellipsoid-shaped failure surface in the $\sigma_1 - \sigma_2 - \tau_{12}$ stress space. Here, σ_1 and σ_2 are the normal stresses in the plane (1-2 directions) of the lamina and τ_{12} is the in-plane shear stress. For bi-directional composites, σ_1 and σ_2 are in the fibre directions; σ_1 is in the fibre direction for a unidirectional lamina. For both of these cases, the failure envelope is symmetrical about the shear plane $\tau_{12} = 0$. This ellipsoidal failure surface is defined by six strength parameters, five of which are calculated from uniaxial and shear tests on a lamina. One remaining constant of the tensorial equation, referred to as the interaction parameter, describes the orientation and length of the major and minor axes of the ellipsoid. It is usually calculated from the results of a biaxial test. However, its value depends on the particular biaxial test selected and consequently, different failure envelopes can be obtained when different tests are employed. This suggests that an ellipsoid may not always provide the best analytical representation of the experimental failure envelope for a fibre-reinforced lamina.

The correlation between experimental data and theoretical failure surfaces can sometimes be improved by including cubic terms in the tensor polynomial equation (e.g. Tennyson et al. (4)). However, the gain in accuracy obtained is accompanied by the significant complexity of determining many additional strength parameters in the failure equation. In plane stress, the number of interaction parameters is increased to five, instead of one with the quadratic theory. Despite this drawback, this approach is still actively being pursued.

As the tensor polynomial theory is tending to become prohibitively complicated and difficult to handle with the inclusion of terms of higher degree, it is perhaps worthwhile to attempt a general formulation of strength criteria from an alternative point of view. In this study, an approach, inspired from recent work done in the area of metal plasticity, is suggested. The unidirectional or bi-directional laminae considered here are orthotropic, with the fibre axes as the axes of orthotropy (1-2 directions). The general ideas presented can be applied to more general anisotropic materials.

The general form of a new plastic yield criterion proposed by Budiansky (5) for transversely isotropic sheet materials will first be discussed. The premises that must be satisfied in developing a failure criterion for orthotropic laminae will then be identified. This will lead to an extension of the basic ideas put forward by Budiansky for orthotropic, fibre-reinforced laminae. An example of application of this approach will be presented in detail.

BUDIANSKY'S YIELD CRITERION FOR PLANE-ISOTROPIC SHEET MATERIALS

The formulation of failure theories for composite materials has closely followed the development of yield criteria for ductile metals. These phenomena are clearly unrelated, but the failure surfaces for fibre-reinforced materials often bear close resemblances to the surfaces which describe the plastic yielding of anisotropic metals. Accordingly, the origin of most existing failure theories for composite materials can be traced back to previously proposed models of ductile plastic flow. Hill's (6) well known yield criterion for anisotropic metals, for example, was one of the first plasticity theories to be adopted as a failure criterion for composites. Here, recent advances in sheet metal plasticity will be used as a basis for the formulation of general failure criteria for fibre-reinforced composite laminae.

In the area of sheet metal forming, the effects of anisotropy are often important. Some sheets are transversely isotropic, with their plane of isotropy in the plane of the sheet metal. Moreover, the applied stresses are in the plane of the element. Various yield criteria especially applicable to these materials and loading conditions have thus been developed. Hill's yield criterion, originally written for anisotropic materials, is often used by reducing the appropriate parameters to the transversely isotropic case. Another criterion subsequently proposed by Hill (7) has also been employed. More recently, a general yield criterion for transversely isotropic sheet metals under in-plane loads was introduced by Budiansky (5). Budiansky showed that all previously proposed criteria for such materials are special cases of his generalized yield function.

Taking a sheet metal, isotropic in its plane and submitted only to in-plane stresses, Budiansky's formulation is written in terms of the two principal stresses in the plane σ_p and σ_q. A general yield locus in the stress space $\sigma_p - \sigma_q$ is shown in Figure 1. Symmetries of the failure envelope are related to material symmetries. For example, the uniaxial tensile and compressive strengths in the plane of isotropy must be identical in all directions. Also, it is assumed to possess equal strengths in uniaxial tension and compression. The failure surface must accordingly be symmetric with respect to the planes $(\sigma_p + \sigma_q)$, and $(\sigma_p - \sigma_q)$. The 45° and -45° axes of symmetry intercept the failure surface at the points of yield under equal biaxial stress σ_b and under pure shear σ_s, respectively. This allowed Budiansky to state that any failure condition can be described parametrically in the following polar coordinate form:

$$(\sigma_p + \sigma_q)/2\sigma_b = g(\mu) \cos \mu$$
$$(\sigma_q - \sigma_p)/2\sigma_s = g(\mu) \sin \mu \tag{1}$$

This yield locus is depicted in Figure 2.

In Equations (1), the failure criterion $g = g(\mu)$ is a function of the parametric angle μ. Budiansky showed that all yield criteria previously proposed for isotropic materials are special cases of Equation (1). Moreover, this parametric form can be used to describe any shape of failure envelope. It makes it theoretically possible to describe given experimental data with any desired degree of accuracy. For example, an experimental yield locus could conceivably be fitted with suitable expressions of the form

$$g(\mu) = \Sigma \ a_n \cos (2n\mu) \tag{2}$$

where the a_n are constants. With such a representation, an accurate curve-fit can be obtained with a sufficient number of terms. Alternatively, different portions of the failure envelope could be fitted with different types of equations.

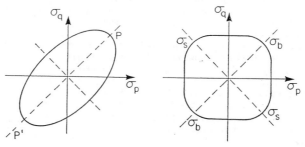

Fig. 1 Isotropic failure criteria.

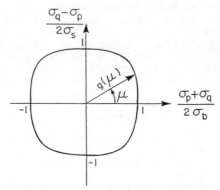

Fig. 2   Budiansky's failure criterion.

CHARACTERISTICS OF A GENERAL FAILURE CRITERION FOR FIBRE-REINFORCED COMPOSITES

In the same way that Hill's yield criterion has been adapted to describe the strength characteristics of fibre-reinforced materials, Budiansky's yield criterion will be extended here to formulate a failure theory for orthotropic laminae. First, it is necessary to identify the characteristics that must be satisfied for fibre-reinforced orthotropic laminae submitted to various combinations of in-plane tensile, compressive, and shear loadings. For these laminae, the orientation of the loading axes with respect to the axes of orthotropy must be taken into account. From the usual transformation rules, any set of applied stresses can be transformed into components σ_1, σ_2, τ_{12} with respect to the material axes. The failure criterion can thus, without loss of generality, be written in terms of the stresses σ_1, σ_2 and τ_{12} , and be represented graphically in this three-dimensional stress space. To ensure that the ultimate strength for any stress combination be finite, the failure surface must be closed.

With unidirectional or bi-directional fibre-reinforced laminae, the tensile strengths in the two directions of orthotropy are generally different. The compressive strengths in these directions are not necessarily equal either. Also, pure shear loadings on laminae with material axes oriented at $\theta = 45°$ or $\theta = -45°$ with respect to the applied loads will give different strengths. Similarly, by considering equal biaxial tension or compression, where the corresponding strengths P and P' generally differ, we see that no inherent symmetry with respect to the planes $(\sigma_1 - \sigma_2)$ and $(\sigma_1 + \sigma_2)$ exists. For a material transversely isotropic with respect to the 1-direction, the usual symmetries for isotropic materials no longer apply.

An orthotropic lamina under pure shear with respect to its principal directions must have the same strength whether the shear is positive or negative. Also, such a lamina with material axes oriented at an angle $+\theta$ and submitted to a positive shear stress will exhibit the same ultimate strength as a lamina with an angle $-\theta$ and negative shear stress. This implies that the failure surface must be symmetrical with respect to the plane $\tau_{12} = 0$. This is the only symmetry inherent to such materials, and it must be reflected in any proposed failure equation.

The number of complete experimental failure envelopes published in the literature is rather limited. However, from the results published by Wu (3) for graphite epoxy, or by Suhling et al. (8) for paperboard, the failure envelopes in the $\sigma_1 - \sigma_2$ plane maintain the same overall shape for different levels of constant shear stress τ_{12} . This observation could be retained as an hypothesis for some materials and eventually be integrated in the failure criterion. Although it has never been explicitly specified in the development of existing failure criteria, this characteristic is implicit in all existing quadratic failure criteria.

In summary, any plane stress failure criterion for orthotropic fibre-reinforced laminae can be written in terms of the stress components σ_1, σ_2 and τ_{12} with respect to the axes (1-2) of orthotropy. It should not exhibit in general any inherent symmetry except for symmetry of the failure surface about the plane $\tau_{12} = 0$. Finally, if the shapes of the projections of the failure envelope in the $\sigma_1 - \sigma_2$ plane at different levels of constant shear stress τ_{12} are similar, this observation should be incorporated in the proposed failure function.

A GENERAL FAILURE CRITERION FOR FIBRE-REINFORCED COMPOSITE LAMINAE

We shall now extend Budiansky's yield criterion and formulate an analogous failure theory for fibre-reinforced orthotropic laminae. As established previously, the failure envelope will be a closed surface in the three-dimensional stress space $\sigma_1 - \sigma_2 - \tau_{12}$, and will be symmetrical with respect to the plane $\tau_{12} = 0$. A general failure surface satisfying these conditions is depicted in Figure 3. Only the upper half is shown since a mirror image gives the form of

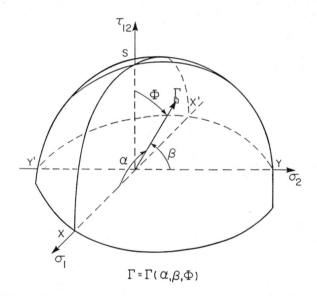

$$\Gamma = \Gamma(\alpha, \beta, \Phi)$$

Fig. 3   Parametric failure envelope for fibre-reinforced materials.

the failure envelope for negative values of τ_{12}. The X, X', Y, Y' and S values are the uniaxial tensile, compressive and shear strengths of the lamina.

The position of any point on this failure surface can be identified in spherical coordinates as follows:

$$\sigma_1 = \Gamma(\alpha, \beta, \phi) \cos \alpha$$

$$\sigma_2 = \Gamma(\alpha, \beta, \phi) \cos \beta \qquad (3)$$

$$\tau_{12} = \Gamma(\alpha, \beta, \phi) \cos \phi$$

where $\Gamma(\alpha, \beta, \phi)$ is the length of the radius vector form the origin to the failure point for a given stress combination. The angles α, β, and ϕ locate the position of this vector with respect to the axes σ_1, σ_2, and τ_{12}, respectively. These parameters can be defined separately as functions of the stress values σ_1, σ_2, and τ_{12} at failure. The values of α, β, and ϕ are given by:

$$\alpha = \cos^{-1}\left[\sigma_1/(\sigma_1{}^2 + \sigma_2{}^2 + \tau_{12}{}^2)^{\frac{1}{2}}\right] \quad (0 < \alpha < \pi)$$

$$\beta = \cos^{-1}\left[\sigma_2/(\sigma_1{}^2 + \sigma_2{}^2 + \tau_{12}{}^2)^{\frac{1}{2}}\right] \quad (0 < \beta < \pi) \tag{4}$$

$$\phi = \cos^{-1}\left[\tau_{12}/(\sigma_1{}^2 + \sigma_2{}^2 + \tau_{12}{}^2)^{\frac{1}{2}}\right] \quad (0 < \phi < \pi)$$

and satisfy $\cos^2\alpha + \cos^2\beta + \cos^2\phi = 1$.

The position of any point on the failure surface is determined uniquely by the angles α, β, and ϕ. Since these angles are sufficient to describe the position of any stress vector (σ_1, σ_2, τ_{12}), its modulus will be a function $\Gamma(\alpha,\beta,\phi)$. The length of the stress vector from the origin to the failure point is:

$$\Gamma(\alpha,\beta,\phi) = (\sigma_1{}^2 + \sigma_2{}^2 + \tau_{12}{}^2)^{\frac{1}{2}} \tag{5}$$

This value must be positive for any combination of angles α, β, and ϕ. Also, symmetry with respect to the plane $\tau_{12} = 0$ requires that the following equality be satisfied for all points:

$$\Gamma(\alpha,\beta,\phi) = \Gamma(\alpha,\beta,\pi-\phi) \tag{6}$$

Equations (3) are a natural extension of Budiansky's yield function and can be considered as general failure equations for orthotropic laminae in plane stress. Once a parametric failure criterion given by $\Gamma = \Gamma(\alpha,\beta,\phi)$ is found to represent satisfactorily the failure surface, it can be used with the classical stress transformation equations to predict the failure conditions for stress combinations defined with respect to any system of reference axes.

GENERALITY OF THE PARAMETRIC FAILURE CRITERION

It has already been demonstrated by Wu (3) that the tensor polynomial criterion encompasses all failure criteria previously proposed for fibre-reinforced laminae. It will now be shown that the tensor polynomial criterion is itself a special case of Equations (3) when plane stress conditions are assumed. The tensor polynomial equation, in the standard reduced notation for stresses, is:

$$F_i \sigma_i + F_{ij} \sigma_i \sigma_j + F_{ijk} \sigma_i \sigma_j \sigma_k + \ldots = 1 \tag{7}$$

where the F_i, F_{ij}, etc. are strength constants. These constants involve the strength values X, X', Y, Y' and S.

Substituting in Equation (7) the values σ_1, σ_2, and τ_{12} ($=\sigma_6$) from Equation (3) gives

$$\sum_{k=1}^{n} \Gamma^k(\alpha,\beta,\phi) \, f_k(\alpha,\beta,\phi) = 1 \tag{8}$$

where n is the degree of the tensor polynomial expression. The coefficients F_i, F_{ij}, etc. are incorporated implicitly in the various functions $f_k(\alpha,\beta,\phi)$. To clarify this, consider the quadratic tensor polynomial, or Tsai-Wu criterion, written as follows:

$$F_1 \sigma_1 + F_2 \sigma_2 + F_{11} \sigma_1{}^2 + F_{22} \sigma_2{}^2 + 2F_{12} \sigma_1 \sigma_2 + F_{66} \tau_{12}{}^2 = 1 \tag{9}$$

Substituting Equations (3) in the above expression gives

$$\Gamma(\alpha,\beta,\phi) \, f_1(\alpha,\beta,\phi) + \Gamma^2(\alpha,\beta,\phi) \, f_2(\alpha,\beta,\phi) = 1 \tag{10}$$

where

$$f_1(\alpha,\beta,\phi) = F_1 \cos \alpha + F_2 \cos \beta$$

$$f_2(\alpha,\beta,\phi) = F_{11}\cos^2\alpha + F_{22}\cos^2\beta + F_{66}\cos^2\phi + 2F_{12}\cos \alpha \cos \beta$$

In this case, the failure function is given by the positive root of

$$\Gamma(\alpha,\beta,\phi) = \frac{-f_1(\alpha,\beta,\phi) \pm [f_1{}^2(\alpha,\beta,\phi) + 4f_2(\alpha,\beta,\phi)]^{\frac{1}{2}}}{2 \, f_2(\alpha,\beta,\phi)} \tag{11}$$

In an analogous fashion, the cubic tensor polynomial used by Tennyson et al. (4) can be written as:

$$\Gamma(\alpha,\beta,\phi) \ f_1(\alpha,\beta,\phi) + \Gamma^2(\alpha,\beta,\phi) \ f_2(\alpha,\beta,\phi) \ ve, \ and$$

$$+ \ \Gamma^3(\alpha,\beta,\phi) \ f_3(\alpha,\beta,\phi) = 1 \qquad (12)$$

$$+ \ 3F_{166} \cos \alpha \cos^2\phi + 3F_{266} \cos \beta \cos^2\phi \qquad (13)$$

In principle, it is possible to express any degree of the tensor polynomial expression in the parametric form $\Gamma = \Gamma(\alpha,\beta,\phi)$. However, for polynomials of degree higher than the quadratic tensor polynomial, the expression $\Gamma = \Gamma(\alpha,\beta,\phi)$ most unlikely cannot be written in explicit form. Nevertheless, these examples do demonstrate the generality of the parametric function $\Gamma = \Gamma(\alpha,\beta,\phi)$ and show the equivalence of the parametric and tensor polynomial approaches. As well, it can be shown that Budiansky's representation, Equations (1), is a special case of the parametric form, Equations (3).

A single parametric strength equation valid for all fibre-reinforced laminae has not yet been developed and, as a result, a particular failure equation must be invented for each material. This is a disadvantage in comparison with Hill-type and tensor polynomial theories in which the strength parameters are defined directly from specific experimental results such as uniaxial strengths in the axes of material symmetry. However, with the previous theories, the failure envelopes all have the same overall shape. For example, all quadratic tensor polynomials are ellipsoids. It is perhaps unrealistic to expect that a single equation can be found to represent all possible failure envelopes for a wide variety of fibre-reinforced laminae. Such an equation would certainly be useful, and it is to a certain extent an interesting feature of the tensor polynomial theory. This aspect has not been investigated in detail here, but there is reason to believe that such a general equation, should it indeed exist, might be easier to obtain in parametric form than with a tensorial equation. The reason for this conjecture is that any parametric equation can be constructed from a series of closed trigonometric surfaces. This technique is more likely to generate more universally applicable general failure surfaces than the higher-degree tensor polynomials, which do not necessarily describe closed surfaces.

As an example of a general parametric form, consider the following series representation:

$$\Gamma(\alpha,\beta,\phi) = \Sigma \ B_m \ [\cos m\alpha \sin m\beta \sin m\phi + \sin m\alpha \cos m\beta \sin m\phi$$

$$+ \sin m\alpha \sin m\beta \cos^2 m\phi] \qquad (14)$$

This form satisfies all of the requirements previously specified for orthotropic laminae. Alternatively, the following series could also be used:

$$\Gamma(\alpha,\beta,\phi) = \Sigma \ B_m \ [1 + \cos m\alpha \sin m\beta \sin m\phi]$$

$$[1 + \sin m\alpha \cos m\beta \sin m\phi] \ [1 + \sin m\alpha \sin m\beta \cos^2 m\phi] \qquad (15)$$

Both of the above trigonometric functions can theoretically describe any failure surface which is symmetric about the plane $\tau_{12} = 0$. However, a large number of terms may be required to describe the real failure surface in question. Other functions requiring less terms may be written. These equations should be more closely scrutinized in attempts to improve the usefulness of the parametric approach for the description of failure surfaces.

The parametric failure criterion presented above was written directly in stress space using spherical coordinates. However, depending on available experimental data, it may be more convenient to use other representations or coordinate systems. For example, it might be preferable to represent the failure envelope in non-dimensional form. We would then use the system of axes $\sigma_1/X - \sigma_2/Y - \tau_{12}/S$. Equations similar to Equations (3) can be written in this normalized space.

EXAMPLE OF APPLICATION

The experimental results used in this section have been obtained by Huang
(9). These are for a porous and brittle graphite, identified as a grade AGOT
graphite. This material has been tested under a wide range of biaxial tensile
and compressive loads. Neither pure shear nor combined shear-tension or shear-
compression tests have been performed. The results are in the plane $\tau_{12} = 0$.
These data show considerable scatter, particularly in the fourth quadrant in
which significantly different failure points along the same loading path have
been measured. This observation is common for this type of material. In the
biaxial compression area, that is the third quadrant, there is also some scatter
but it is relatively less important than that for the fourth one.

In order to fit the experimental failure data with a single failure equation, Huang suggests the following cubic tensorial failure criterion:

$$(F_1\sigma_1 + F_2\sigma_2)^K + (F_{11}\sigma_1^2 + F_{22}\sigma_2^2 + 2F_{12}\sigma_1\sigma_2 + F_{66}\tau_{12}^2)^L$$

$$+ (F_{111}\sigma_1^3 + F_{222}\sigma_2^3 + F_{112}\sigma_1^2\sigma_2 + F_{122}\sigma_1\sigma_2^2 \qquad (16)$$

$$+ F_{266}\sigma_2\tau_{12}^2 + F_{66}\sigma_1\tau_{12}^2)^M = 1$$

Since no shear stress τ_{12} is applied, this can be further reduced to

$$(F_1\sigma_1 + F_2\sigma_2)^K + (F_{11}\sigma_1^2 + F_{22}\sigma_2^2 + 2F_{12}\sigma_1\sigma_2)^L + (F_{111}\sigma_1^3$$

$$+ F_{222}\sigma_2^3 + F_{112}\sigma_1^2\sigma_2 + F_{122}\sigma_1\sigma_2^2)^M = 1 \qquad (17)$$

This equation contains nine constants, but the author gives no indication about
the method used to calculate these in order to fit the experimental data. In
the work by Huang two quadratic failure surfaces as well as Equation (17) were
illustrated and are reproduced here in Figure 4. All three curves give a fairly
good representation of the experimental data in the first three quadrants.
Unfortunately, the expansion of the failure equation to cubic form did not
really improve the simulation where it was the most important to obtain a better
curve-fit, namely in the fourth quadrant. In view of the considerable effort

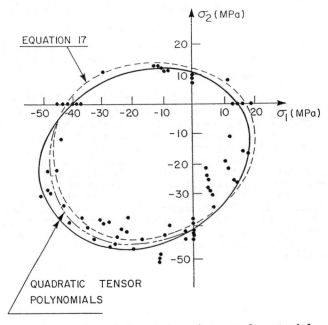

Fig. 4   Tensor polynomial equations for example material.

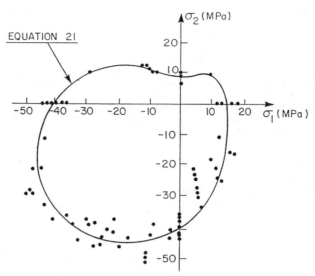

Fig. 5 Parametric solution for example material.

One additional advantage of the parametric failure criterion is that the number of parameters can easily be adjusted as a function of the experimental data available and the desired degree of accuracy. In contrast, with the tensor polynomial formulation, new values for the strength parameters F_{ij} and F_{ijk} are required, and identifying these values involves a relatively cumbersome procedure.

CONCLUSION

In recent years, many failure criteria have been proposed for composite materials. Among these, the tensor polynomial theory has received the most attention. This is due to many factors, the most important being that it is a single continuous equation that can be easily manipulated using the rules of tensor transformation. Also, it encompasses every criterion proposed previously. However, the quadratic form of the tensor polynomial expansion does not always provide an accurate representation of failure data for composites. Higher expansions require a much higher number of parameters and do not always describe closed surfaces. Furthermore, determining the failure stresses for a given loading path entails the sometimes cumbersome task of solving for the roots of a cubic equation.

In view of the ongoing discussions concerning the merits of the tensor polynomial theory, it seemed timely to explore a new approach for predicting failure of fibre-reinforced laminae. Ideas recently brought forward by Budiansky for predicting plastic yielding in sheet materials were adapted here to describe the strength of orthotropic laminae. It was suggested that a general parametric representation of the failure surface be explored.

The parametric approach has the following advantages. First, it was shown that the tensor polynomial theory is a special case of the proposed parametric criterion, and that all existing criteria could be written in parametric form. Second, the failure surface will always be closed if appropriate functions are selected. Third, failure surfaces of irregular shapes can be described with few parameters. This can be achieved through the use of higher order equations, the use of non-integer exponents, or by fitting different portions of the failure surface with different equations. In fact, the flexibility of the parametric approach is its greatest advantage. Finally, applying the parametric criterion to find the failure strength for any proportional loading path is a straightforward matter, in contrast to the involved methods usually required with the higher order forms of the tensor polynomial theory.

121

On the other hand, a single parametric function, applicable to all fibre-reinforced laminae, has not yet been postulated. It was shown that series of trigonometric functions can describe all failure surfaces that can be obtained experimentally. However, a new equation $\Gamma = \Gamma(\alpha,\beta,\phi)$ has to be invented for each material, depending on the available experimental data. When only the uniaxial strengths in the axes of symmetry are known, there are no grounds for rejecting the quadratic tensor polynomial equation in favour of the parametric approach because all experimental data are well reproduced with this theory. However, if additional results cannot be reconciled with the quadratic form, using a parametric approach can be a useful alternative to higher expansions of the tensor polynomial equation.

ACKNOWLEDGEMENTS

This work was supported by the Natural Sciences and Engineering Research Council of Canada (NSERC) and the Government of the Province of Québec (Programme FCAR).

REFERENCES

1. R.S. Soni, J. Reinf. Plast. Compos., 1, 34 (1983).

2. S.W. Tsai and E.M. Wu, J. Compos. Mater., 5, 58 (1971).

3. E.M. Wu, "Composite Materials, Vol. 2: Mechanics of Composite Materials," p. 353, Academic Press, New York (1974).

4. R.C. Tennyson, D. MacDonald and A.P. Nanyano, J. Compos. Mater., 12, 63 (1978).

5. B. Budiansky, "Mechanics of Material Behavior," p. 15, Elsevier Science, Amsterdam (1984).

6. R. Hill, "The Mathematical Theory of Plasticity," p. 318, Oxford University Press, London (1950).

7. R. Hill, Math. Proc. Camb. Phil. Soc., 85, 179 (1979).

8. J.C. Suhling, R.E. Rowlands, M.W. Johnson and D.E. Gunderson, Exp. Mech., 25, 75 (1985).

9. C.L.D. Huang, Proc. 10th Can. Congr. Appl. Mech., London, Ont., p. A45, (1985).

MICRO/MACRO MECHANICS FOR COMPOSITE FRACTURE CRITERIA

J. J. Kollé, K. Y. Lin,* C. C. M. Eastland, and A. C. Mueller
Flow Research, Incorporated
Kent, Washington

*Department of Aeronautical Engineering
University of Washington
Seattle, Washington

ABSTRACT

This paper describes the results of an integrated microscopic/macrosopic analysis of fracture in continuous fiber-reinforced composite. A macroscopic analysis of a composite double-cantilever-beam fracture toughness test specimen was carried out using a singular element technique. The shear component of the strain energy release rate becomes important for some fiber orientations. Results from this analysis were input as boundary conditions to a microscopic model used to calculate J-integral values in the crack tip region. The plasticity of the matrix in the crack tip region dissipates the fracture energy. The constitutive equation chosen for the matrix plasticity was shown to have an important effect on the J-integral value, while the crack tip geometry had little effect.

NOMENCLATURE

d	distance	T	traction vector
E	Young's modulus	U	boundary displacement conditions
G	defined in Equation (6)	u	displacement vector
G_T	total strain energy release rate	$\tilde{u}$	interelement boundary displacement
G_I	opening mode strain energy release rate	W	strain energy density
G_{II}	shear mode strain energy release rate	α	scalar constant
H	defined in Equation (6)	β	defined in Equation (7)
I_m	hybrid functional for m^{th} element	Γ	line integral curve
J_I	J-integral	ϵ_{max}	maximum strain contour level
K_I	stress intensity factor	θ	ply angle
k	element stiffness matrix	μ	internal coefficient of friction
L	shape function along interelement boundaries	ν	Poisson's ratio
		σ	internal stress tensor
N	number of terms used	τ_o	strength in simple shear
n	component of unit vector normal to the element boundary	ϕ	yield criterion
P	load applied to sample specimen		
P	boundary stress conditions		
q	nodal displacement vector		
R	Boundary traction conditions		
r_y	radius of plastic zone		
S_m	interelement boundary		

INTRODUCTION

Advanced composite materials offer unsurpassed stiffness-weight ratios coupled with high strength. These materials are now being considered as cost-effective replacements for applications requiring high stiffness, strength and a low coefficient of thermal expansion (CTE). The Army, Navy and Air Force have identified a number of structural components that could benefit from discontinuous silicon-carbide-whisker- or particle-reinforced aluminum (SiC_w/Al and SiC_p/Al) (DiCarlo, 1985). NASA has identified continuous boron-fiber-reinforced aluminum (B/Al) composite as a possible replacement for titanium in advanced gas turbine engines (Logsdon and Liaw, 1986). Graphite-reinforced aluminum has applications in space structures, which require high rigidity when exposed to temperature extremes, as it combines the high strength of aluminum with the low CTE of graphite (Chawla, 1987).

In all of these applications, the reinforcing fiber and matrix material both contribute important properties to the composite. For example, McDanels (1985) has shown that the ultimate elongation and strength of a 20-percent SiC_w/Al composite strongly depends on the choice of matrix alloy. The geometric arrangement of the fibers also contributes to the macroscopic properties. Cooper and Kelly (1967) have demonstrated that the work of fracture in a plane normal to reinforcement in a metal matrix composite is proportional to the fiber diameter for a fixed fiber volume fraction. This work led to the development of larger-diameter boron fibers for B/Al composites, which have relatively low toughness (DiCarlo, 1985). The increase in diameter resulted in an increase in strength, although not proportional to the increase in fiber diameter. Failure in the material studied was transferred to the boron fibers.

The characterization of composite materials requires a practical means of deducing the macroscopic properties of a finished part from the microscopic properties of its component fibers and matrix. In this paper, we evaluate the feasibility of applying microscopic/macroscopic design techniques developed for the elastic structural analysis of composites to the analysis of fracture. A preprocessor code using the method of mixtures or shear-lag models of the matrix/reinforcement behavior is used to define the elastic properties of a single ply. The elastic properties of the composite laminate can then be determined from a combination of the ply properties using laminated plate theory. A structural analysis using plate theory or finite element analysis provides the state of stress in a given ply, and failure criteria can be applied. Examples of this integrated approach include the limit-load analysis technique presented by Tsai (1987) and the integrated composites analyzer code (ICAN) for structural analysis of advanced composites developed by NASA (Hopkins, 1987).

Analysis of other types of mechanical behavior, such as thermal loading, fatigue or fracture, can also be approached by relating microscopic and macroscopic properties. In this work, we combine a microscopic J-integral evaluation of the strain energy release rate near a crack tip with a macroscopic finite element analysis of the displacement field in a laminated composite test specimen to obtain fracture propagation criteria.

Figure 1 shows the type of microscopic failure region modeled in this work. The figure shows several cracks propagating through a unidirectional ply transversely. This is typical of the initial damage observed during the loading of composite materials. This type of damage can have a large effect on the ultimate strength of the composite structure. Unfortunately, fracture initiation and propagation in uniaxial composite plies cannot be predicted from the fracture toughness of the matrix. The situation is even more complicated in materials reinforced with brittle fibers, such as boron, which may also fracture.

The microscopic and macroscopic analyses are presented separately in this paper. The macroscopic analysis was used to evaluate mode I and II strain energy release rates for a double-cantilever beam (DCB) fracture toughness test specimen. The analysis was carried out for laminated composite DCB specimens consisting of four plies with varying fiber orientations. The results show the importance of the mode II strain energy release rate for some ply configurations.

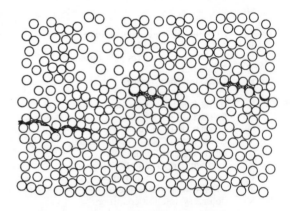

Fig. 1. Microscopic failure associated with intralaminar delamination

The microscopic analysis was carried out with the FEAP finite element analysis code coupled with commercial pre- and post-processing software. This code was modified to calculate the strain energy release rate in the crack tip region using the J-integral approach. Boundary conditions for the microscopic analysis were provided by an orthotropic material finite element code with an embedded singular element. The analysis shows the importance of matrix plasticity in providing composite toughness and a surprising insensitivity of strain energy release rate to crack tip geometry.

MACROSCOPIC FRACTURE ANALYSIS

In macroscopic analysis, each layer in a composite laminate is assumed to be homogeneous and orthotropic. Interlaminar failure usually occurs in the form of delamination between adjacent layers when a fracture parameter, such as the strain energy release rate, G, reaches its critical value. Thus, an accurate determination of the strain energy release rate as a function of crack geometry and laminate orientation is important in developing an appropriate macroscopic fracture criterion.

In the present study, the G values associated with a delamination crack are calculated using the singular finite element method. In this approach, singular elements incorporating the exact displacement and stress fields from the closed-form solution are employed near the crack tip region. These singular elements are then combined with the regular isoparametric elements in the surrounding region. Standard finite element procedures are then used to obtain the displacement at each node, after which the stress intensity factor or strain energy release rate can be calculated from the displacement solution. Since the singular element contains the exact form of the stress and displacement fields, this method yields extremely accurate solutions, typically within 1 or 2 percent.

Formulation

The development of singular elements is based on a hybrid functional derived by Tong et al. (1973). This hybrid functional was used by Lin and Mar (1976) for the study of bi-material crack problems, by Lin and Tong (1978) for crack problems in orthotropic materials and recently by Aminpour and Holsapple for the dynamic analysis of cracks between two anisotropic materials (Aminpour, 1986). The formulation of a singular element for interlaminar fracture problems is briefly described in the following. Details of this hybrid finite element technique for the study of fracture problems in composites can be found in the references.

For a two-dimensional continuum divided into individual elements, the hybrid functional for the mth element is given by

$$I_m = \int_{S_m} (n\sigma)^T \tilde{u} \, ds - \frac{1}{2} \int_{S_m} (n\sigma)^T u \, ds \qquad (1)$$

where u and n are interior displacements and stresses, respectively. The term $\tilde{u}$ is the interelement boundary displacement, n is the component of the unit vector normal to the element boundary, and S_m is the interelement boundary.

In the above functional, the interior stresses and displacements are required to satisfy all the elasticity equations and stress-free boundary conditions over the crack surface. The Euler equations for the given functional are

$$\tilde{u} - u = 0 \qquad \text{and} \qquad \Sigma(n\sigma) = 0 \qquad (2)$$

on the interelement boundary.

Making use of the interior stresses and displacements from the closed-form solution, we can write

$$\sigma = P\beta \, , \qquad u = U\beta \qquad \text{and} \qquad n\sigma = R\beta \qquad (3)$$

The interelement boundary displacements are assumed to be related to the nodal displacements q by

$$u = Lq \qquad (4)$$

where L is the shape function along the interelement boundaries and q is the nodal displacement vector.

Substituting Equations (3) and (4) into (1), and taking the variation of the functional with respect to β_i, the element stiffness matrix $\mathbf{k}$ is obtained as

$$\mathbf{k} = \mathbf{G}^T \mathbf{H}^{-1} \mathbf{G} \tag{5}$$

where

$$\mathbf{G} = \int_{S_m} \mathbf{R}^T \mathbf{L} \ ds \ , \qquad \mathbf{H} = \frac{1}{2} \int_{S_m} (\mathbf{R}^T \mathbf{U} + \mathbf{U}^T \mathbf{R}) \mathbf{L} \ ds \tag{6}$$

and

$$\beta = \mathbf{H}^{-1} \mathbf{G} \mathbf{q} \tag{7}$$

It must be noted that the evaluation of the stiffness matrix given above requires integration only along the boundary of the crack element, thus avoiding the stress singularity. Solution of the global equations results in the nodal displacements $\mathbf{q}$, from which β can be calculated using Equation (7). The stresses and displacements at any point in the element can then be obtained.

Numerical Results

The singular element method described above is especially useful for direct evaluation of mixed-mode stress intensity factors (K_I, K_{II}) or strain energy release rates (G_I, G_{II}). Since the K and G values are related, only the strain energy release rate G will be reported in this study. For a general two-dimensional crack problem, the strain energy release rate can be expressed by the following crack closure integral (Paris and Sih, 1965):

$$G_T = \lim \frac{1}{d} \int_0^d \frac{1}{2} \left[\sigma_z(x,0) \ w(x-d,0) + \tau_{xz}(x,0) \ u(x-d,0) \right] dx \tag{8}$$

The first term in the integral can be defined as G_I (opening mode) and the second term as G_{II} (shear mode). The stresses are evaluated at a distance x ahead of the crack tip, and the corresponding displacements are calculated at a distance d-x behind the tip of the crack.

The model problem considered in the present analysis is a DCB specimen, which has been widely used in interlaminar fracture toughness tests. As shown in Figure 2, the specimen consists of a delamination crack between plies and is subjected to a pair of concentrated loads, P. The material properties and specimen dimensions used are given in Table 1. These properties are typical of graphite/epoxy composites. The condition of plane strain is assumed in the analysis.

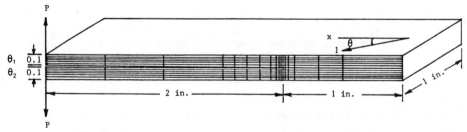

Fig. 2. Finite element mesh pattern used for macroscopic analysis

Table 1. Material properties used in macroscopic model*

Property	E_{11}	E_{22}	E_{33}	G_{12}	G_{23}	G_{31}	ν_{12}	ν_{23}	ν_{31}
Value	18.0	1.4	1.4	0.9	0.5	0.9	0.34	0.40	0.34

*E_{ii} and G_{ii} are expressed in 10^6 psi.

Numerical results on strain energy release rates have been obtained for both symmetric and unsymmetric laminates. In the symmetric case, the laminate is made of an angle ply of uniform orientation. The strain energy release rates G_I for various values are given in Table 2 and plotted in Figure 3. In these calculations, 96 Gaussian integration points were used to calculate the integral in Equation (8) numerically. It is seen that the G_I value increases with orientation. This is because the stiffness of the laminate decreases as θ increases, resulting in a larger crack opening displacement and thus a larger G_I value. However, the critical G_{Ic} value or fracture toughness for different laminate orientations may not be the same. Therefore, experimental determination of G values is required before a fracture criterion can be established.

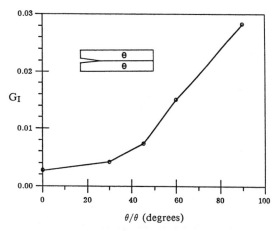

Fig. 3. Mode I strain energy release rate versus orientation angle for symmetric (θ/θ) laminate DCB specimen

Table 2. G_I values for (θ/θ) laminates, P = 1 lb

Rate (in-lb/in²) \ θ	0°	30°	45°	60°	90°
$G_I \times 10^3$	2.6677	4.1695	7.3316	14.962	28.238

In the case of unsymmetric laminates, both tensile and shear stress fields are induced near the crack tip under a load P due to differences in layer properties. As a result, the mode II strain energy release rate G_{II} exists in addition to the mode I value. Table 3 shows the results of G_I and G_{II} calculations for laminates with $(0/\theta)$ orientations. The total strain energy release rate G_T is also given. These results are also plotted in Figures 4 and 5 as a function of θ. Similar to the symmetric case, the strain energy release rate increases with θ. Also, it is noted that the G_{II} values are smaller than the corresponding G_I values for all but the $(\theta/60)$ laminate. Again, test data for critical G_{Ic} and G_{IIc} toughness values are needed to develop a mixed-mode fracture criterion.

Table 3. G_I and G_{II} values for $(0/\theta)$ laminates, P = 1 lb

Rate (in-lb/in²) \ θ	30°	45°	60°	90°
$G_I \times 10^3$	3.2330	3.8274	4.4537	5.3704
$G_{II} \times 10^3$	0.1953	1.2697	4.8632	1.1531
$G_T \times 10^3$	3.4283	5.0971	9.3169	6.5235

The order of stress singularity for a crack between the 0 and θ plies $(0/\theta)$ is a complex number and, thus, the stresses are oscillatory near the crack tip (Williams, 1959). The G_I and G_{II} values were shown to be oscillating with the distance d in Equation (8). However, the total G value is independent of the d value (Lin and Walker, 1988). Therefore, in this case, it is more meaningful to use the total G value to characterize fracture behavior rather than its components G_I and G_{II}.

Further parametric studies on the effects of crack geometry and laminate configuration can be conducted using the present analysis technique. These analysis results coupled with experimental

127

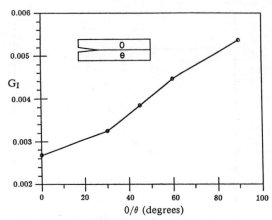

Fig. 4. Mode I strain energy release rate versus orientation angle for unsymmetric $(0/\theta)$ laminate

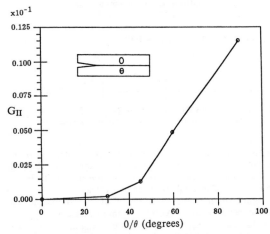

Fig. 5. Mode II strain energy release rate versus orientation angle for unsymmetric $(0/\theta)$ laminate

data on toughness values and an appropriate interlaminar failure criterion can then be developed for the characterization of composite materials.

MICROMECHANICAL FRACTURE ANALYSIS

It is possible to determine the load at which a fracture will propagate from a pre-existing flaw in metals by determining the value of the stress intensity factor from the theory of linear elastic fracture mechanics (LEFM) and comparing this value with the fracture toughness of the material. In isotropic, nonlinear materials, such as ceramics or polymers, LEFM theory no longer holds, but fracture predictions can still be made by calculating the strain energy release rate G, which would be the energy released were the fracture to propagate a small distance. The value of G can be obtained from the J-integral around the region where the nonlinear behavior occurs.

We are interested here in the propagation of fractures on a microscopic scale through a fiber-reinforced matrix. In this case, the stress intensity factor may be expected to vary considerably as the fracture passes the reinforcing fibers, and the meaning of the stress intensity factor becomes ambiguous. The strain energy release rate fracture criterion retains its physical meaning in the case of material inhomogeneity, large strain and material plasticity. In this study, we determine whether the value of G determined from the macroscopic anisotropic model described above corresponds to the G value obtained from a detailed analysis of the stress field near the crack tip.

J-Integral Formulation

The J-integral calculation provides a convenient means of calculating strain energy release rates for both elastic and nonlinear materials because of its path invariance in the linear region. It is usually calculated indirectly by crack closure techniques, which amount essentially to a direct calculation of the strain energy release rate. These techniques are convenient and are presently available for straight cracks in isotropic nonlinear materials on a variety of codes, including MARC, Abaqus and Ansys. However, in a nonhomogeneous material, the fracture interacts with the reinforcement. The tip of the crack may be curved around a fiber or may even be discontinuous due to debonding. These considerations would complicate a crack closure technique, so we calculate the line integral directly.

The integral is calculated on an arbitrary boundary, Γ, connecting the upper and lower surfaces of the crack as shown in Figure 6. While this integral can take a variety of forms (Nishioka and Atluri, 1984), we use the one given by Rice (1968):

$$J = \int_{\Gamma} W \, dy - \mathbf{T} \cdot \partial\mathbf{u} \, \partial x \, ds \qquad (9)$$

where W is the strain energy density:

$$W = \int_0^\epsilon \sigma_{ij} \, de_{ij} \qquad (10)$$

and T and u are the traction and displacement vectors on the boundary. This integral is path independent for any path that ends on the straight portion of the crack. The path used must be defined for the finite element program. This is accomplished in the preprocessor code.

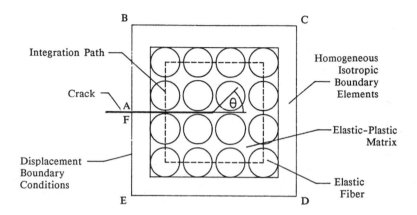

Fig. 6. Microscopic region near crack tip showing J-integral path

Finite Element Model

The microscopic analysis was carried out using FEAMOD, which is a modified version of the FEAP finite element analysis code originated by R. L. Taylor (see Zienkiewicz, 1977). This code has a modular structure designed for the easy introduction of new element types and analysis procedures by introducing new subroutines without modifying the code structure (Mueller, 1986). Analysis procedures are called, using a macro-level command language, that read input data, form the stiffness matrix, increment loads, iterate nonlinear solutions and calculate desired outputs. At present, the code calculates the J-integral when the macro post-processing command to calculate and print stress is issued.

Elastic calculations were carried out using a plane-strain linear elastic formulation with linear triangular elements. The plastic solutions were obtained with a plane-strain elastic-perfectly plastic finite strain formulation (Pritchard, 1986). This plasticity formulation allows the selection of any desired isotropic flow rule including associative and nonassociative flow.

Mesh generation was carried out using a commercial preprocessor. This is a menu-driven code designed to create input decks for a variety of finite element codes. The user first generates a geometry that describes the problem at hand. In our case, we are interested in evaluating the state of stress and the stress singularity in an area of unidirectional composite near the tip of an intralaminar crack. The near-tip region can be defined as the area within which nonlinear energy dissipative mechanisms such as plasticity are active.

It has been shown in elasticity studies (Chen et al., 1970; Parks, 1974) that the path-independent J-integral provides accurate estimates of stress intensity even when the crack tip region is modeled with a relatively coarse grid. The mesh density was chosen to represent the curvature of the fiber and matrix geometry accurately but not the stress singularity at the crack tip. This model represents the type of fracture observed in the early stages of composite damage and can be responsible for severe degradation of component stiffness and ultimate strength.

The model consists of 16 fibers in a square array with a volume fraction of 62 percent. It is surrounded by a boundary layer of elements with elastic moduli determined from a phase-averaged method of mixtures technique based on the fiber and matrix moduli (Tsai, 1987). The matrix material is assumed to act either as an elastic material or as an elastic-perfectly plastic material.

The J-integral path is chosen to pass through the outer ring of fibers and to have sides parallel with the principal axes of the model to simplify calculations.

Boundary Conditions

Displacement boundary conditions were derived from the elastic orthotropic plane strain model of a (90/90) composite laminate DCB fracture toughness sample subject to opening-mode loading. Displacements from this model were calculated for the small region surrounding the crack tip and used as boundary conditions for the microscopic finite element analysis. The displacement of the boundary of the model thus corresponds to the analytic solution. These boundary conditions were applied to a homogeneous, isotropic boundary layer surrounding the inhomogeneous region to avoid inconsistencies of applying homogeneous boundary conditions to an inhomogeneous material. In the case of a (90/90) laminate, the elastic orthotropic properties reduce to the isotropic case used as the boundary layer of the model.

Model Verification

The technique was validated by applying displacement boundary conditions from the analytic solution to a mode I opening crack to the boundary of a homogeneous, isotropic model of a crack. The strain energy release rate G in an isotropic, linear elastic medium is equal to the value of the J-integral:

$$J_I = G_I = (1 - \nu^2)K_I^2/E \tag{11}$$

where E is Young's modulus.

The analytic solution for displacements along the boundary was calculated to two significant figures at 80 points along the boundary of the model and input as displacement boundary conditions. Figure 7 gives the displacements on this boundary at an arbitrary value of the mode I stress intensity factor, assuming isotropic elastic moduli from a mixtures estimate and plane strain conditions. A similar distribution could be obtained from a macroscopic calculation of the mode II stress intensity factor.

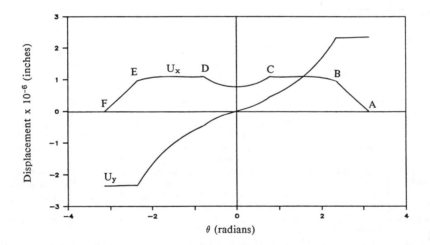

Fig. 7. Analytic displacements for mode I crack on J-integral boundary

The comparison between the analytic and finite element solutions is given in Table 4 for two values of Young's modulus. The model gives a value of G_I that is 6.4 percent greater than the analytic solution. This is remarkably accurate, given the limited accuracy of the displacement boundary conditions. When a higher stiffness is used, the accuracy declines slightly because the displacements are smaller. These results confirm the accuracy of the J-integral technique for stress intensity factor calculations even when no attempt is made to account for the singularity at the crack tip.

Table 4. Validation of J-integral calculation in an isotropic medium

K_I (psi $\sqrt{\text{in}}$)	Analytic Solution			Model	
	$E \times 10^{-6}$ (psi)	$G_I \times 10^3$ (in–lb/in^2)		$J_I \times 10^3$ (in–lb/in^2)	$(G_I - J_I)/G_I$
60.64	1.29	2.6677		2.84	6.5%
226.5	18.0	2.6677		2.88	8.0%

Finite Element Analysis

Three different crack geometries were investigated using this technique. The straight crack geometry was the same as used in the homogeneous calculation. The other two cases simulate debonding failure at the tip of the crack (Debond 1) and with a debond around a fiber ahead of the crack tip (Debond 2). These geometries are shown along with their finite element meshes in Figure 8. The analytic displacement field is applied to the outer edge of a set of boundary elements with moduli determined by the Halpin-Tsai equations (Tsai, 1987).

The material properties used in these calculations are given in Table 5. In the non-homogeneous case, the interior is represented by a two-phase region with high-modulus fibers embedded in a low-modulus matrix. Calculations were also made for the homogeneous case where all the material properties were set equal to that determined by the Halpin-Tsai equations.

Table 5. Material properties used in finite element analysis

Property	Fiber	Matrix	Boundary
$E \times 10^{-6}$ (psi)	19.1	0.26	1.4
ν	0.22	0.35	0.27

Elastic Strain Energy Release Rate

The results of the J-integral calculations are given in Table 6. The J_I values given should be compared with $G_I = 28.2 \times 10^{-3}$ in-lb/in^2 from the macroscopic analysis results for a (90/90) laminate given in Table 2. The microscopic calculations of the strain energy release rate are reasonably close to the macroscopic value for the straight crack and the Debond 1 geometry. The stiffness of the Debond 2 configuration is significantly greater, and the J_I value is consequently greater. It is interesting to note that the nonhomogeneous J_I is significantly lower than that for the homogeneous case, indicating that this is a lower-energy configuration. The difference in these two values may also indicate that the value used for the homogeneous modulus is high by as much as 20 percent.

Table 6. Results of J-integral calculations

Configuration	Homogeneous		Nonhomogeneous	
	$J_I \times 10^3$ (in–lb/in^2)	$(J_I - G_I)/G_I$	$J_I \times 10^3$ (in–lb/in^2)	$(J_I - G_I)/G_I$
Straight Crack	30.5	8.2%	25.7	-8.9%
Debond 1	31.5	11.7%	26.2	-7.1%
Debond 2	55.3	96.1%	35.2	24.8%

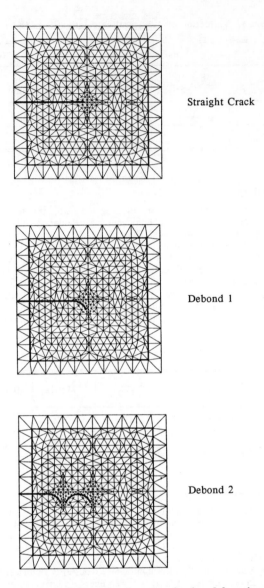

Straight Crack

Debond 1

Debond 2

Fig. 8. Crack configurations and finite element meshes developed for microscopic analysis

In Figure 9, the octahedral shear strain near the tip of the Debond 1 crack tip is plotted for three cases. This is a measure of maximum shear strain directly analogous to octahedral shear stress. Case 1 is a curved crack tip in a homogeneous material. Curvature of the crack tip distorts the strain field slightly but does not have a significant effect on the stress intensity factor as derived from J_I. In the second case shown, the fibers and matrix are assumed to have different elastic moduli, with the fiber being much stiffer than the matrix. This severely distorts the strain field with the apparent effect of reducing the effective stress intensity factor for this crack.

Plasticity Effects

In the third case shown in Figure 9, the matrix is allowed to yield plastically assuming a Von Mises yield surface and associative flow. This type of plasticity is generally adequate for yield in metals. The yield criterion is

$$\phi = \tau - \tau_0 \tag{12}$$

132

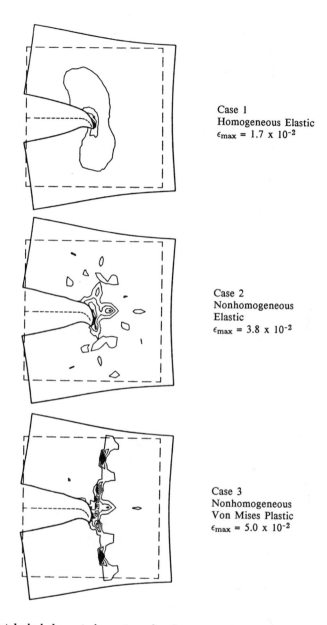

Case 1
Homogeneous Elastic
$\epsilon_{max} = 1.7 \times 10^{-2}$

Case 2
Nonhomogeneous
Elastic
$\epsilon_{max} = 3.8 \times 10^{-2}$

Case 3
Nonhomogeneous
Von Mises Plastic
$\epsilon_{max} = 5.0 \times 10^{-2}$

Fig. 9. Octahedral shear strain contours for three cases of the Debond 1 geometry

where τ is the octahedral shear and τ_o is the shear strength. The plastic strain rate is normal to the yield surface

$$d\epsilon/dt = \alpha \quad \partial\phi/\partial\sigma \qquad (13)$$

where α is a scalar constant that is adjusted to restrain the stresses to lie within the yield surface. In this case, the highest strains are no longer found at the crack tip but are displaced to the matrix near the column of fibers immediately ahead of the crack tip.

Figure 10 shows the effect of changing the yield criterion on the strain field near the tip of the Debond 2 crack configuration. The first plot shows the strain field for the elastic case. Maximum strains are found in the ligament between the tip of the crack and the debond around the fiber ahead of the crack. When this material is allowed to yield according to the Von Mises

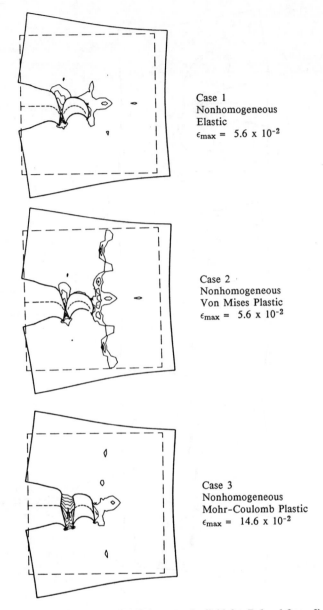

Case 1
Nonhomogeneous
Elastic
$\epsilon_{max} = 5.6 \times 10^{-2}$

Case 2
Nonhomogeneous
Von Mises Plastic
$\epsilon_{max} = 5.6 \times 10^{-2}$

Case 3
Nonhomogeneous
Mohr-Coulomb Plastic
$\epsilon_{max} = 14.6 \times 10^{-2}$

Fig. 10. Effects of plasticity on octahedral shear strain field for Debond 2 configuration

criterion, yielding is concentrated at the ligament as expected; it then begins to transfer to the column of fibers ahead of the debond. Once the ligament fails, the Debond 2 configuration becomes the same as Debond 1 as in Case 3 of Figure 9. If the flow rule is changed, the deformation pattern in the ligament becomes quite different.

In Case 3 yield was assumed to follow a Mohr-Coulomb criterion

$$\phi = \tau - \tau_0 + \mu\sigma \tag{14}$$

where μ is the internal coefficient of friction and σ is the mean stress. The Mohr-Coulomb yield criterion is probably more appropriate for an epoxy matrix, which is stronger in compression than tension. Because of tensile yielding, the strain in the ligament is quite high, and yielding is transferred to the column of fibers ahead of the debonded fiber. The contour increment is different for Case 2 and Case 3, and the strain level in the column ahead of the crack tip is actually similar.

The results of the J-integral calculation for the plastic case are given in Table 7. Plasticity significantly reduces the strain energy release rate in all cases, because some of the elastic energy that would be released by crack propagation is dissipated by yielding. The type of yielding also has a large effect. Because the Mohr-Coulomb criterion allows for tensile yield, the strain levels are higher and the energy dissipated plastically is greater. Strain levels were so high in the Debond 1 configuration that the Mohr-Coulomb case would not converge.

Table 7. Results of J-integral calculations with plasticity

Configuration	Von Mises		Mohr-Coulomb	
	$J_I \times 10^3$ (in-lb/in^2)	$(J_I - G_I)/G_I$	$J_I \times 10^3$ (in-lb/in^2)	$(J_I - G_I)/G_I$
Debond 1	19.5	-31%	NC*	--
Debond 2	32.4	15%	12.4	-56%

*NC = no convergence.

The finite element analysis indicated significant levels of plastic strain at a relatively low load. Because we assume that the line integral completely encloses the plastic region, it is important to determine its size and the limits on material type for which this analysis is valid.

An estimate of the size of the plastic region can be obtained by comparing the material yield stress and the stress at some radius, r, from a crack in an equivalent linear elastic material subject to a critical stress intensity factor. In our model, only the matrix material is subject to yielding, and the radius of the plastic zone is found from

$$r_y = (1/2\pi)(K_I/\sigma_{ys})^2 \tag{15}$$

For example, a high-strength aluminum alloy, 7178, has a yield strength of 78 x 10^3 psi with a fracture toughness of 25 x 10^3 psi $\sqrt{\text{in}}$. The yield radius is about 0.016 inch. A typical Al/B metal matrix composite contains boron fibers with a diameter of 0.005 inch, so it is necessary to include a number of fibers in the model if the J-integral at fracture is desired. A typical epoxy material will have a yield stress of 25 x 10^3 psi and a toughness of 10 x 10^3 psi $\sqrt{\text{in}}$ for a yield radius of 0.025 inch. The radius of graphite fibers is only 0.0005 inch, so the yielded area will include over a thousand fibers.

An appropriate model for such a large number of fibers will require a phase-averaged flow rule that can be verified by the type of detailed finite element analysis presented here. In this study, we have focused on the metal matrix problem where the plastic region around the crack tip interacts with a few fibers and averaging properties no longer make sense.

CONCLUSION

The results obtained show the feasibility of applying micromechanical material properties to the analysis of fracture in composite materials using existing numerical techniques.

The macromechanical analysis showed the importance of layup angle and mode II stress intensity factors on calculating the strain energy release rate for composite laminates. The micromechanical analysis showed the importance of fiber geometry and microscopic properties of fiber and matrix. These types of analysis will be critical to the development and selection of high-performance metal matrix and ceramic matrix materials and to demonstrating that significant performance improvements can be achieved by proper geometry and materials selection.

Both the fiber reinforcement configuration and the material plasticity have been shown to reduce strain energy release rates for the material properties studied. These reductions would lead to a prediction of greater toughness for the reinforced material than would be expected in an unreinforced matrix test if the critical strain energy release rate of the reinforced material is the same as that of the unreinforced. The reduced strain energy release rate is the result of a high strain energy density due to stress concentrations near fibers in the crack tip region. The effect is multiplied when plasticity occurs, because much of the strain is concentrated away from the crack tip reducing the energy available to drive crack propagation.

The J-integral path used for this work included a relatively large region with both fiber reinforcement and matrix material near the crack tip. Insensitivity of this integral to crack tip geometry does not imply that the fracture propagation criterion will be insensitive to geometry. The prediction of fracture propagation may require consideration of the strain energy release rate in a very small region of matrix material at the crack tip. An experimental program is now underway to determine whether either of these fracture criteria are appropriate for metal matrix composites.

REFERENCES

Aminpour, M. A., 1986, "Finite Element Analysis of Propagating Interface Cracks in Composites," Ph.D. Dissertation, University of Washington, Seattle, WA.

Chawla, K. K., 1987, *Composite Materials Science and Engineering*, Springer Verlag, New York.

Chen, S. K., Tuba, I. S., and Wilson, W. K., 1970, "On the Finite Element Method in Linear Fracture Mechanics," *Eng. Fract. Mech.*, Vol. 2, pp. 1-17.

Cooper, G. A., and Kelly, A., 1967, "Tensile Properties of Fibre Reinforced Metals: Fracture Mechanics," *Journal of the Mechanics and Physics of Solids*, Vol. 15, pp. 279-297.

DiCarlo, J. A., 1985, "Fibers for Structurally Reliable Metal and Ceramic Composites," *Journal of Meteorology*, June, pp. 44-49.

Hopkins, D. A., 1987, "Integrated Analysis and Applications," *Aeropropulsion '87, Session 2 - Aeropropulsion Structures Research*, NASA Conference Publication 10003, NASA Lewis Research Center, pp. 37-48.

Lin, K. Y., and Mar, J. W., 1976, "Finite Element Analysis of Stress Intensity Factors for Cracks at a Bi-material Interface," *Int. J. Fract.*, Vol. 12, pp. 521-531.

Lin, K. Y., and Tong, P., 1978, "A Hybrid Crack Element for the Fracture Mechanics Analysis of Composite Materials," *Proceedings, 1st International Conference on Numerical Methods in Fracture Mechanics*, A. R. Luxmoore and D. R. J. Owen, eds., Noordhoff International, Leyden.

Lin, K. Y., and Walker, T., 1988, "A Singular Finite Element Analysis of Delamination Fracture in Composite Materials," to be published.

Logsdon, W. A., and Liaw, P. K., 1986, "Tensile, Fracture Toughness and Fatigue Crack Growth Rate Properties of Silicon Carbide Whisker and Particulate Reinforced Aluminum Metal Matrix Composites," *Eng. Fract. Mech.*, Vol. 24, pp. 737-751.

McDanels, D. L., 1985, "Analysis of Stress-Strain, Fracture and Ductility Behavior of Aluminum Matrix Composites Containing Discontinuous Silicon Carbide Reinforcement," *Met. Trans. A*, Vol. 16, pp. 1105-1115.

Mueller, A. C., 1986, "FEAMOD: A Modifiable Finite Element Analysis Program," Flow Research Report No. 381, Flow Research, Inc., Kent, WA.

Nishioka, T., and Atluri, S. N., 1984, "Path Independent Integral and Moving Isoparametric Elements for Dynamic Crack Propagation," *AIAA Journal*, Vol. 22, pp. 409-414.

Paris, P. C., and Sih, G. C., 1965, "Stress Analysis of Cracks," *Symposium on Fracture Toughness Testing and Its Applications*, ASTM Special Technical Publication No. 381, pp. 30-83.

Parks, D. M., 1974, "A Stiffness Derivative Finite Element Technique for Determination of Crack Tip Stress Intensity Factors," *Int. J. Fracture*, Vol. 10, pp. 487-502.

Pritchard, R. S., 1986, "Elastic-Plastic Material Model," Technical Report No. 86-01, IceCasting Inc., Seattle, WA.

Rice, J. R., 1968, "A Path Independent Integral and the Approximate Analysis of Strain Concentration by Notches and Cracks," *Journal of Applied Mechanics*, June, pp. 379-386.

Tong, P., Pian, T. H. H., and Lasry, S. J., 1973, "A Hybrid-Element Approach to Crack Problems in Plane Elasticity," *Int. J. Num. Methods Eng.*, Vol. 7, pp. 297-308.

Tsai, S. W., 1987, *Composites Design*, Think Composites, Dayton.

Williams, M. L., 1959, "The Stresses Around a Fault or Crack in Dissimilar Media," *Bulletin of the Seismological Society of America*, Vol. 49, pp. 119-204.

Zienkiewicz, O. C., 1977, *The Finite Element Method*, 3rd Edition, McGraw-Hill, London.

MODELING FRACTURE OF RESIN MATRIX COMPOSITES

C. T. Herakovich
University of Virginia
Charlottesvile, Virginia

ABSTRACT

Characteristics of the fracture of resin matrix fibrous composites are discussed with emphasis on the physical parameters influencing fracture and the requirements of a suitable model for predicting crack growth. The influence of notch geometry, fiber orientation, specimen geometry, and far field loading on crack growth in unidirectional resin matrix composites is discussed. The normal stress ratio theory is reviewed and its applicability for crack initiation in unidirectional resin matrix composites are considered. Experimental results from a variety of tests on resin matrix composites are reviewed and compared with theoretical predictions. It is shown that the normal stress ratio theory correctly predicts the initiation site and direction of crack growth from an existing notch. Predictions of critical stress are less consistent being quite accurate in some cases, but less accurate in others.

INTRODUCTION

Fracture mechanics is concerned with the prediction of failure in the presence of existing flaws. The foundation of linear elastic fracture mechanics for isotropic materials was laid by the original work of A. A. Griffith (1921). This work was refined and expanded by G. R. Irwin (1957) to provide present-day *fracture mechanics*. The theory has been studied extensively and is now in general use for isotropic materials. Textbook accounts of the subject can be found, for example, in the book by Broek (1983).

There is no such universally accepted theory for crack growth in anisotropic materials such as fibrous composites which are orthotropic on the macro scale and heterogeneous on the micro scale. One primary reason for the lack of an acceptable fracture theory for composites is the fact that linear elastic fracture mechanics relys on the assumption that cracks grow in a self-similar manner (i.e. from the tip of an existing sharp crack and in a direction parallel to the axis of the crack, Fig. 1a). While this may be a good assumption for isotropic materials under a fairly broad class of loading conditions, it is not a good assumption for unidirectional resin matrix composites, except for a *very limited* class of loading conditions.

A recent study by Beuth and Herakovich (1989) showed that cracks in unidirectional graphite-epoxy grow parallel to the fibers under a wide variety of loading and crack orientation conditions (Fig. 1). Further, they found that all attempts to grow a crack in a direction which would cause fiber breakage were unsuccessful. Similar results were reported by Hidde and Herakovich (1988) for unidirectional aramid-epoxy. Unidirectional resin matrix composites have a preferred crack growth direction parallel to the fibers because they are highly orthotropic with the axial to transverse strength ratios typically being thirty or more. Wu (1968) argues that flaws in resin matrix composites will tend to be aligned parallel to the fibers and, therefore, study of notches parallel to the fibers is of most practical interest.

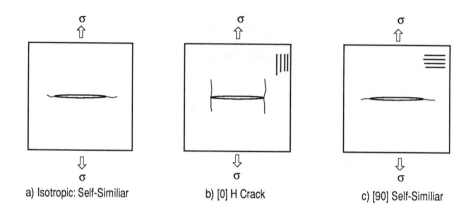

Fig. 1. Crack Growth Directions

Because cracks generally do not grow in a self-similar manner in unidirectional composites, the problem of predicting crack growth is significantly more complicated in these materials than in homogeneous isotropic materials. In addition to the variable direction of crack growth, experimental evidence has shown that cracks do not generally grow from the "tip" of an existing, *rounded* notch in unidirectional composites (Beuth and Herakovich, 1989). Thus the problem of crack growth in unidirectional composites requires the prediction of:

- the location of crack growth from an existing flaw
- the direction of crack growth
- the critical far-field stresses for initiation of crack growth
- the stability of crack growth

After a model to predict the critical features listed above has been verified experimentally, it is then necessary to extend the model for laminates where the macroscopic state of stress is truly three-dimensional in the vicinity of a flaw.

The choice of a fracture model for composites is further complicated by the fact that the material is heterogeneous at the level of the fiber and matrix (micro scale). A three-dimensional analysis of the microscopic stresses in the vicinity of a crack tip where extremely high stress gradients may be present is a formidable task even for supercomputers. Thus it seems desirable to develop an approach which permits stress analysis at the macrolevel while still incorporating the most important features of the microlevel heterogeneity.

A model which incorporates both macroscopic stress analysis and microscopic failure is discussed in this paper.

PREVIOUS STUDIES

One of the earliest papers concerned with cracks in anisotropic materials was that by Sih, Paris, and Irwin (1965). They modeled a line crack with an infinitely sharp tip in an anisotropic material using the complex variable approach of Lekhnitskii (1963) to derive expressions for the near-crack-tip stresses.

A combined theoretical - experimental study on crack growth in unidirectional fiberglass reinforced Scotch-ply was reported by Wu (1968). Wu also used Lekhnitskii's formulation to describe the stress state. He considered the limited class of problems in which the existing flaw was parallel to the fiber direction. His experimental results indicated that the crack growth was self-similar under a variety of loading conditions ranging from pure tension to pure shear. He determined critical stress intensity factors k_{1c} for this limited class of problems and found that k_{1c} was a constant equal to -0.49 for uniaxial tension, -0.46 for pure shear, and -0.50 for combined tension and shear. Thus the results (for the limited class of crack orientations) were reasonably consistent with linear elastic fracture mechanics which predicts $k_{1c} = -0.50$.

While it is true that cracks in unidirectional graphite-epoxy tend to grow parallel to the fibers, this is certainly not generally the case for laminates. A clear demonstration of this fact is shown in Fig. 2 which shows a failed tensile coupon for a $[(\pm30)_2]_s$ angle-ply laminates (Herakovich, 1982). This figure shows that the fracture surface is parallel to the fiber direction in half the layers, but clearly breaks the fibers in the remaining layers. As was discussed in the previous cited paper, the fracture surfaces for angle-ply laminates are a function of both the laminate stacking sequence and the fiber orientations. Differences in the ultimate stress of as much as 30% were reported for laminates with identical fiber orientations, but different stacking sequences. In some cases the fracture consisted of entirely matrix and fiber/matrix bond failure as opposed to the fiber breakage evident in Fig. 2 These variable results for crack growth in laminae and laminates show that it is imperative that a general purpose model be developed which is capable of predicting crack growth in composites under arbitrary conditions of stacking sequence and far-field loading.

Fig. 2. Failed $[(\pm30)_2]_s$ Angle-ply Laminate

THE NORMAL STRESS RATIO THEORY

In an effort to model the fracture response of composites, Buczek and Herakovich (1985) proposed the normal stress ratio theory. This model is based upon knowledge of the local state of stress near the tip of a notch (crack) as obtained from a macroscopic stress analysis with the material modelled as homogeneous and anisotropic. The theory assumes that a crack will grow from an existing flaw at the site and in the direction of the maximum value of the ratio of normal tensile stress $\sigma_{\phi\phi}$ to tensile strength $T_{\phi\phi}$ (Fig. 3). This model was proposed as a natural extension of the generally accepted belief that for isotropic materials cracks grow in the direction of maximum normal stress. The observed behavior in isotropic materials was modified for anisotropic materials by taking into account the anisotrophy of strength through $T_{\phi\phi}$. For the parameters as defined in Fig. 3, the mathematical statement of the normal stress ratio theory is that crack growth will occur when

$$R(r,\phi)=1.0 \tag{1}$$

where

$$R(r,\phi) = \sigma_{\phi\phi}/T_{\phi\phi} \tag{2}$$

and

$$T_{\phi\phi} = X_T \sin^2\beta + Y_T \cos^2\beta \qquad (3)$$

where β is the angle from the ϕ plane to the fiber direction (Fig. 3), X_T and Y_T are the axial and transverse strength, respectively, and r is the distance from the crack surface (which may be zero).

The definition of $T_{\phi\phi}$ used above satisfies the three conditions:

1. for isotropic materials, $T_{\phi\phi}$ is independent of ϕ

2. for crack growth parallel to the fibers in a unidirectional composite, $T_{\phi\phi}$ must equal the transverse strength Y_T

3. for crack growth perpendicular to the fibers in a unidirectional composite, $T_{\phi\phi}$ must equal the axial strength X_T

A definition of $T_{\phi\phi}$ (as opposed to experimental results) is used because it has not been possible to grow a crack in a unidirectional resin matrix composite along a prescribed direction. Thus it has not been possible to experimentally determine $T_{\phi\phi}(\phi)$. This definition does satisfy the three limiting conditions indicated above.

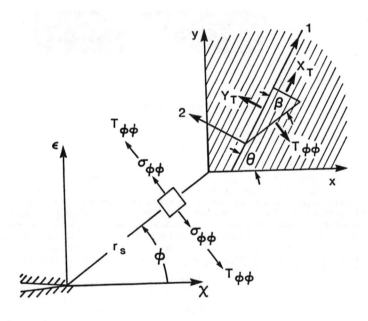

Fig. 3. Normal Stress Ratio Parameters

STRESS ANALYSIS

Infinite Anisotropic Plate with a Sharp Crack

Lekhnitskii (1963) outlines a complex variable plane elasticity solution for an infinite homogeneous anisotropic plate with an elliptical flaw at its center (Fig. 4). By reducing the minor axis dimension to zero and evaluating the stress potential functions in the neighborhood of the crack tip, the solution can be used to analyze the center-cracked infinite homogeneous anisotropic plate problem. This procedure models a crack in an actual material as a line crack with an infinitely sharp tip. Sih, Paris, and Irwin (1965) use the complex variable approach of Lekhnitskii to derive expressions for the near-crack-tip stresses which are of the form

$$\sigma_{\chi\chi} = \frac{\sigma^\infty\sqrt{a}}{\sqrt{2r}}\mathrm{Re}\left\{\frac{S_1S_2}{(S_1-S_2)}\left[\frac{S_2}{\psi_2^{1/2}}-\frac{S_1}{\psi_1^{1/2}}\right]\right\} + \frac{\tau^\infty\sqrt{a}}{\sqrt{2r}}\mathrm{Re}\left\{\frac{1}{(S_1-S_2)}\left[\frac{S_2^2}{\psi_2^{1/2}}-\frac{S_1^2}{\psi_1^{1/2}}\right]\right\} \qquad (4)$$

$$\sigma_{\epsilon\epsilon} = \frac{\sigma^\infty\sqrt{a}}{\sqrt{2r}}\mathrm{Re}\left\{\frac{1}{(S_1-S_2)}\left[\frac{S_1}{\psi_2^{1/2}}-\frac{S_2}{\psi_1^{1/2}}\right]\right\} + \frac{\tau^\infty\sqrt{a}}{\sqrt{2r}}\mathrm{Re}\left\{\frac{1}{(S_1-S_2)}\left[\frac{1}{\psi_2^{1/2}}-\frac{1}{\psi_1^{1/2}}\right]\right\}$$

$$\tau_{\chi\epsilon} = \frac{\sigma^\infty\sqrt{a}}{\sqrt{2r}}\mathrm{Re}\left\{\frac{S_1S_2}{(S_1-S_2)}\left[\frac{1}{\psi_1^{1/2}}-\frac{1}{\psi_2^{1/2}}\right]\right\} + \frac{\tau^\infty\sqrt{a}}{\sqrt{2r}}\mathrm{Re}\left\{\frac{1}{(S_1-S_2)}\left[\frac{S_1}{\psi_1^{1/2}}-\frac{S_2}{\psi_2^{1/2}}\right]\right\},$$

where σ^∞ and τ^∞ are the applied far-field stresses in the crack coordinate system, S_1 and S_2 are the roots of the characteristic equation for a plane linear elastic anisotropic material, and

$$\psi_1 = \cos\phi + S_1\sin\phi; \quad \psi_2 = \cos\phi + S_2\sin\phi. \qquad (5)$$

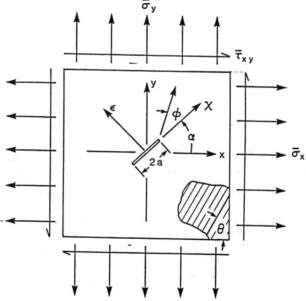

Fig. 4. Notched Anisotropic Plate

The angle ϕ is measured with respect to the crack and a is one half of the crack length (Fig. 4). These expressions are used in this study to model the near-crack-tip stress field for the problem illustrated in Fig. 4 by using the crack angle, α, to rotate the far-field stresses into the crack coordinate system.

Infinite Anisotropic Plate with an Elliptical Flaw

The stresses given in Eq. 4 are for the special case of an elliptical flaw having a zero minor axis dimension. The more general case of an elliptical flaw of arbitrary shape in an infinite anisotropic plate is also of interest as a model for near-notch-tip stresses. This stress solution is taken from Savin (1961). The analysis in [11] yields stresses of the form

$$\sigma_{\chi\chi} = \sigma_{\chi\chi}^\infty + 2\text{Re}\left[S_1^2 \phi_o'(z_1) + S_2^2 \psi_o'(z_2) \right] \tag{6}$$

$$\sigma_{\varepsilon\varepsilon} = \sigma_{\varepsilon\varepsilon}^\infty + 2\text{Re}\left[\phi_o'(z_1) + \psi_o'(z_2) \right]$$

$$\tau_{\chi\varepsilon} = \tau_{\chi\varepsilon}^\infty - 2\text{Re}\left[S_1\phi_o'(z_1) + S_2\psi_o'(z_2) \right] ,$$

where

$$\phi_o(z_1) = -\frac{i(a - iS_1 b)}{2(S_1 - S_2)}\left\{ \frac{\sigma_{\chi\chi}b + \sigma_{\varepsilon\varepsilon}(iaS_2) + \tau_{\chi\varepsilon}(bS_2 + ia)}{z_1 + \sqrt{(z_1^2 - (a^2 + S_1^2 b^2)}} \right\} \tag{7}$$

$$\chi_o(z_2) = \frac{i(a - iS_2 b)}{2(S_1 - S_2)}\left\{ \frac{\sigma_{\chi\chi}b + \sigma_{\varepsilon\varepsilon}(iaS_1) + \tau_{\chi\varepsilon}(bS_1 + ia)}{z_2 + \sqrt{(z_2^2 - (a^2 + S_2^2 b^2)}} \right\} .$$

In these expressions, a and b are, respectively, the major and minor semi-axes of the ellipse (Fig. 4), $\sigma_{\chi\chi}$, $\sigma_{\varepsilon\varepsilon}$, and $\tau_{\chi\varepsilon}$ are the far-field stresses in the crack coordinate system, $z_1 = x + S_1 y$, and $z_2 = x + S_2 y$.

CRACK GROWTH INITIATION SITE AND DIRECTION

Experimental studies on crack growth in unidirectional composites require that a flaw or notch be introduced into the specimen. Thus far it has not been possible to introduce "sharp" cracks as the flaw in a composite. Hence slots with finite radius at the "tip" have been used. Figure 5 shows a typical notch in graphite-epoxy obtained using a wire saw (Beuth and Herakovich, 1989). The notch consists of a slot with semi-circular ends or "tips".

Experimental results have shown that the crack initiation site is a function of notch geometry, fiber orientation relative to the notch, and far-field loading. The dependence of crack initiation site on these parameters was studied analytically in a paper by Gurdal and Herakovich (1987). They used the normal stress ratio theory to predict the crack initiation site from elliptical notches and circular holes in an infinite, anisotropic layer. The elliptical notch was used as an approximation to a slot because of the availability of an analytical solution which facilitated a parametric study. The ellipse was sized so that the radius at the tip of the ellipse was the same as the radius of the actual notch in the specimens of Beuth and Herakovich (1989).

Figures 6-8 show typical predictions for the crack initiation site (and the direction of crack growth) in unidirectional graphite-epoxy for: a) off-axis tensile coupons with slots perpendicular to the loading axis (Fig. 6), b) off-axis tensile coupons with circular holes (Fig. 7), c) pure shear loading of off-axis slotted lamina (Fig. 8). With the exception of one idealized case, crack growth direction was predicted to be parallel to the fiber direction. The one exception was the case of an ideally sharp crack in a [0] lamina under tensile loading. In this case, it was predicted that the crack grows in a self similar manner; however, the results were very unstable and any slight deviation from the 0 ° fiber orientation or "sharp" crack geometry resulted in crack growth parallel to the fibers.

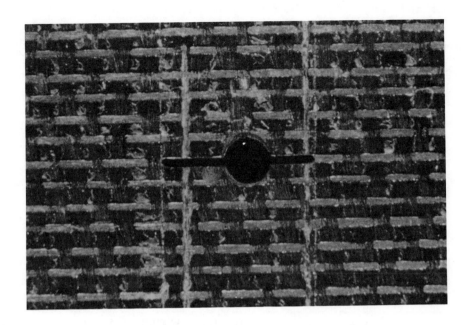

Fig. 5. Typical Notch in Graphite-Epoxy

The crack initiation site predictions depended on all three parameters: notch geometry, fiber orientation, and far-field loading. For tensile loading of slotted coupons, the initiation site angle Γ was predicted to be nonzero for fiber orientations up to 60°. For fiber orientations larger than 60° the initiation site was predicted to be at the tip of the ellipse ($\Gamma = 0$). As clearly indicated in Fig. 7, the location of the crack initiation site for tensile coupons with circular holes was very dependent on fiber orientation. For the case of 0° fibers the initiation site was predicted to be 10° from the horizontal. This angle compares very closely with a measurement of 13° obtained by Post and Czarnek (1986) using moire' interferometry. The predicted angle Γ corresponding to the crack initiation site for circular holes ranged from $\Gamma = 10°$ for 0° fiber orientation to greater than 30° for a fiber orientation of 22° . For larger fiber orientations the initiation site angle decreased with increasing fiber orientation, approaching $\Gamma = 0°$ in the limit of 90° fiber orientation.

Experimental results providing information on the site location of crack initiation from a flaw were presented by Beuth & Herakovich (1989). They are in generally good agreement with the experimental results.

CRITICAL FAR-FIELD STRESSES

Theoretical-experimental comparisons for the critical stress for crack initiation are very difficult to make to a high degree of accuracy. This is because it is difficult to model the exact shape of the notch and also because it can be very difficult to experimentally identify the exact stress level corresponding to crack initiation. In general, the 90° tension test produces accurate results for crack initiation, but for most other configuration there is some degree of uncertainty.

A comparison of theory and experiment for the far-field stresses initiating crack growth in unidirectional graphite-epoxy was presented by Beuth and Herakovich (1989) for far-field tensile and shear loading. Theoretical comparisons were made both on the bases of a sharp crack and an elliptical notch. The 90° tension test was used as the baseline for determining a critical distance from the sharp notch to be used in the predictions. This critical distance was assumed to be a material constant for all other cases. The results of that investigation are summarized in Table 1.

Table 1 Theory & Experiment for Crack Initiation

Loading	Fiber	Notch	Experiment	NSR Theory	
	θ	α	σ_{cr}	Crack	Ellipse
	(deg)	(deg)	ksi(MPa)	ksi(MPa)	ksi(MPa)
Tension	0	0	38.2(26.3)	20.9(144.0)	22.8(157.0)
Tension	45	0	4.12(28.4)	5.02(34.6)	4.83(33.3)
Tension	90	0	2.81(19.4)	Baseline	Baseline
Shear	0	90	4.02(27.8)	3.10(21.4)	6.51(44.9)
Shear	15	90	3.50(24.1)	2.88(19.9)	5.12(35.3)
Shear	30	90	4.13(28.5)	3.01(20.8)	4.56(31.4)
Shear	45	90	4.08(28.1)	3.80(26.2)	4.70(32.4)

These results show that the normal stress ratio does quite well in predicting the critical stress for crack initiation. While there is still room for improvement in the comparisons, it is noteworthy that a theory which is based solely on normal stress does quite well in predicting the critical stress for a pure shear far field loading configuration.

EFFECTS OF CRACK LENGTH

As indicated previously, Wu (1968) reported experimental results for the critical stress as a function of crack length for center notched fiberglass. He considered tensile, shear and combined loading of unidirectional laminae with notches oriented parallel to the fiber direction. He reported that stress intensity factors k_{1c} were constant in all cases as evidenced by straight line plots of log σ^∞ vs log a with slopes very nearly equal to the theoretical value of -0.5.

Recent results for ln-ln plots of critical stress vs half-crack length are presented in Figs. 9 & 10 for aramid-epoxy and graphite-epoxy, respectively. Both figures show results for [0] and [90] laminae. The aramid-epoxy results are taken from reference [5] and the graphite-epoxy results were reported by Rydin (1989). There are several interesting aspects of these results. The aramid-epoxy results (Fig. 9) fall along straight lines. For the 90 ° tests the slope is -0.34, and for the 0 ° tests the slope is -0.75. These results were obtained from 2" wide specimens with a maximum crack length of 0.5".

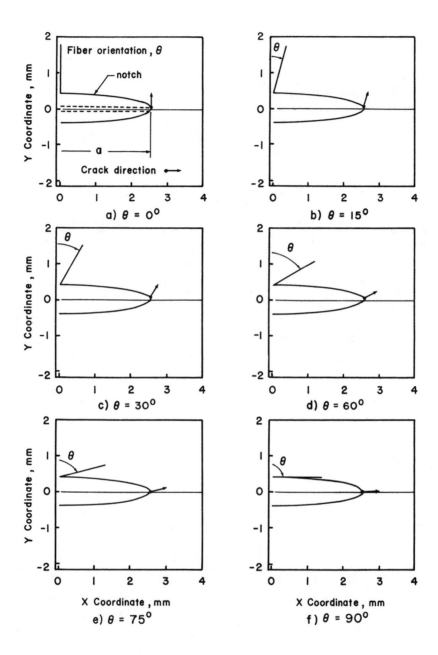

Fig. 6. Crack Growth in Off-axis Slotted Coupons

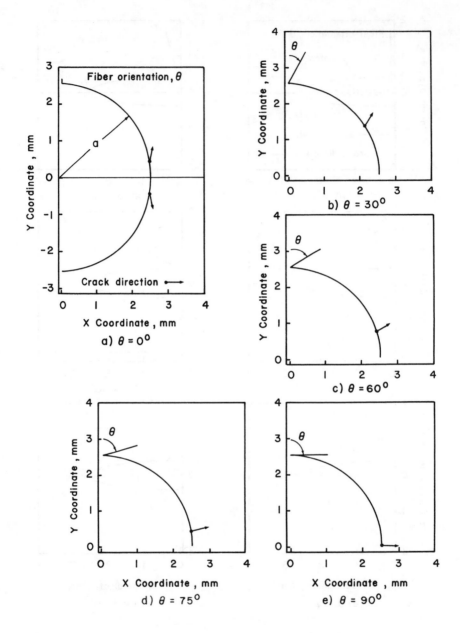

Fig. 7. Crack Growth in Off-axis Coupons with Circular Holes

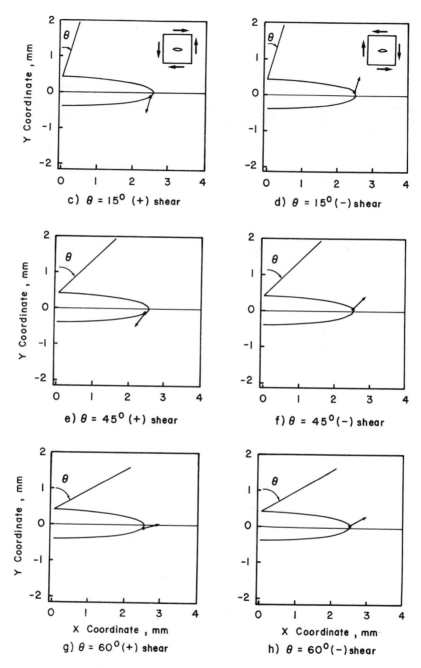

Fig. 8. Crack Growth in Shear Loaded Slotted Lamina

The graphite-epoxy results in Fig. 10 do not exhibit the same straight line character. These results were obtained from 1" wide specimens with the maximum crack length being 0.7". Thus it is entirely possible that specimen width is affect the results for larger crack lengths. It is interesting to note that if only the two smallest crack lengths are considered for the [90] lamina, the slope is -0.49. Further testing is required to fully assess the significance of these results.

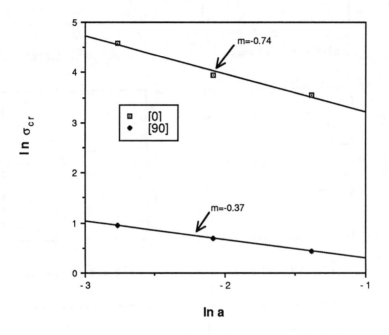

Fig. 9. Effect of Crack Length for Aramid-Epoxy

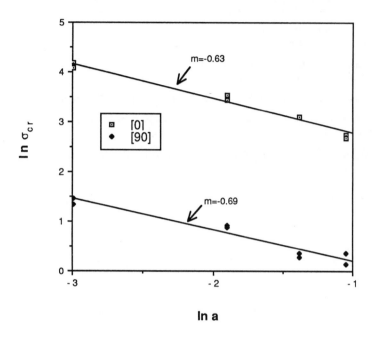

Fig. 10. Effect of Crack Length for Graphite-Epoxy

SUMMARY

Important features of crack growth in unidirectional composites have been discussed. The normal stress ratio theory has to shown to be quite accurate for predicting the initiation site and direction of crack growth from an existing notch. Critical stresses predicted by the normal stress ratio theory are good in some cases but not in others. Additional work is necessary for accurate comparison of theory and experiment for critical stresses.

ACKNOWLEDGEMENT

This work was supported by Hercules Incorporated, Aluminum Company of America, and the Center for Innovative Technology.

REFERENCES

Beuth, J. L. and Herakovich, C. T., 1989, "Analysis of Crack Extension in Anisotropic Materials Based on Local Normal Stress", *Theoret. Appl. Fracture Mech.,* ., 11, pp. 27-46.

Broek, D., 1982, *Elementary Engineering Fracture Mechanics*, Martinus Nijhoff, Boston.

Buczek M. B., and Herakovich, C. T., 1985, "A Normal Stress Criterion for Crack Extension Direction in Orthotropic Composite Materials," *Journal of Composite Materials*, Vol. 19, pp. 544-578.

Griffith, A. A., 1921, "The phenomena of rupture and flow in solids", *Phil. Trans. Roy. Soc. of London, A 221, pp. 163-197.*

Gurdal, Z., and Herakovich, C. T., 1987, "Effect of Initial Flaw Shape on Crack Extension in Orthotropic Composite Materials," Theoretical and Applied Fracture Mechanics, Vol. 8, pp. 59-75.

Herakovich, C.T., 1982, "Influence on Layer Thickness on the Strength of Angle-Ply Laminates", *Journal of Composite Materials*, Vol. 16, pp. 216-227.

Hidde, J. S. and Herakovich, C. T., 1988, "An Experimental Study of the Notch Sensitivity of Aramid/Epoxy", SEAS Report, Univ. Virginia, April.

Irwin, G. R., 1958, "Fracture", *Handbuch der Physik VI*, Flugge, Ed. pp. 558-590, Springer.

Lekhnitskii, S. G., 1963, *Theory of Elasticity of an Anisotropic Body*, English translation by Brandstatton, Holden-Day Inc., San Francisco.

Post, D. and Czarnek, R., 1986, Private Communication, Engineering Science & Mechanics Department, VPI&SU, Blacksburg, VA,.

EVALUATION OF THE STRENGTH AND MODE OF FRACTURE OF STANDARD LAP SHEAR JOINTS WITH AND WITHOUT INITIAL LARGE DEBONDS

P. R. Borgmeier, K. L. DeVries, and J. K. Strozier
University of Utah
Salt Lake City, Utah

G. P. Anderson
Morton Thiokol
Brigham City, Utah

ABSTRACT

In this study standard lap shear joints with and without half area debonds are investigated both analytically (numerically), and experimentally. The regions of debond varied from the center of the overlap to the bond termini. Linear elastic fracture mechanics, in conjunction with finite element methods is used to determine the mode of fracture and energy release rates for the joints. Finite element methods are also used to compute the stress distribution, displacements, and energy release rates for the joints. Analytical investigation shows that the geometry without a debond initially experiences large cleavage stresses near the bond termini resulting in a Mode I failure while the specimen with the half area debond may (depending on the exact location of the debond) initially experience large shear stresses at the bond termini resulting in a predominantly Mode II failure. The finite element analysis, using an energy release rate approach, predicts that the load required for failure of the joint with the half area debond should be approximately 58% of the load required for failure of the joint without a debond. Experimental results show that the load required for failure of the joint with the debond to be in excess of 75% of the load required for failure of the joint without an initial debond. Based on these observations, the authors suggest that the total energy release rate determined for the different geometries is not constant but rather mode dependent.

INTRODUCTION

In recent years there has been an increasing interest in implementing adhesives into design. Traditionally, the leading adhesives used by man from prehistoric times to the present were animal glues usually made from cooking down collagenous materials such as hides, hooves or bones of animals. Even these early adhesives have enjoyed continual use and improvement. These adhesives were used for bonding wooden structures such as furniture and wood laminates. These animal glues eventually found their way into everything from book bindings to the paper on which the books were printed on. In the early twentieth century these same adhesives were used in constructions such as automobiles, aircraft, etc.

For the first half of the twentieth century, adhesives were used for their ease of application which saved time joining materials. The need for machining parts to accommodate fasteners was eliminated and fewer parts needed to be fabricated. In addition to the above was the added bonus of the reduction of stress concentrations due to the applied load being distributed over a larger area. Although the use of adhesives has continually increased throughout history, recent technology has caused an enormous increase in both the interest and use of adhesives today.

With the many advances in adhesive chemistry and aerospace technology, adhesives are being used in more complex and critical applications than ever before. The reasons adhesives are such an attractive method for joining materials are different than those for earlier use. Many of today's adhesives are expensive to manufacture and time consuming to lay up. However, high costs are offset by their ability to reduce weight compared to traditional fastening methods as well as reduce critical stress concentrations. Due to the critical and complex nature of advanced adhesive use, accurate design and analysis methods for strength and reliability have to be made to insure success. Most often, current design techniques consist of a "trial and error" type analysis. Each specific geometry considered must be tested separately due to the fact that data obtained from standardized testing methods is hard (or virtually impossible) to correlate between different geometries. With this method, optimizing designs becomes very time consuming and expensive due to the number of tests required for the many geometries that are possible. This results in fewer geometries being tested, which in turn, increased the risk that the optimum design is not discovered. However, in order to use adhesives effectively in critical applications today and in the future, accurate strength and failure predictions must be able to be made prior to testing. One method that shows promise for accomplishing this task is the use of fracture mechanics. But before fracture mechanics methods can be used to predict failure of complex designs, much more research is needed to assure success.

As mentioned earlier, standardized tests are most often used for determining the strength properties of adhesives. Of these tests, the most widely used test is probably the lap shear test. The lap shear test usually consists of two substrates bonded in an overlapping fashion. The applied failure load for the test specimen is divided by the overlap area resulting in a so-called failure stress.

In the study reported here, standard lap shear joints with and without half area debonds are investigated in detail. Finite element methods are used to examine and obtain an understanding of the stress state throughout the lap shear joint overlap. Finite element methods were again used in combination with fracture mechanics methods to predict failure loads for the lap shear joint using two different methods for determining energy release rates.

TECHNICAL DISCUSSION

Of the many lap shear joints, the single lap shear joint is by far the most commonly used joint for design as well as for testing adhesives. In reporting the results of the simple lap shear joint test, the failure load is usually divided by the area of the overlap producing an average shear stress. This average shear stress is often used as a design parameter for other applications. Sample fabrication and testing procedures are covered in testing standards such as ASTM D1002-72, titled "Strength Properties of Adhesives in Shear by Tension Loading."

The simplest analysis of this joint would model the adherends as rigid and allow the adhesive to deform only in shear. The average shear stress would follow as

$$\sigma_{xy} = \frac{F}{b \cdot l} \tag{1}$$

where F is the failure load, b is the joint width and l is the length of the overlap, see figure 1. It should be noted that this simple analysis predicts the maximum (average) shear stress from the ASTM standard referenced above.

Although there are at least two different approaches to using fracture mechanics, the energy method formulated by Griffith will be used because it requires less knowledge of the details of the stress state at the crack.[1] Griffith, considered accounting for energy changes in an isotropic elastic plane space (extending to infinity in all directions) of thickness b with a line crack or flaw of length 2a in it. See figure 2. The elastic plane space is axially stressed in the direction normal to the crack at "infinity" by an amount σ. The input energy or external work E_1, produced by the stress σ for this system, must be converted to energy associated with deformation or strain energy E_2, dissipated energy E_3 or kinetic energy E_4. Since we are considering elastic bodies where plastic deformation does not exist, E_3 is only due to the "creation" of new surface. Summing the energies, it is easily seen that

$$E_1 = E_2 + E_3 + E_4 \tag{2}$$

Considering the first law of thermodynamics, the system of figure 2 is not concerned with how the energy E_1 is distributed among E_2, E_3 and E_4. For instance, if no energy is dissipated (no crack growth), then E_1 would equal E_2 while E_3 and E_4 would be zero. Or as E_1 increases, E_2 can increase while E_3 and E_4 are zero. Then at a point in time energy will be dissipated through E_3. As can be seen, E_2, E_3 and E_4 can vary as long as they sum to E_1, or in essence a constant.

Griffith hypothesized that the total energy of the crack system, E_1, was constant as the crack extended an infinitesimal distance, da, creating new surface, dA, where

$$dA = da \times b \tag{3}$$

Differentiating equation (1) with respect to the new crack area, dA, leads to

$$-\frac{dE_2}{dA} = \frac{dE_3}{dA} + \frac{dE_4}{dA} \tag{4}$$

But since the crack of figure 2 only extended an incremental amount, the change in E_4 is assumed zero. Therefore equation (4) reduces to

$$-\frac{dE_2}{dA} = \frac{dE_3}{dA} \tag{5}$$

Using the sign convention that the energy stored in the system is negative and the energy released from the system is positive, the stain energy for the system in figure 2 can be shown by elastic strain analysis to be equal to

$$E_2 = -\frac{\pi a^2 \sigma^2}{E} \tag{6}$$

for plane stress conditions and

$$E_2 = -\frac{\pi a^2 \sigma^2 (1 - v^2)}{E} \tag{7}$$

for plane strain conditions. Griffith hypothesized that the energy needed to create new surface can be represented as

$$E_3 = 4aT \tag{8}$$

where T is the fracture energy for the material. Griffith termed this quantity the surface energy and for the glass he studied initially it was very near the surface tension. For most structural materials this energy (required to create a unit of surface) is much larger than the surface tension as we will next consider.[2] Substituting E_2 and E_3 into equation (5) and solving for the critical stress, σ_c, needed to create new surface leads to

$$\sigma_c = \sqrt{\frac{2TE}{\pi a}} \tag{9}$$

for plane stress and

$$\sigma_c = \sqrt{\frac{2TE}{\pi a (1 - v^2)}} \tag{10}$$

for plane strain.

In a more general sense, consider a similar system that is not necessarily linear or homogeneous. In addition, assume that the total energy of the system does not have to be constant during crack extension da. See figure 3. Equation (2) must still hold for this general system since it is a statement of the first law of thermodynamics. It should be noted for example, that the dissipated energy E_3 can be almost totally due to energy associated with plastic deformation (rupture), totally due to energy associated with formation of fracture surface (brittle fracture) or a combination of both (fracture with plastic deformation). Again, differentiating equation (2) with respect to dA gives

$$\frac{dE_1}{dA} - \frac{dE_2}{dA} = \frac{dE_3}{dA} + \frac{dE_4}{dA} \qquad (11)$$

or letting

$$G = \frac{dE_1}{dA} - \frac{dE_2}{dA}$$
$$R = \frac{dE_3}{dA} \qquad (12)$$

where G then represents the amount of energy available to create new surface. R represents the material's ability to resist dissipating energy, and is assumed to be a material constant. Substituting equation (12) into equation (11) gives

$$G = R + \frac{dE_4}{dA} \qquad (13)$$

It can now be seen that when G is less than R no fracture can occur. If G equals R, the amount of energy available to create new surface just equals the amount of energy needed to create new surface so that fracture is just initiating. And if G is larger than R, fracture will occur. The amount of kinetic energy is equal to G minus R.

Considering the case where G equals R leads us to the critical point where there is just enough energy available to create new surface. The critical energy release rate, where G equals R is represented as

$$G_c = R = \frac{dE_3}{dA} \qquad (14)$$

In cohesive fracture mechanics, G_c is assumed to be a material constant. After determining G_c for a certain material, determining the point of crack initiation can be determined for any geometry of this material by accounting for the energy of the system by means of stress analysis and an energy balance.

The development of equation (11) above was based on energy considerations of the body and in no way limits its use to only homogeneous systems. Therefore, equations (11), (12), (13), and (14) will apply equally well to a bonded joint. However, R for the adhesive system is not a material property in the same sense that R for the cohesive system was. R for the bonded joint can depend on many variables such as material type, surface preparation, curing conditions, humidity, etc. Therefore, R for the adhesive system shall be represented by R_a. R_a can still be used in the sense of a material property as long as the variables mentioned above are kept constant from joint to joint. Making this substitution into equation (14) leads to

$$G_c = R_a = \frac{dE_3}{dA} \qquad (15)$$

We see therefore that, with the exception of the definition of R_a, adhesive systems can be treated in an identical way that cohesive systems mechanics are treated. However, accounting for the energy in a bi-material system is a much more difficult task and has been accomplished, in closed form, for only a limited number of rather ideal systems. The use of numerical methods makes this task in principle possible for almost any geometry. Both the adherends and adhesive were specifically selected such that they exhibit a nearly linear elastic stress-strain relation. The research for this paper will, therefore, be limited to linear elastic fracture mechanics (LEFM) analysis. As mentioned earlier, in an elastic system, there is no plastic deformation and the dissipated energy, E_3, is due totally to the creation of new surface. LEFM applies to both adhesive and cohesive systems.

Consider the condition where the crack of figure 3 extends from an initial length a to a final length a+da and the input energy is caused by the force F. The crack could grow from length a to a+da in three different manners. The crack could grow assuming a constant force throughout the crack extension, assuming a constant displacement throughout the crack extension or a mixture of changed loads and displacements throughout the crack extension.

154

The change in input energy or external work dE_1, can be represented in general as

$$dE_1 = \left(\frac{F^{(a + da)} + F^{(a)}}{2} \right)(u^{(a + da)} - u^{(a)})$$

(16)

where F is the input force on the body and u is the displacement of the body. This is represented graphically in figure 4. The change in strain energy dE_2, can be represented in general as

$$dE_2 = \frac{1}{2}F^{(a + da)}u^{(a + da)} - \frac{1}{2}F^{(a)}u^{(a)}$$

(17)

This is represented graphically in figure 5. Rearranging G from equation (12) leads to

$$G = \frac{dE_1}{dA} - \frac{dE_2}{dA} = \frac{d(E_1 - E_2)}{dA} \rightarrow$$

$$G \cdot dA = d(E_1 - E_2)$$

(18)

Substituting equations (16) and (17) into (18) reduces to

$$G \cdot dA = \frac{1}{2}\left(F^{(a)}u^{(a + da)} - F^{(a + da)}u^{(a)} \right)$$

(19)

or rearranging

$$G = \frac{\frac{1}{2}\left(F^{(a)}u^{(a + da)} - F^{(a + da)}u^{(a)} \right)}{dA}$$

(20)

Graphically, equation (18) is represented in figure 6. Figure 7 graphically shows the limiting case of equation (18) where the crack is assumed to grow under constant stress assumptions. Figure 8 graphically shows the limiting case of equation (18) where the crack is assumed to grow under constant displacement assumptions.

It should be noted that the compliance curves for crack length of a and a+da, respectively, are identical for figures 6, 7, and 8.

The limiting case of crack growth under constant stress is of special interest in this paper. Not only does this case predict the largest or most conservative energy release value G, but can also be used to solve for the energy release rate for the limiting case of constant displacement crack growth. Now, compliance curves will be used for evaluating energies. The strain energy stored in the sample created from the input energy is

$$E_2 = \frac{1}{2}Fu$$

(21)

The compliance C for the system is

$$C = \frac{u}{F}$$

(22)

Rearranging equation (22) and substituting into equation (21) leads to

$$E_2 = \frac{1}{2}F^2C$$

(23)

155

For the limiting case of constant stress crack extension

$$\frac{dE_2}{dA} = \frac{1}{2}F^2\frac{dC}{dA} + \frac{1}{2}C\frac{d(F^2)}{dA} = \frac{1}{2}F^2\frac{dC}{dA}$$

$$\frac{dC}{dA} = \frac{d\left(\frac{u}{F}\right)}{dA} = \frac{1}{F}\frac{du}{dA}$$

$$\therefore$$

$$\frac{dE_2}{dA} = \frac{1}{2}F\frac{du}{dA} = \frac{1}{2}F^{(a)}\left(u^{(a+da)} - u^{(a)}\right) \tag{24}$$

which is identical to equation (17) for constant F. And the limiting case of equation (19) for constant F is

$$G \cdot \Delta A = \frac{dE_2}{dA} = \frac{1}{2}F^{(a)}\left(u^{(a+da)} - u^{(a)}\right) \tag{25}$$

Now the limiting case of constant displacement crack growth can be determined using information from above. The load $F^{(a+da)}$ for constant displacement crack growth is derived from

$$C^{(a+da)} = \frac{u^{(a+da)}}{F^{(a)}}$$

$$C^{(a)} = \frac{u^{(a)}}{F^{(a)}}$$

$$\therefore$$

$$C^{(a+da)} = \frac{u^{(a+da)}}{F^{(a)}} = \frac{u^{(a)}}{F^{(a+da)}}$$

$$\therefore$$

$$F^{(a+da)} = \frac{u^{(a)}}{u^{(a+da)}}F^{(a)} \tag{26}$$

Rearranging equation (22) and substituting into equation (21) leads to

$$E_2 = \frac{1}{2}\frac{u^2}{C} \tag{27}$$

For the limiting case of constant displacement crack extension

$$\frac{dE_2}{dA} = \frac{1}{2}\left[\frac{C \cdot 0 - u^2\frac{dC}{dA}}{C^2}\right] = -\frac{1}{2}\frac{u^2}{C^2}\frac{dC}{dA}$$

$$\frac{dC}{dA} = \frac{d\left(\frac{u}{F}\right)}{dA} = \left[\frac{F \cdot 0 - u\frac{dF}{dA}}{F^2}\right] = -\frac{u}{F^2}\frac{dF}{dA}$$

$$\therefore$$

$$\frac{dE_2}{dA} = \frac{1}{2}\frac{u^2}{C^2}\frac{u}{F^2}\frac{dF}{dA} = \frac{1}{2}u\frac{dF}{dA} = \frac{1}{2}u^{(a)}\left(F^{(a+da)} - F^{(a)}\right) =$$

$$\frac{1}{2}u^{(a)}\left(\frac{u^{(a)}}{u^{(a+da)}}F^{(a)} - F^{(a)}\right) \tag{28}$$

which again is identical to what equation (17) predicts for constant displacement growth. And the limiting case of equation (19) for constant u is

$$G \cdot dA = -\frac{dE_2}{dA} = \frac{1}{2}u^{(a)}(F^{(a)} - F^{(a+da)}) = \frac{1}{2}u^{(a)}\left(F^{(a)} - \frac{u^{(a)}}{u^{(a+da)}}F^{(a)}\right)$$

(29)

These relationships will be used in the numerical analysis of the lap shear joint.

The software choice for this research was the ADINA finite element code available on a Digital VAX 11/750 computer. The current research was limited to two-dimensional analyses with plane strain assumptions. Quadrilateral elements were used throughout the analyses of the different modified lap shear specimens. The analyses were also limited to static assumptions.

Two methods were used to determine energy release rates for crack extensions. The first and simplest of the two methods calculates only the total energy release rate. The second method calculates the energy release rate for each mode of fracture as well as the total energy release rate.

The basic consideration for calculating energy release rates is accounting for the energy before and after crack extension dA. The basic equation for balancing energy for crack extension was shown to be

$$\frac{dE_1}{dA} - \frac{dE_2}{dA} = \frac{dE_3}{dA}$$

(30)

for the case of no kinetic energy for a continuum system. Using a finite elements representation of the continuum, the crack extension dA can be represented by a finite extension ΔA. Equation (30) can be approximated in the following form

$$\frac{\Delta E_1}{\Delta A} - \frac{\Delta E_2}{\Delta A} = \frac{\Delta E_3}{\Delta A}$$

(31)

where

$$\Delta E_1 = E_1^{(a+\Delta a)} - E_1^{(a)}$$
$$\Delta E_2 = E_2^{(a+\Delta a)} - E_2^{(a)}$$
$$\Delta E_3 = E_3^{(a+\Delta a)} - E_3^{(a)}$$
$$\Delta A = b(a + \Delta a - a) = A^{(a+\Delta a)} - A^{(a)}$$

(32)

The energy release rate, G, using a finite element representation can be approximated in a similar fashion as that of the continuum representation from above as

$$G = \frac{\Delta E_1}{\Delta A} - \frac{\Delta E_2}{\Delta A} = \frac{\Delta E_1 - \Delta E_2}{\Delta A}$$

(33)

The energy release rate, G, for the infinitesimal crack extension da was for a crack length of a. But in the finite element approximation Δa has finite length. Therefore the approximated energy release rate of equation D is for an average crack length of $(a+(a+\Delta a))/2$.

The compliance method is easily done using ADINA. This method requires only two computer runs for each energy release rate desired. The energy release rate calculated is for the assumption of constant force crack extension. The first run is used to calculate the total relative displacement, $u^{(a)}$, of the sample for the crack length of a, under a constant force $F^{(a)}$. In the second run, a node is released extending the crack length to a length of $a+\Delta a$. The second run is therefore used to calculate the total relative displacement, $u^{(a+\Delta a)}$, of the sample for a crack length of $a+\Delta a$ under the same constant load $F^{(a)}$. Using equation (25) and substituting in the values of $F^{(a)}$, $u^{(a+\Delta a)}$ and $u^{(a)}$ leads to

157

$$G = \frac{F^{(a)}(u^{(a + \Delta a)} - u^{(a)})}{2\Delta A}$$

(34)

which is the energy release rate under constant stress assumptions for a crack length of $(a+(a+\Delta a))/2$. The energy release rate under constant displacement assumptions for the crack length of $(a+(a+\Delta a))/2$ is easily determined from equation (29) as

$$G = \frac{F^{(a)} u^{(a)}\left(1 - \frac{u^{(a)}}{u^{(a + \Delta a)}}\right)}{2\Delta A}$$

(35)

It should be noted that using a third computer run the crack can now be extended by releasing the next node increasing the crack length to $a+2\Delta a$. Letting $u^{(a+\Delta a)}$ equal the new total relative displacement and $u^{(a)}$ equal the total relative displacement from run two, the energy release rate for the new crack length can be computed using equations (34) and (35) with only the addition of one computer run.

The idea behind the second method, the crack closure integral method (CCI), is that the energy is calculated that is needed to close the crack extension.[3] It is assumed that this energy is equivalent to the energy that was needed to create the crack's new surface.

The CCI method is also easily accomplished using ADINA. The energy release rate is again for constant stress crack extension assumptions when using the following method. In general, this method requires four computer runs for each energy release rate required. The first run is used to calculate the displacements of the nodes at the crack tip of the sample for the crack length of a, under a constant load $F^{(a)}$. Let u^{BX1} and u^{BY1} represent the displacements of the bottom node in the X and Y directions from run 1 respectively. Let u^{TX1} and u^{TY1} represent the displacements of the top node in the X and Y directions from run 1 respectively.

In the second run, a node is released extending the crack length to a length of $a+\Delta a$. The second run is therefore used to calculate the new nodal displacements of the nodes that were released from run 1 still under the constant load $F^{(a)}$. Let u^{BX2} and u^{BY2} represent the displacements of the bottom node in the X and Y directions from run 2 respectively. Let u^{TX2} and u^{TY2} represent the displacements of the top node in the X and Y directions from run 2 respectively. The difference between the displacements from run 1 and run 2 determine how far each node must be moved to close the crack.

In the third run, vertical unit forces are applied to the released nodes in the direction needed to bring them back to their original vertical positions determined from run 1. Again the nodal displacements of these released nodes of the sample for a crack length of $a+\Delta a$ under the same constant load $F^{(a)}$ are noted. Let u^{BX3} and u^{BY3} represent the displacements of the bottom node in the X and Y directions from run 3 respectively. Let u^{TX3} and u^{TY3} represent the displacements of the top node in the X and Y directions from run 3 respectively. The difference between the node displacements from run 3 and run 2 determine how far the vertical unit force moved each node towards its original position from run 1.

In the forth run, horizontal unit forces are applied to the released nodes in the direction needed to bring them back to their original horizontal positions determined from run 1. Again the nodal displacements of these released nodes of the sample for a crack length of $a+\Delta a$ under the same constant load $F^{(a)}$ are noted. Let u^{BX4} and u^{BY4} represent the displacements of the bottom node in the X and Y directions from run 4 respectively. Let u^{TX4} and u^{TY4} represent the displacements of the top node in the X and Y directions from run 4 respectively. The difference between the node displacements from run 4 and run 2 determine how far the horizontal unit force moved each node towards it's original position from run 1.

Since the material is modeled as linear elastic, the force required to move each node back to its original position from run 1 can be calculated from

$$F^{BX} = abs\left[\frac{\left(u^{BX2} - u^{BX1}\right)}{\left(u^{BX4} - u^{BX2}\right)}\right]$$

$$F^{TX} = abs\left[\frac{\left(u^{TX2} - u^{TX1}\right)}{\left(u^{TX4} - u^{TX2}\right)}\right]$$

$$F^{BY} = abs\left[\frac{\left(u^{BY2} - u^{BY1}\right)}{\left(u^{BY3} - u^{BY2}\right)}\right]$$

$$F^{TY} = abs\left[\frac{\left(u^{TY2} - u^{TY1}\right)}{\left(u^{TY3} - u^{TY2}\right)}\right] \tag{36}$$

where F^{BX}, F^{TX}, F^{BY} and F^{TY} are the forces required to move the nodes back to their original position from run 1 in the direction the unit forces were applied. The energies required to close the crack can now be calculated from the following equations

$$\Delta E^X = \frac{F^{BX} \cdot abs\left(u^{BX4} - u^{BX2}\right)}{2} + \frac{F^{TX} \cdot abs(u^{TX4} - u^{TX2})}{2}$$

$$\Delta E^Y = \frac{F^{BY} \cdot abs\left(u^{BY3} - u^{BY2}\right)}{2} + \frac{F^{TY} \cdot abs(u^{TY3} - u^{TY2})}{2} \tag{37}$$

where ΔE^X is the energy required to close the crack in the X direction and ΔE^Y is the energy required to close the crack in the Y direction. Knowing that the energy required to close the crack was equal to that available for crack extension, and making the following substitutions

$$\Delta E^X_1 - \Delta E^X_2 = \Delta E^X$$

$$\Delta E^Y_1 - \Delta E^Y_2 = \Delta E^Y \tag{38}$$

into equation (33), the energy release rates can be calculated from

$$G_I = \frac{\Delta E^Y_1 - \Delta E^Y_2}{\Delta A} = \frac{\Delta E^Y}{\Delta A}$$

$$G_{II} = \frac{\Delta E^X_1 - \Delta E^X_2}{\Delta A} = \frac{\Delta E^X}{\Delta A} \tag{39}$$

Therefore, the total energy release rate is

$$G = G_I + G_{II} \tag{40}$$

where G represents the energy release rate under the constant stress crack extension assumptions for an average crack length of $(a+(a+\Delta a))/2$.

Two variations of the modified lap-shear test were manufactured. Some of these specimens were bonded with intentional half area debonds in different locations while others were bonded without debonds. The following gives specific details on the various modified lap-shear specimens.

The modified lap shear joints were constructed per procedures of ASTM D 3165-73. The adherend material used in these specimens was a 4340 hardened steel. Hysol® 934-NA was used as the adhesive for these specimen groups.

The lap shear specimens were modelled in an identical way, except where noted, to maintain consistency between analyses. Isoparametric four node two-dimensional plane strain quadrilateral elements were used throughout the analysis.

Figure 9 shows a schematic of the generalized finite element grid used for all the analyses. The boundary conditions for the analyses are as follows. The node at point B was constrained from moving in both the X and Y directions. All the nodes along the line between point B up to point A were constrained as to only allow displacement in the Y direction. All the nodes on the line from point

C to point D were rigidly constrained in the X direction. In other words, all the nodes from point C to point D displaced an equal amount in the X direction. Two equal concentrated forces were applied to two nodes near location F as shown in figure 9.

The adherend material used in all the models was modelled as steel. This material had a Young's modulus of 200×10^9 Pascals (29×10^6 psi) and a Poisson's ratio of .3. The adhesive material used in all the models was modelled as HYSOL® 934-NA epoxy. This material had a Young's modulus of 4.6×10^9 Pascals ($.67 \times 10^6$ psi) and a Poisson's ratio of .37.

Square elements with a side length of .0127 mm (.0005 in.) were used near point E of figure 9 for both stress state and energy release rates calculations for all the specimen groups.

Figure 10 shows the exact finite element grid used for this analysis as well as blown up regions of the area around the singularity. In the smallest region of figure 10, the adhesive is shown as shaded. The bond line between the adhesive and adherend is the horizontal line right of point E.

Figure 11 shows a plot of the normalized cleavage and shear stresses versus the normalized distance down the bond line horizontal from point E with no crack or debond. See figure 10. The stresses were normalized by dividing the calculated stresses by the average shear stress. The distance was normalized by dividing by the total linear length of the overlap. It should be noted that the stresses in the Y direction (cleavage stresses) in figure 11 are higher than the shear stresses.

EXPERIMENTAL RESULTS

Two series of tests were conducted. In the first, modified lap shear specimens with a 12.7 mm x 25.4 mm (.5 in x 1 in.) overlap with a geometry similar to that shown schematically in figures 12 and 13 but with no initial debonds were pulled to failure in an Instron Testing Machine at a constant loading rate of .5mm/min (.02 in/min). In the second series of tests, the samples had regions of initial debond, as shown in Figures 12 and 13 and were pulled to failure at the same displacement rate . In each of these cases, the "bonded" area is half that of the first series of specimens. If there is any validity at all to an average shear stress criteria, one would expect the failure load to be approximately half that of the original specimens. From consideration of the anticipated high stress concentration at reentrant corners such as in the samples of Figures 12 and 13, one might, on the basis of a stress criteria expect the load carrying capability to be even further reduced for these samples. The experimentally determined average failure load for the first series of tests was 4841 Newtons (1088 pounds) with a standard deviation of 1299 Newtons (292 pounds). The samples with the initial half area debonds failed at an average load of 4067 Newtons (914 pounds) with a standard deviation of 1143 Newtons (257 pounds). In this data, the samples with the debond starting at the bond termini had an average failure load of 3630 Newtons (816 pounds). The samples with the debond centered in the adhesive bond experienced an average failure load of 4561 Newtons (1025 pounds). This is a reduction of only 25% for the samples with the edge debond and a 6% reduction for the samples with the debond centered. These value are less than the 50% or more reduction in load carrying capability that one would anticipate based on maximum stress criteria.

The authors feel that application of the principals of fracture mechanics (including the mode dependence of the energy release rate) provides insight into the causes of the apparent anomaly as described in the next section.

FRACTURE MECHANICS ANALYSIS

A finite element analysis of the original lap shear test specimens was conducted assuming an inherent flaw size (debond) of 0.09 mm (0.0035 in) and a load of 4830 newtons (1088 pounds). This inherent debond was selected as typical for this adhesive surface preparation and fabrication methods as typical from the tests conducted in this laboratory and at Thiokol Chemical Corp in which data was extrapolated from tests on samples with varying sizes of introduced debonds. The total energy release rate determined by the compliance method described earlier was 118 J/m^2 (0.67 lb-in/in^2). Using this value of the energy release rate and finite element analysis in the compliance methods for the 50% initial edge debond specimens, one predicts a total load carrying capability of 2840 Newtons (640 pounds). This is well below the experimentally determined value of 3520 newtons (816 pounds) reported in the last section. Stated another way, the effective total energy release rate for these initially 50 % debond specimens is 192 J/m^2 (1.09 lb-in/in^2). To explain the difference, we must look at the fracture mechanics analysis. As noted earlier, the energy release rate can be partitioned into various modes. One associated with the crack opening displacement at the crack (debond) face and one associated with the shear displacements perpendicular and parallel to the crack face.

We would like to note that we are reporting preliminary results on ongoing research. The results in this phase are based on idealized models of the joint. For these mode calculations, it is assumed that the adherends are rigid (the actual\ ratio of the moduli is on the order of 50) and that only the adhesive deforms. Perhaps by the time of the presentation at the ASME winter meeting a more complete physically realistic model will be completed. The problems we are currently having appear to be related to the utility of applying some principles of superposition in the crack closure method for these rather complex asymmetric cracks. We are hopeful this will soon be resolved. In any case, we feel even the idealized model provides interesting insight to explain the observed behavior reported in these last two sections.

The finite element analysis of the crack closure behavior of the original lap joint indicates the energy release rate is of the order of 99% Mode I (crack opening), while for the initial 50% debond specimen geometry, the total energy release rate is 95% Mode II. Several researchers, including some in this laboratory, have previously reported that the energy release rate is mode dependent.[4-7] For a number of adhesives, the energy release rate under Mode II loading is significantly higher than for Mode I, completely consistent with the results reported here.

The application of fracture mechanics to adhesive problems has been an active area of research for several years. See references 5 and 8 for a review of some aspects. Research on the use of finite element methods has been and continues to be important in expanding the use of these methods. There are problems in treating problems of stresses at and near interfaces, numerical problems (including the treatment of singularities, stress oscillations near the point of debonding, etc). Researches in this laboratory and elsewhere have worked on, and hopefully have partially solved some of these problems.[5,8]

CONCLUSIONS

The experimental results reported here clearly demonstrate the difficulty in using average stress criteria to predict strength of lap shear joints. Adhesive fracture mechanics with the aid of finite element analysis is more amenable to the the treatment of such problems that must be cognizant of the dependence of the energy release rate (adhesive fracture energy) on loading mode.

A finite element analysis of the mode of failure is currently being completed for the specimens that have half area debonds in the center region. Current studies are also in progress to determine the effects of each mode of fracture in the fracture process using some recently acquired advanced finite element software. In both cases, the overlap bond area was reduced in half. However, the required load for failure was well above half the load for the specimens with twice the area.

ACKNOWLEDGMENTS

Much of this research was supported by Morton Thiokol Corporation, with equipment made available by NSF Grant # DMR 85-087175.

REFERENCES

1.   Griffith, A. A., *Proc. 1st Int. Congr. Appl. Mech.*, 55 (1924).

2.   Orowan, E., *Rep. Prog. Phys.*, 12, 185 (1949).

3.   Rybicki, E. F. and Kanninen, M. F., A finite element calculation of stress intensity factors by a modified crack closure integral, *Eng. Fract. Mech.* 9, 921 (1974).

4.   Anderson, G. P., DeVries, K. L., and Williams, M. L., The influence of loading direction upon the character of adhesive debonding, *J. Colloid Interface Sci.*, 47 (3), 600 (1974).

5.   Anderson, G. P., Bennett, S. J., and DeVries, K. L., "Analysis and Testing of Adhesive Bonds," Academic Press, New York, 1977, p. 223.

6.   Trantina, G. C., Combined mode crack extension in adhesive joints, *J. Compos. Mater.*, 6, 371 (1972).

7.   Johnson, W. S. and Manzalgire, P. D., Influence of the resin on interlaminar mixed mode fracture, NASA technical memorandum 87571 (1985).

8.   Kinloch, A. J., "Adhesion and Adhesives," Chapman and Hall, New York, 1987.

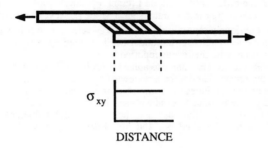

σ$_{xy}$

DISTANCE

FIGURE 1   Shows a schematic of the rigid adherend model of a single lap shear joint
                 with a plot of the theoretical shear stress distribution below.

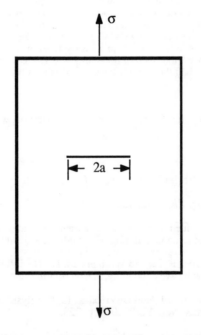

FIGURE 2   An isotropic elastic plane space with a crack length of 2a. The plane space
                 is also stressed by an amount σ.

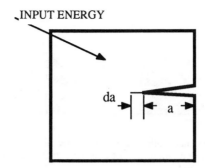

FIGURE 3   Shows the model used for the derivation of the energy release rate used in this paper.

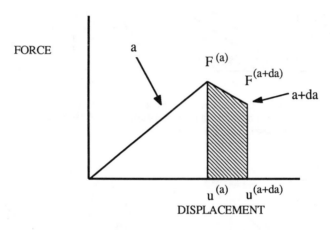

FIGURE 4   General case of the change in energy as the crack extends from a length a to a length a+da.

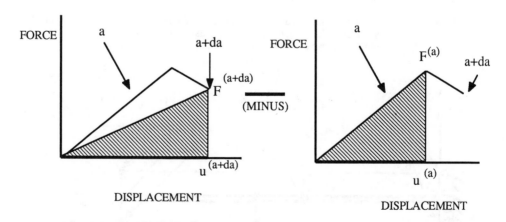

FIGURE 5   Shows the change in energy under the displacement curve as the crack extends from a length a to a length of a+da.

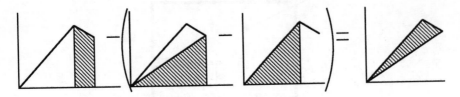

FIGURE 6   Shows the net result of energy under the displacement curve that is available
                to create new surfaces. The axis have the same labels as those used in
                figures 4 and 5.

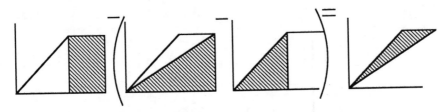

FIGURES 7   Show the limiting cases of the energy available for the creation of new crack
                surfaces under constant stress assumptions. Axis have the same labels as
                those used in figures 4 and 5.

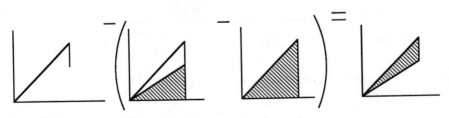

FIGURES 8   Show the limiting cases of the energy available for the creation of new crack
                surfaces under constant displacement assumptions. Axis have the same
                labels as those used in figures 4 and 5.

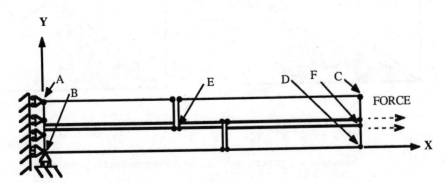

FIGURE 9   Shows an outline of the finite element grid used for the numerical analysis.

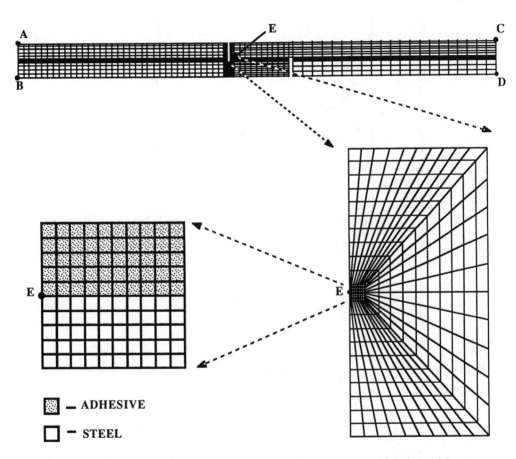

FIGURE 10   The finite element grid used for the analysis with expanded views of the
                   areas of interest.

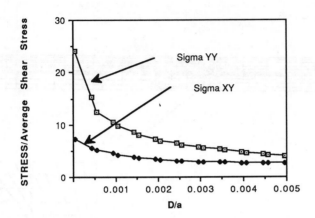

FIGURE 11   Shows the relationship of the normalized stress versus the normalized distance. Notice that the cleavage stress is higher than the shear stress near the bond termini.

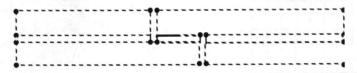

FIGURE 12   Shows the location of the half area debond that starts at the bond termini. The solid line represents the debond.

FIGURE 13   Shows the location of the half area debond that is in the center of the bond area. The solid line represents the debond.

166

AN INTEGRATED METHODOLOGY FOR OPTIMIZING STRUCTURAL COMPOSITE DAMPING

D. A. Saravanos and C. C. Chamis
Structures Division, MS 49-8
NASA Lewis Research Center
Cleveland, Ohio

ABSTRACT

A method is presented for tailoring plate and shell composite structures for optimal forced damped dynamic response. The damping of specific vibration modes is optimized with respect to dynamic performance criteria including placement of natural frequencies and minimization of resonance amplitudes. The structural composite damping is synthesized from the properties of the constituent materials, laminate parameters, and structural geometry based on a specialty finite element. Application studies include the optimization of laminated composite beams and composite shells with fiber volume ratios and ply angles as design variables. The results illustrate the significance of damping tailoring to the dynamic performance of composite structures, and the effectiveness of the method in optimizing the structural dynamic response.

INTRODUCTION

Fiber composite materials are broadly utilized in light-weight structures, as they readily provide superior specific modulus and strength. In addition to stiffness and strength, polymer-matrix composite materials provide higher material damping than most metals because of their "viscoelastic" matrix and heterogeneity. High specific stiffness and strength are sufficient conditions for improved static performance, but they do not always ensure improved dynamic performance. Passive structural damping is also a crucial dynamic property in vibration and sound control, as it generally improves resonance phenomena, settling times, and fatigue life. Composite materials are primarily targeted for structures requiring good dynamic performance, such as engine, aircraft, and space structures. Therefore, the inherent damping capacity of composites becomes a significant design factor, making polymer-matrix fiber composites even more attractive as structural materials.

Research on the damping capacity of composite materials, laminates, and beams (Adams and Bacon, 1973; Siu and Bert, 1974; Ni and Adams, 1984; Suarez et al, 1986; Saravanos and Chamis, 1989a; Saravanos and Chamis, 1989b), has shown that laminate damping is highly tailorable with respect to constitutive properties, volume fractions, and ply orientation angles. The same work suggests that composite structures should be tailored for optimal combinations

167

of damping and stiffness in order to obtain improved dynamic performance. To the authors' best knowledge, formal methods for tailoring general plate/shell composite structures for optimal damping and optimal damped dynamic response are not presently available. Research has been reported on the tailoring of plate/shell composite structures for optimum static and/or undamped dynamic performance. Composite plates have been optimized (Schmit and Farhsi, 1977; Schmit and Mehrinfar, 1982; Tauchert and Adibhatla, 1984; Eschenauer and Fuchs, 1985) based on static performance criteria. Composite plates have been also optimized either with a constraint on the first natural frequency (Soni and Iyengar, 1983), or with performance criteria based on the undamped frequency response (Adali, 1983). Liao et al. (1986), and Sung and Thompson (1986) presented the optimal tailoring of composite beams and links for maximum damping capacity based on static constraints. Most reported methods can produce designs with improved integrity and stiffness, having natural frequencies within desirable ranges. The inclusion of the structural composite damping into the present method provides a new dimension regarding the optimal tailoring of composite structures, as trade-offs between damping, weight, stiffness, and placement of natural frequencies are now possible. Furthermore, to the authors' best knowledge, damping micromechanics have not been included so far into the optimal design of composite structures, and in most reported cases the analysis starts from the laminate level. The present method incorporates such a damping micromechanics theory, hence, it provides the capability for tailoring also the constituent materials and fiber volume ratios.

This paper presents an integrated computational methodology for simulating the dynamic response of composite structures, and optimizing structural damping in conjunction to other static and dynamic design criteria. The method is targeted for composite structures under forced excitation. The dynamic performance criteria are based on modal damping capacities, resonance dynamic amplitudes, and natural frequencies. The method can produce designs having natural frequencies inside the desirable frequency domain, and also minimized resonance amplitudes. The optimal tailoring includes tailoring of the basic composite material(s), and tailoring of the laminate configuration. The modeling of composite damping and other elastic properties is based on micromechanics and laminate theories (Saravanos and Chamis, 1989a; Saravanos and Chamis, 1989b; Murthy and Chamis, 1984). The global damping capacity of composite structures is simulated based on a specialty finite element. Finite element damping matrices have been developed, based on previous theories, for a triangular plate element. The optimal design problem is formulated, and solved with the feasible-directions non- linear programing algorithm. The method is demonstrated by applying it to a composite beam and a composite cylindrical shell.

METHOD

This section summarizes the methodology. The method includes: (1) simulation of the structural composite damping, (2) simulation of the damped frequency response of composite structures, and (3) formulation and solution of the optimal design problem.

Structural Composite Damping

Local Laminate Damping

Approximate micromechanics equations are utilized, which have been derived based on hysteretic damping. On-axis specific damping capacities (SDC's) are represented in terms of elastic and dissipative properties of the fibers and matrix, interface properties, temperature, and moisture. Off-axis SDC's, ie. the SDC's of a ply loaded at an off-axis angle, are related to the on-axis SDC's with proper transformations, and subsequently, the specific damping capacity (SDC) of the composite laminate is synthesized. The laminate SDC is related to constituent properties, volume fractions, interface properties, hygro-thermal parameters, fiber orientation, laminate configuration, and local deformation. The micromechanics and laminate damping theories were developed by Saravanos and Chamis (1989a), and Saravanos and Chamis (1989b). Additional micromechanics and laminate theories (Murthy and Chamis, 1984) are utilized for other mechanical properties of the composite material.

Structural Damping

The laminate damping capacity is a local structural property. The global damping capacity of the structure at a given deformation shape would be the integrated action of local laminate damping over the structural volume. For plate and shell structures, the structural SDC ψ_s would be the ratio of the integrals of the local laminate damping energy δW_L and local strain energy W_L respectively, over the structural area A,

$$\psi_s = \frac{\int_A \delta W_L \, dA}{\int_A W_L \, dA} \tag{1}$$

Hence, the local laminate damping can only provide a rough estimate of the global structural damping of simple structural components subjected to forced vibration loading conditions. In case of complex structures or structural components, global structural damping is recommended. Numerical integration of laminate damping over the structural volume is performed based on finite element discretization. The developed procedure is summarized in the following paragraphs.

The local laminate damping and strain energies are first integrated over the volume of the finite element based on numerical quadrature. The finite element damping and stiffness matrices, $[C_e]$ and $[K_e]$ respectively, are obtained from the laminate damping matrices ($[A_d]$, $[C_d]$, $[D_d]$) (Saravanos and Chamis, 1989b) and stiffness matrices ($[A_s]$, $[C_s]$, $[D_s]$) as follows:

$$[C_e] = \int_{A_e} [B]^T \begin{bmatrix} [A_d] & [C_d] \\ [C_d] & [D_d] \end{bmatrix} [B] \, dA_e \tag{2}$$

$$[K_e] = \int_{A_e} [B]^T \begin{bmatrix} [A_s] & [C_s] \\ [C_s] & [D_s] \end{bmatrix} [B] \, dA_e \tag{3}$$

where [B] is the strain shape function matrix, and A_e the element area. After the integration, the damping and strain energies of the finite element, δW_e and W_e, are directly related to the respective element matrices and nodal displacements $\mathbf{u_e}$.

$$\delta W_e = \frac{1}{2} \mathbf{u_e}^T [C_e] \mathbf{u_e} \tag{4}$$

$$W_e = \frac{1}{2} \mathbf{u_e}^T [K_e] \mathbf{u_e} \tag{5}$$

Summation of the damping and strain energies of the individual elements provides the global damping and strain energies of the structure, δW_s and W_s respectively, for the specific deformation state.

$$\delta W_s = \sum_{i=1}^{nel} \delta W_{e,i} \tag{6}$$

$$W_s = \sum_{i=1}^{nel} W_{e,i} \tag{7}$$

The SDC of the structure at a specific deformation state is the ratio of damping and stored strain energies.

$$\psi_s = \frac{\delta W_s}{W_s} \tag{8}$$

In order to establish a coherent measure regarding the damping capacity of a structure, specification of representative deformation shapes is required. The mode shapes of the structure are the natural choice for such representative deformation shapes, since any small vibrational deformation can be expressed as a linear combination of mode shapes. The structural damping associated with an individual mode shape is defined as modal structural damping.

The previously described structural damping theory has been incorporated into a triangular plate element with three nodes, six degrees of freedom per node (3 deflections u_x, u_y and u_z; and 3 rotations ϕ_x, ϕ_y and ϕ_z), linear shape functions for the in-plane (membrane) deformations, and cubic shape functions for the out-of-plane (flexural) deformations. The same element is also utilized for modal analysis and simulation of the dynamic response.

Structural Dynamic Response

Main emphasis is focused on the forced (frequency) response of composite structures. Using finite element discretization, the dynamic response of the composite structure may be approximated by the following system of n discrete dynamic equations,

$$[M]\{\ddot{u}\} + [C]\{\dot{u}\} + [K]\{u\} = \{F(t)\} \tag{9}$$

where [M], [C], and [K] are the mass, damping and stiffness matrices respectively, and $\{u\}$ is the vector of the n discretized degrees of freedom.

A typical excitation force $\{F(t)\}$ would involve: surface tractions, body forces, and hygrothermal forces. In the present study, the effects of body forces and hygrothermal forces have been neglected. The study is limited to structures operating in room temperature dry environment. Hygrothermal effects may be significant for the design of composite structures, and the theory has provisions to model them. However, the design problem is complicated beyond the scope of the current paper, because the damping and stiffness of the composite depend also on temperature and moisture variations (Saravanos and Chamis, 1989b).

The dynamic system in eq. (9) is transformed to the $m \times m$ modal space through the linear transformation:

$$\{u\} = [\Phi]\{q\} \tag{10}$$

where q is the modal displacement vector, and $[\Phi]$ is ensembled from the first m normalized undamped eigenvectors of eq. (9). Assuming proportional damping, the transformation yields the reduced uncoupled dynamic system,

$$[m]\{\ddot{q}\} + [c]\{\dot{q}\} + [k]\{q\} = \{f(t)\} \tag{11}$$

where,

$$[m] = [\Phi]^T[M][\Phi]$$

$$[c] = [\Phi]^T[C][\Phi]$$

$$[k] = [\Phi]^T[K][\Phi]$$

$$\{f(t)\} = [\Phi]^T\{F(t)\}$$

The damping matrix [c] is formulated from the first m modal SDC's in accordance to eqs. (2-8).

Forced Dynamic Response

The frequency response of the j-th nodal displacement would be,

$$u_j(\omega) = \sum_{k=1}^{m} \Phi_{jk} q_k(\omega) \tag{12}$$

where $q_k(\omega)$ is the frequency response of the k-th mode in modal space. In general, $u_j(\omega)$ would be a complex number describing both amplitude and phase. The dynamic amplitude would be:

$$U_j(\omega) = ||u_j(\omega)|| \tag{13}$$

The resonance amplitude of the k-th mode, would be:

$$U_{jk}^r = ||u_j(\omega_{d,k})|| \tag{14}$$

where $\omega_{d,k}$ is the damped natural frequency of the k-th mode. The resonance amplitudes U_{jk}^r are used as dynamic performance measures.

Optimal Design

The structural damping methodology aims to optimize the forced response of composite structures. The objective is to tailor the composite materials and the laminate configurations for optimal combination of stiffness and damping such that selected natural frequencies are placed within desirable frequency ranges, and selected resonance peaks are minimized. Compared to other methods which only ensure proper placement of natural frequencies, the present method is far more powerful, as it drives all resonance frequencies in the feasible frequency domain, and concurrently minimizes the resonance peaks of selected modes at selected structural sites. Hence, the present methodology is even applicable when placement of all resonant frequencies into the desirable frequency domain is infeasible. The later problem is frequently encountered in design and is further complicated by the fact that the frequency bounds are usually stiff, in that, we have no control on them during the design of a given structural component. The present method readily includes the capacity to identify the most critical natural frequencies within the infeasible frequency range and to minimize their resonance amplitudes by increasing the respective modal damping values. The optimization problem is formulated in non-linear programming form, as described bellow.

Design Vector: The design vector may include: (1) the ply orientation angles θ_i of each sublaminate (group of $\pm\theta$ angle plies), and (2) the fiber volume ratios (FVR's) $k_{f,i}$ of each sublaminate.

The structural dimensions are not altered. The effect of shape optimization on the damped dynamic response of composite structures and its interaction with material tailoring will be addressed in a future study.

Objective Function: In general, the method provides the flexibility to minimize selected resonance amplitudes, based on the particular design requirements. The following objective function will minimize the maximum resonance amplitude of the first m vibration modes for q_d displacements:

$$min(max\{U_{jn}^r(\theta_i, k_{f,i})\}) \qquad j = 1, ..., q_d, \quad n = 1, ..., m \qquad (15)$$

Constraints: The minimization of the objective function in Eq. (15) is subject to the following constraints:

Upper and lower bounds on design variables:

$$-90.0° \leq \theta_i \leq 90.0° \qquad (16)$$

$$0.0 \leq k_{f,i} \leq 0.70 \qquad (17)$$

Upper bounds $U_j^{s,U}$ on q_s static displacements u_j^s:

$$u_j^s \leq U_j^{s,U} \qquad j = 1, ..., q_s \qquad (18)$$

Frequency constraints described in general by k upper and lower bounds, Ω_j^U and Ω_j^L respectively, on m natural frequencies:

$$\omega_n \leq \Omega_j^U, \qquad \omega_n \geq \Omega_j^L \qquad (19)$$

$$n = 1, ..., m \quad and \quad j = 1, ..., k$$

Upper bounds $U_{jn}^{r,U}$ on m resonance amplitudes for q_d displacements :

$$U_{jn}^r \leq U_{jn}^{r,U} \qquad j = 1, ..., q_d, \quad and \quad n = 1, ..., m \qquad (20)$$

171

It is pointed out that the objective function (15) and constraints (18) and (20) may include either deflections or rotations. Stress failure constraints are not included in the present paper, but they will be included in future work.

Optimization Algorithm

As already mentioned, the optimization problem has been formulated in non-linear programming form and is solved with the method of feasible directions. The feasible directions algorithm is a primal optimization method, performing a direct search in the feasible design space based on first order sensitivity for the objective function and active constraints. Primal methods are more suitable for non-linear programing problems which have computationally expensive constraints, since they typically require fewer constraint evaluations than do the penalty transformation methods. The feasible directions method used in this study incorporated active constraint set and line-search strategies for improved computational efficiency.

APPLICATION STUDIES

Assumptions

The present section presents applications of the previously described method on the two composite structures shown in Fig. 1:
(1) a cantilever laminated composite beam, 6in (152.4mm) long, 1in (25.4mm) wide, and 0.2in (5.08mm) uniform thickness, and
(2) a cantilever cylindrical laminated composite shell, 16in (406.4mm) long, 16in (406.4mm) wide, of 10in (254mm) radius, and uniform 0.2in (5.08mm) thickness.

Both structures were modeled with 80 triangular plate elements. The basic composite material system for each case was HM-S (high modulus surface treated) graphite fiber in an epoxy matrix. The mechanical properties of the fiber and matrix are shown in Table 1. Each ply is 0.01in (0.254mm) thick (equal to 2 preimpregnated tapes of 0.005in), hence, the composite laminate has 20 plies through the thickness. The applications were limited to symmetric laminate configurations, shown in Fig. 2, consisting by either one or three sublaminates at each symmetric side. As seen in Fig. 2, each sublaminate is a set of regular $\pm\theta$ angle-plies. The fiber orientation angle and the fiber volume ratio (FVR) of each sublaminate were considered as design variables. The effects of multiple sublaminates with different fiber orientation angles and FVR's are investigated with the following 4 laminates incorporating:
(a) one sublaminate of fixed 50% FVR with the ply angle as design variable.
(b) one sublaminate with both the ply angle and FVR as design variables.
(c) three sublaminates of fixed 50% FVR with their ply angles as design variables.
(d) three sublaminates with the ply angles and FVR's as design variables.

Table 1. Mechanical properties of HM-S/epoxy system.

Epoxy	HM-S Graphite	50% HM-S/Epoxy (Ni and Adams, 1984)
$E_m = 0.500$ Mpsi (3.45 GPa)	$E_{f11} = 55.0$ Mpsi (379.3 GPa)	$\psi_{l11} = 0.45$ %
$G_m = 0.185$ Mpsi (1.27 GPa)	$E_{f22} = 0.9$ Mpsi (6.2 GPa)	$\psi_{l22} = 4.22$ %
$\psi_{mn} = 10.30$ %	$G_{f12} = 1.1$ Mpsi (7.6 GPa)	$\psi_{l12} = 7.05$ %
$\psi_{ms} = 11.75$ %	$\nu_{f12} = 0.20$	
	$\psi_{f11} = 0.4$ %	
	$\psi_{f22} = 0.4$ %	
	$\psi_{f12} = 0.4$ %	

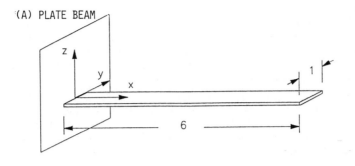

(A) PLATE BEAM

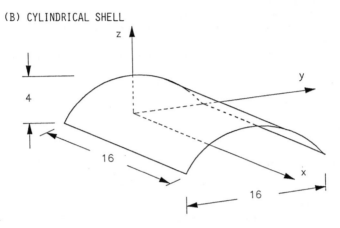

(B) CYLINDRICAL SHELL

Fig. 1 Candidate composite structures. (a) Beam. (b) Open cylindrical shell. Dimensions are in inches.

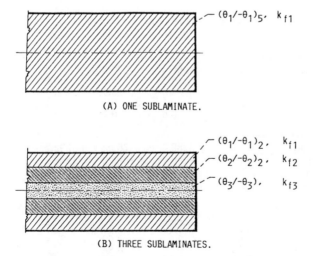

$(\theta_1/-\theta_1)_5$, k_{f1}

(A) ONE SUBLAMINATE.

$(\theta_1/-\theta_1)_2$, k_{f1}
$(\theta_2/-\theta_2)_2$, k_{f2}
$(\theta_3/-\theta_3)$, k_{f3}

(B) THREE SUBLAMINATES.

Fig. 2 Candidate laminate configurations. (a) with 1 angle-ply sublaminate; (b) with 3 angle-ply sublaminates.

173

Case 1: Beams

The assumed loading conditions of the beam were a static load of 5 lb/in (876 N/m) applied along the free edge, and a cyclic load of 0.1 lb/in (17.5 N/m) also applied along the free edge. The design objective was to minimize the maximum resonance (z-axis) amplitude of the first five modes at the middle of the tip. The first natural frequency was constrained to be less than 350 Hz. An upper bound of 0.050in (1.27mm) was imposed on the z-axis static deflection (u_z) of the free edge.

For comparison purposes, an additional optimization study was performed for one sublaminate with varying ply orientation angle and FVR without optimizing damping. In the later case, the maximum static deflection of the beam was minimized subject to an identical frequency constraint. The performance criteria did not include any damped resonance amplitudes, hence, the effect of composite damping on the design was neglected.

The optimum designs for each of the previously described subcases are presented in Table 2. The same table presents the reference design of a 50% FVR unidirectional composite beam. The unidirectional beam was selected as a reference design, because is known to exhibit the maximum static bending stiffness. As seen in Table 2, the first mode has the higher resonance amplitude, consequently, this resonance peak was minimized. The present method

Table 2. Optimum designs for the beam (Case 1).

	Unidirectional Design	Optimum Designs (Case 1a)	(Case 1b)	(Case 1c)	(Case 1d)	(w/o damping optimization)
Ply Angles, degrees						
θ_1	0.0	26.69	26.09	26.46	25.97	5.66
θ_2	0.0	–	–	26.60	26.18	–
θ_3	0.0	–	–	0.02	29.99	–
Fiber volume ratios						
k_{f1}	0.50	–	0.700	–	0.700	0.408
k_{f2}	0.50	–	–	–	0.700	–
k_{f3}	0.50	–	–	–	0.495	–
Max. z-axis Static deflection (tip), (10^{-3} in)						
u_z^s	19.99	48.94	34.56	47.54	34.49	25.56
Resonance z-axis Amplitudes (tip), (10^{-3} in)						
U_{z1}^r	426.9	235.1	209.4	237.3	209.3	481.9
U_{z2}^r	0.0	0.0	0.0	0.3	0.0	0.1
U_{z3}^r	0.0	5.4	4.8	5.5	4.8	0.0
U_{z4}^r	9.3	0.1	0.1	0.1	0.1	10.2
U_{z5}^r	0.0	0.7	0.6	0.7	0.5	0.0
Natural Frequencies, (Hz)						
ω_1	383.2	254.1	289.2	257.5	292.2	350.0
ω_2	1088.7	1369.5	1530.9	1558.6	1492.5	1219.9
ω_3	1565.1	1587.1	1802.2	1608.0	1820.7	1577.7
ω_4	2395.5	2945.6	3272.5	2924.0	3293.6	2193.2
ω_5	3854.0	5072.6	5770.2	5130.9	5828.2	4167.6
Modal SDC's, (%)						
ψ_1	0.57	2.55	2.02	2.45	2.01	0.65
ψ_2	5.57	3.85	3.12	2.28	3.28	4.15
ψ_3	3.43	2.69	2.15	2.60	2.14	2.05
ψ_4	0.66	0.83	0.70	0.84	0.70	0.73
ψ_5	4.54	2.19	1.74	2.13	1.74	3.68

has produced optimum designs (Cases 1a, 1b, 1c, and 1d) having lower resonance amplitudes for the bending modes by factor of 2, compared to the unidirectional design. In addition, the unidirectional design violates the frequency constraint, while all optimum designs have a first natural frequency less than 350 Hz. The trade-off for the improvements in dynamic performance was a reduction in static stiffness.

The superiority of the current method to other methods which do not optimize damping is also demonstrated in Table 2. The optimum design without considering damping in the performance criteria satisfies the frequency constraint marginally, but has a higher first resonance amplitude than the reference design. In contrast, all optimum designs produced by the present method have reduced first resonance amplitudes by factor of 2, and have first natural frequencies substantially below the upper bound constraint.

Additional comparison of the optimum designs produced by the current methodology for various laminate configurations indicates that the optimization of FVR has produced an additional 11% improvement in the objective function. This seems a rather insignificant improvement compared to the higher manufacturing costs related with the production of composites with customized FVR. Laminates with multiple sublaminate systems have virtually produced no additional reductions in the objective function.

The frequency response functions (FRF's) of the optimum design for Case 1d (three sublaminates with varying fiber volume), and the reference design are shown in Fig. 3. The z-axis dynamic deflection at the center of the tip induced by the previously described cyclic load is plotted as a function of frequency. Clearly, the optimum design has a better frequency response.

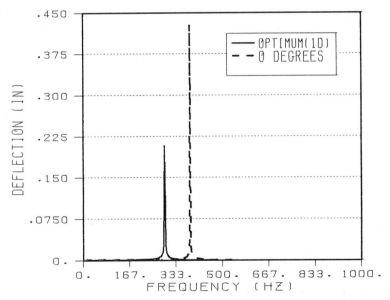

Fig. 3 Frequency response function of the composite beam (Case 1). Unidirectional and optimum (Subcase 1d) designs.

Case 2: Open Cylindrical Shells

The current section covers applications on a more complicated structure, such as the cylindrical laminated composite shell shown in Fig. 1b. The assumed loading conditions are: (1) a uniform static force of 0.31 lb/in (54.3 N/m) applied along the free-edge in the z-direction, and (2) a uniform cyclic force of 0.0625 lb/in (10.9 N/m) also applied along the free-edge in the z-direction. The objective function was set to minimize the maximum z-axis resonance amplitude of the first 5 natural frequencies at 3 sites (the middle and two

Subcase 2a: The optimum design for the simplest laminate configuration has produced a 39% decrease in the maximum resonance amplitude, in addition to a 59% decrease in static deflection, and has natural frequencies within the feasible frequency domain. The resultant optimum design is a ±23.3 degrees angle-ply symmetric laminate.

Subcase 2b: The introduction of the FVR into the design vector has produced an optimum design of ∓25.8 degrees angle-plies with 0.62 FVR. With respect to the reference design, the optimum design has reduced the maximum resonance peak by 50%, the maximum static deflection by 69% and satisfies all frequency constraints. The optimum design has reduced the modal SDC's of the shell, hence, the additional improvements were mostly accomplished by the increased stiffness due to the higher FVR.

Subcase 2c: Multiple sublaminates produced further reductions in the objective function. In contrast to the beam in Case 1, the open cylindrical shell is a three-dimensional structure, therefore, multiple sublaminate systems are expected to provide better tailoring capacity. Indeed, the consideration of three sublaminates with equal and fixed fiber volume ratios has produced an optimum design with 53% reduction in objective function, 66% reduction in static deflections, and natural frequencies within the feasible frequency domain.

Subcase 2d: Three sublaminates with varying FVR's have produced the best improvement in the objective function, as the resultant optimum design exhibits a 64% reduction in the maximum resonance amplitude, decreased static compliance by 63%, and satisfies all frequency constraints. Interestingly, sublaminate 2 has been reduced to a passive damping layer. The outer sublaminate with ∓26.5 degrees angle-plies and 0.691 fiber volume ratio provides most of the stiffness. Sublaminate 2 with 0.010 FVR is virtually pure matrix and provides most of the damping. The inner sublaminate 3 with ∓12.8 degrees angle-plies and 0.245 FVR provides both in-plane stiffness and damping. This optimum design produced a constrained layer damping structure. The high first modal SDC indicates that the reduction in the dynamic amplitude was mostly accomplished due to increased damping.

In all previous subcases, both maximum dynamic and static deflections occurred at the corners of the free-edge. In contrast to the unidirectional design, all optimal designs have asymmetric bending modes due to coupling between torsion and bending. The mode shapes of the first four modes for the reference and optimum (subcase 2d) designs are shown in Figs. 4 and 5 respectively. The first mode in both designs is the first transverse bending mode. The second bending and twisting modes of the optimum design have been switched. Coupling between torsion and flexure was observed in the modes of the optimum designs.

The FRF's of the reference and optimum (Subcase 2d) designs are shown in Fig. 6. The z-axis dynamic deflections of the corner with the maximum resonance amplitude are plotted. The superior frequency response of the optimal design is apparent.

SUMMARY

An integrated formal method for the tailoring of structural composite damping was described. The method is based on static and dynamic performance criteria and its primary objective is optimization of the forced damped dynamic response of the candidate composite structure. The damping capacities of individual modes were optimized such that selected resonance amplitudes were minimized subject to constraints on static and dynamic deflections, and natural frequencies. The method incorporates unified micromechanics and laminate theories for composite damping and other mechanical properties, therefore, can be utilized for the simultaneous tailoring of basic composite systems and laminate configurations. The simulation of structural composite damping is based on finite-element analysis, for this reason, the method is applicable to a wide array of plate and shell composite structures.

Applications of the method included: (1) the structural tailoring of a composite beam, and (2) the structural tailoring of a composite shell. The effect of laminates with multiple sublaminate systems on the optimal designs was also investigated. The more important conclusions are summarized in the following paragraphs.

1. Both application cases have illustrated the effectiveness of the method. In both cases

corners of the free-edge). The dynamic deflections of both corners were incorporated into the design criteria, in addition to the deflections of the midpoint, in order to take into account the bending in transverse direction (y-axis) and the material coupling between bending and torsion. Frequency constraints were imposed on the first 5 natural frequencies such that:

$$\omega_n \leq 150Hz, \quad or \quad \omega_n \geq 350Hz, \qquad n = 1, ..., 5 \qquad (21)$$

The z-axis static deflections along the free-edge were restricted to be less that 0.050in (1.27mm). The z-axis resonance amplitudes of the first five modes at the three sites were also restricted to be less than 0.050in (1.27mm). The reference design was a unidirectional shell $(0)_{20}$ of 50% FVR.

The resultant optimum designs for the 4 different laminate configurations are shown in Table 3. Table 3 also shows the maximum z-axis static deflection along the tip, the z-axis resonance dynamic amplitudes at the site with the maximum resonance peak, the natural frequencies, and the modal SDC's. The unidirectional reference design is also shown in Table 3. The reference design violates the static displacement constraint and the upper frequency bound of 350 Hz, since the 3rd and 4th natural frequencies are less than 350 Hz. All optimum designs have produced significant improvements to all performance measures, as they are within the feasible design space and exhibit superior dynamic and static stiffness. Contrary to Case 1 (beam), the consideration of multiple sublaminates and varying FVR's had a definite impact on the resultant optimal designs.

Table 3. Optimum designs for the cylindrical shell (Case 2).

	Unidirectional Design	Optimum Designs (Case 2a)	(Case 2b)	(Case 2c)	(Case 2d)
Ply Angles, degrees					
θ_1	0.0	23.28	-25.81	-30.96	-26.55
θ_2	0.0	–	–	20.79	-42.16
θ_3	0.0	–	–	0.61	-12.79
Fiber volume ratios					
k_{f1}	0.50	–	0.620	–	0.691
k_{f2}	0.50	–	–	–	0.010
k_{f3}	0.50	–	–	–	0.245
Max. z-axis Static deflection (tip), $(10^{-3}$ in)					
u_z^s	7.980	3.240	2.475	2.715	2.934
Resonance z-axis Amplitudes (tip), $(10^{-3}$ in)					
U_{z1}^r	396.4	240.6	196.4	183.9	143.6
U_{z2}^r	5.5	28.8	33.8	24.9	18.9
U_{z3}^r	27.6	11.2	9.0	3.4	5.3
U_{z4}^r	3.6	25.9	30.8	32.7	24.3
U_{z5}^r	0.8	0.1	0.0	0.1	0.2
Natural Frequencies, (Hz)					
ω_1	86.0	128.5	142.0	138.1	139.8
ω_2	98.7	135.1	150.0	146.8	150.0
ω_3	308.4	405.6	444.6	421.5	438.8
ω_4	316.7	417.6	455.9	428.6	447.8
ω_5	344.8	609.7	646.9	611.8	550.1
Modal SDC's, (%)					
ψ_1	2.85	2.12	1.98	2.23	3.10
ψ_2	2.82	2.54	1.65	2.03	2.79
ψ_3	4.36	1.29	1.22	1.44	2.03
ψ_4	4.02	1.54	1.00	1.35	1.67
ψ_5	3.30	2.20	1.66	2.00	2.60

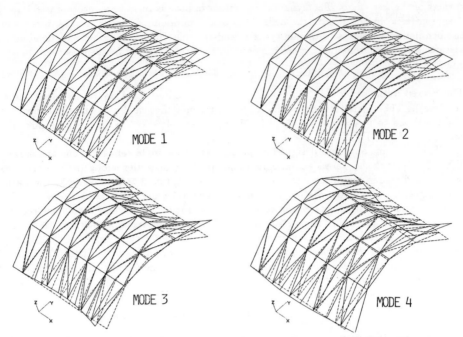

Fig. 4 Mode shapes of the unidirectional composite shell (Case 2).

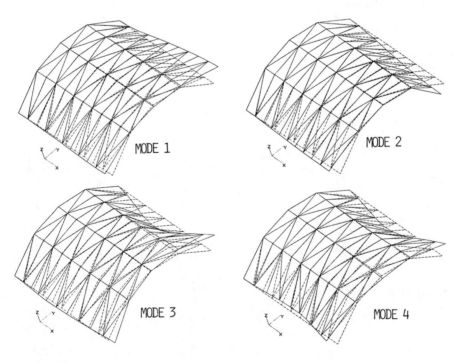

Fig. 5 Mode shapes of the optimized (Subcase 2d) composite shell.

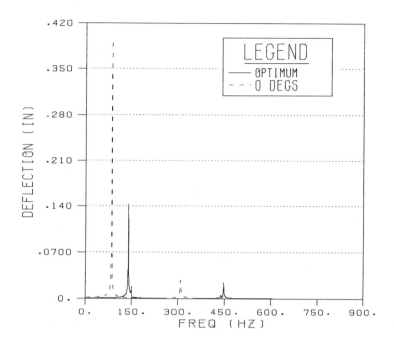

Fig. 6 The frequency response of the composite shell at the site of maximum resonance amplitude. Unidirectional and optimum (Subcase 2d) designs.

the tailoring of structural composite damping produced optimum designs with superior dynamic performance.

2. Tailoring of composite structures based only on static performance criteria and/or frequency constraints, and neglecting the effects of composite damping, may produce designs with poor dynamic performance.

3. The optimum tailoring of composite beams with laminates of one or multiple angle-ply sublaminates produced optimum designs with virtually equivalent dynamic performance. Compared to a unidirectional beam, the optimum designs have reduced the maximum resonance amplitudes by factor of 2, and have natural frequencies in the feasible frequency domain.

4. In the structural tailoring of the composite shell, laminates with multiple angle-plied sublaminates and variable fiber content produced optimum designs with significantly improved dynamic performance. In view of the attained improvements, any higher costs that may result in the fabrication of more complex laminate configurations seem justified. With respect to a unidirectional shell, the optimum designs exhibited lower resonance amplitudes by factor of 2.76, lower static compliance by factor of 2.70, and natural frequencies within the feasible frequency domain.

5. The optimization of shells with multiple sublaminates of variable fiber volume ratio produced laminates with passive damping layers of pure matrix. These laminate configurations have provided the best dynamic performance of all other optimum designs, and resemble a constrained layer damping structure.

179

REFERENCES

Adams, R. D., and Bacon, D. G. C., 1973, "Effect of Fibre Orientation and Laminate Geometry on the Dynamic Properties of CFRP," *Journal of Composite Materials*, Vol. 7, pp. 402-428.

Adali, S., 1983, "Multiobjective Design of an Antisymmetric Angle-Ply Laminate by Nonlinear Programming," *ASME Journal of Mechanisms, Transmissions, and Automation in Design,* Vol. 105, pp 214-219.

Eschenauer, H.A., and Fuchs, W., 1985, "Fiber-Reinforced Sandwich Plates Under Static Loads - Proposals for their Optimization," *ASME paper 85-Det-82.*

Liao, D.X., Sung, C.K., and Thompson, B.S., 1986, "The Optimal Design of Laminated Beams Considering Damping," *Journal of Composite Materials*, Vol. 20, pp. 485-501.

Murthy, P.L.N., and Chamis, C.C., 1984, "ICAN: Integrated Composite Analyzer," *AIAA Paper 84-0974.*

Ni, R.G., and Adams, R.D., 1984, " The Damping and Dynamic Moduli of Symmetric Laminated Composite Beams – Theoretical and Experimental Results," *Journal of Composite Materials*, Vol. 18, pp. 104-121.

Saravanos, D.A., and Chamis, C.C., 1989a, "Unified Micromechanics of Damping for Unidirectional Fiber Composites," *NASA TM-102107.*

Saravanos, D. A., and Chamis, C. C., 1989b, "Mechanics of Damping for Fiber Composite Laminates Including Hygro-Thermal Effects," *30th Structures, Structural Dynamics, and Materials Conference*, Paper No. 89-1191-CP, Mobile, Alabama, Apr. 3-5.

Schmit, L.A., and Farhsi, B., 1977, "Optimum Design of Laminated Fibre Composite Plates," *International Journal for Numerical Methods in Engineering*, Vol. 11, pp. 623-640.

Schmit, L.A., and Mehrinfar, M., 1982, "Multilevel Optimum Design of Structures with Fiber-Composite Stiffened-Panel Components," *AIAA Journal*, Vol. 20, No. 1, pp. 138-147.

Siu, C. C., and Bert, C. W., 1974, "Sinusoidal Response of Composite-Material Plate with Material Damping," *ASME Journal of Engineering for Industry*, pp. 603-610.

Soni, P.J., and Iyengar, N.G.R., 1983, "Optimal Design of Clamped Laminated Composite Plates," *Fibre Science and Technology*, Vol. 19, pp. 281-296.

Suarez, S. A., Gibson, R. F., Sun, C. T., and Chaturvedi, S. K., 1986, "The Influence of Fiber Length and Fiber Orientation on Damping and Stiffness of Polymer Composite Materials," *Experimental Mechanics*, Vol. 26, No. 2, pp. 175-184.

Sung, C.K., and Thompson, B.S., 1986, "A Methodology for Synthesizing High - Performance Robots Fabricated with Optimally Tailored Composite Materials," *ASME Paper No. 86-DET-8.*

Tauchert, T.R., and Adibhatla, S., 1984, "Design of Laminated Plates for Maximum Stiffness," *Journal of Composite Materials*, Vol. 18, pp. 58-69.

FRICTION AND WEAR BEHAVIOR OF GRAPHITE FIBER REINFORCED COMPOSITES

H. H. Shim, O. K. Kwon, and J. R. Youn
Korea Advanced Institute of Science and Technology
Seoul, Korea

ABSTRACT

Friction and Wear behavior of continuous graphite fiber composites was studied for different fiber orientations against the sliding direction. The effect of fiber orientation on friction and wear of the composite and on the deformation of the counterface was investigated experimentally. Pin on disk type testing machine was built and employed to generate the friction and wear data. A graphite fiber composite plate was produced by the bleeder ply molding in an autoclave and machined into rectangular pin specimens with specific fiber orientations, i.e., normal, transverse, and longitudinal directions. Three different wear conditions were employed for two different periods of time, 24 and 48 hours. The wear track of the worn specimens and the metal counterface was examined with a scanning electron microscope (SEM) to observe the damaged fibers on the surface and wear film generation on the counterface. Wear mechanism of the composite during sliding wear is proposed based on the experimental results.

INTRODUCTION

Polymeric composites are nowadays used extensively for tribological applications. Various fillers and reinforcements have been added to polymeric substances to improve strength and wear characteristics. For example, a wide variety of polymeric composites are available today for use as bearing materials, including self-lubricating reinforced plastics which contain solid lubricant such as Teflon, MoS_2, or graphite powders.

Experimental and theoretical studies on wear behavior of fiber composites under various conditions have been reported, e.g., sliding wear, abrasive wear, erosive wear, and fretting. Tsukizoe and Ohmae [1-3] and Sung and Suh [4] studied the effect of fiber orientation on sliding wear of fiber composites. Tsukizoe and Ohmae reported that the high modulus (type I) carbon fiber composite has the lowest wear in the transverse sliding direction and the high strength (type II) carbon fiber composite in the longitudinal direction.

The maximum wear occured in the longitudinal direction for the high modulus fiber composite and in the transverse direction for the high strength fiber composite. They also claimed that both high modulus and high strength carbon fiber composites experienced seizure in the normal sliding direction. According to Sung and Suh, the minimum wear of the graphite fiber composite was obtained in the normal sliding direction and the maximum wear in the transverse direction. Any seizure has not been reported in the normal direction. Tsukizoe and Ohmae selected fully annealed 0.25% carbon steel (equivalent to AISI 1025) as the counterface material and Sung and Suh selected AISI 52100 bearing steel.

It seems that the different behavior in normal direction is caused by the different counterface. Although it was not mentioned in the paper, Sung and Suh may have used high strength carbon fiber composite since it showed the largest wear in the transverse direction and its tensile modulus was reported to be 124 GPa (18 x 10^{6}psi).

There are some references [5,6] which state that the fiber orientation or matrix resin does not have any effect on wear behavior of the continuous carbon fiber composite. Giltrow and Lancaster claimed that the wear film formation on the counterface, the type of fibers in the composite, the composition and hardness of the counterface material, and wear conditions influence the wear rate significantly. Type I fiber composites transferred wear film to the counterface but type II fiber composites transferred little or no wear film. The wear film were reportedly composed of extensively degraded carbon fibers and matrix resin debris. It was also reported that any large broken fiber debris was not observed in the wear film by the electron microscope with high magnification. It seemed that the fractured fiber debris by the asperity interaction was broken and ground further between the two sliding surfaces yielding very fine particles or powders. It was also found that the stainless steel was the best counterface for high modulus graphite fiber composites. Although some conflicting experimental results were reported regarding the effect of fiber orientation on wear properties of the carbon fiber composites as reviewed above, Fahmy and West [7] recently studied wear behavior of high strength graphite fiber composites. They showed that the wear rate of the composite was strongly dependent upon the fiber orientation. The least wear occurred when the fibers make an angle of 45 deg with the wear plane and the sliding direction is along the fiber projection on the plane.

Wear mechanisms of composites in different sliding directions or fiber orientations must be understood well in order to design a good wear resistant composite structure for specific sliding applications. The high modulus carbon fiber composites showed good wear resistance because the graphitized high modulus carbon fiber reinforced composite generates surface wear film which reduces friction and wear owing to the self-lubricating nature of graphite powders. Effects of fiber orientation on sliding wear of unidirectional carbon fiber composites vary depending upon the reinforcing fiber, the counterface material, and the sliding condition.

The high modulus carbon fiber composite was selected for investigation of wear behavior in different sliding conditions, since it is most suitable for tribological applications. Stainless steel disks were prepared as the counterface because it was reported to be the best counterface material for the lowest wear.

EXPERIMENTAL

A unidirectional high modulus graphite fiber composite plate was produced by laminating the prepreg which contains high modulus carbon fibers of tensile modulus, 392 GPa. The laminated composite was cured in an autoclave with a vacuum bag surrounding the composite. The plate was cut into rectangular pin specimens of size 4.5 mm x 5.5 mm x 10 mm with three fiber orientations, i.e., normal, longitudinal, and transverse directions. The sliding surface of the specimen was carefully prepared by grinding and polishing followed by cleaning with a mild detergent, water, and hexane. A pin-on-disk type experimental setup shown in Fig.1 was built. The frictional force was measured by a load cell and recorded continuously in a micro-computer (IBM PC/AT) using an A/D converter and an amplifier. As the counterface, stainless steel (SUS 304) disks were prepared by grinding and polishing with a fine abrasive paper to give surface roughness of 0.02 μm. Three wear test conditions were employed; in condition I sliding speed of 0.5 m/s and normal load of 19.6 N was employed, in condition II 0.5 m/s and 49 N was employed, and in condition III 2.5 m/s and 19.6 N was employed. The friction and wear test was conducted for 24 hours and 48 hours for each specimen in each condition. The frictional force was measured continuously and the weight loss was measured after the test had been completed. Wear track of the worn specimen was carefully examined with a SEM to observe the wear film formation and damaged fibers which were exposed at the surface. The wear track on the counterface was also observed with the microscope to examine the wear film formation or groove formation. The hardness of the stainless steel counterface was measured with the Vicker's hardness tester.

RESULTS AND DISCUSSION

When the specimens were tested under the condition I, the weight loss was too small to

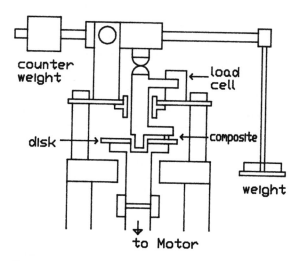

Fig.1. Schematic diagram of experimental Set-up

measure after 24 hour sliding. It means that the applied stresses on the surface was not large enough to cause considerable fiber damage in this wear condition. This result means that the high modulus graphite fiber composite will make an excellent wear resistant material if used with light wear conditions. For example the condition I employed here can be considered as a light wear condition. The friction coefficient was the highest in the normal direction (0.27), the lowest in the longitudinal direction (0.18), and middle in the transverse direction (0.23).

When tested under more severe wear conditions, small amount of wear was obtained. Average friction coefficients and wear factors for each condition after 48 hour sliding are listed in Table 1. When tested under the condition II, i.e., the normal force was increased to 49 N, the friction coefficient in the longitudinal direction was varied as a function of time as shown in Fig.2. The firction coefficient was increased to a high value and then decreased gradually to lower value. The wear after 24 hours of longitudinal sliding was the largest among the three directions but it was not after 48 hours. The friction coefficient in the transverse direction was the minimum and fluctuated as shown in Fig.3. The wear rate in the transverse direction was also the minimum among the three directions after 48 hours. The wear behavior in transverse direction is remarkable since the wear does not increase considerably once it reaches some value. It is believed that the wear film formation both on the specimen and on the counterface reduces the wear significantly. When slid in the normal direction, Fig.4 shows that the wear was increased continuously as it was slid against the counterface and the friction coefficient stayed almost constant at about 0.29. Micrographs of the worn surface of the specimens are shown in Fig.5. The polished cross-section of the fiber was observed on the worn surface in the normal direction. In the transverse direction, wear film adhered to the surface and fibers broken by the counterface asperity motion were found on the wear track as shown in the picture (b). In the case of the longitudinal sliding, carbon fibers are polished to show flat side surface and some fibers are missing owing to counterface interaction. In normal and longitudinal sliding, the fibers exposed to the surface are not covered by the wear film while transversly oriented fibers are covered by the wear film.

When the sliding condition III, the most severe wear condition, was employed, frictional behavior in the longitudinal and transverse directions changed significantly. There was a sharp increase in frictional forces at the beginning of sliding as indicated in Fig.6 and 7. Once the wear film was formed at the counterface the friction coefficients dropped to the lowest value and stayed stable. The lowest wear was obtained in the transverse direction both in 24 hours and in 48 hours. The friction coefficient in the normal direction (Fig.8) was high at the beginning and then reached lower steady value. The wear in normal direction was increased continuously as the sliding continued for 48 hours, resulting in the highest wear. The observation of the counterface slid against the normally oriented composite revealed that there was little wear film generation and grooves

Table 1. Average friction coefficients and wear factors for each sliding condition after 48 hour wear test

Condition	Fiber orientation	Wear factor ($\times 10^{-11}$ g/Nm)	Friction coefficient
Load=49N Speed=0.5m/s	Longitudinal	4.79	0.30
	Transverse	3.54	0.22
	Normal	5.36	0.29
Load=19.6N Speed=2.5m/s	Longitudinal	9.48	0.17
	Transverse	4.07	0.13
	Normal	22.56	0.18

were created by the plowing of the graphite fibers. However, wear film was observed on the counterface in the case of longitudinal and transverse sliding.

The microscopic observation of the worn surface in the transverse direction after tested with the wear condition III disclosed that the wear film covered the composite surface as shown in Fig.9. In the case of longitudinally slid specimens the traces of missing fibers and polished flat fibers were observed. As in the case of condition II the wear track of normally oriented specimen in condition III showed the polished cross-section of the fibers (Fig.9). It must be mentioned that there existed severely damaged regions within the wear track on the normally oriented composite. These regions are not continuous and appear to be like small craters where many debonded and broken fibers and wear debris were found under the SEM (Fig.10). Usually these massive failure regions were created when the friction coefficient was decreased suddenly with a large noise.

The frictional properties of materials were well explained by Suh and Sin[8]. They proposed that the friction coefficient is composed of three components; one due to the deforming asperities, another due to plowing by wear particles entrapped between the sliding surfaces and hard asperities, and the other due to adhesion. It was also claimed that the contribution by the plowing could become greater than that by other components. In the case of condition I which is the lightest condition among the three, wear due to fiber breakage must be small and the contribution of the plowing by wear particles to the friction coefficient should be small. In the longitudinal sliding, the counterface slides continuously on the graphite fiber which has low friction. Without entrapped fiber debris between the surfaces the friction coefficient in the longitudinal direction must be small. In fact, the lowest friction was obtained in longitudinal direction under the condition I. In the severe wear conditions the normally oriented fibers will plow the counterface. The hardness of the stainless steel was measured to be 2.29 GPa (234 kg/mm²) and the tensile strength of the graphite fiber employed was listed as 2.25 GPa (230 kg/mm²). Therefore the hardness of the fiber will be larger than the metal counterface. As the grooves found on the counterface in the case of normal fiber orientation indicate, the plowing action should be the reason why the friction in the normal direction is high. At the beginning of the longitudinal sliding the friction coefficient was increased to 0.35 in the case of condition III and then decreased to steady value of about 0.17. It is believed that the initial friction was determined by the asperity deformation and adhesion components and then the friction increases due to the contribution of plowing by the wear particles until sufficient wear film is transferred to the counterface. Once the wear film is established at the counterface the friction coefficient will be determined mostly by flow of wear film and adhesion component. In the case of transverse sliding, the friction coefficient was increased to 0.46 and then dropped to 0.13 as shown in Fig.7 for the wear condition III. The high initial friction is generated by the interaction between the counterface asperities and the transversely oriented fibers. As the wear film is formed both on the counterface and the specimen, the friction will be reduced to the low value. The wear film adhered to the specimen is especially important in the transverse case because the wear film fills the grooves between fibers as in Fig.9. The graphite fiber against metal sliding is changed to wear film against wear film sliding to cause the lowest friction.

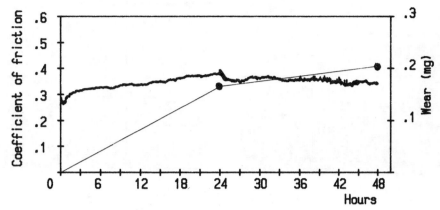

Fig.2. Friction coefficient and wear of the composite tested in longitudinal direction. (normal load = 49 N, sliding speed = 0.5 m/s)

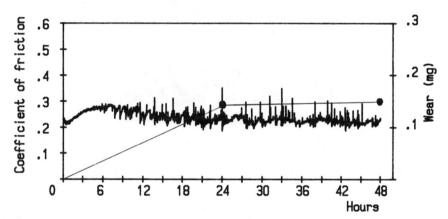

Fig.3. Friction coefficient and wear of the composite tested in transverse direction. (normal load = 49 N, sliding speed = 0.5 m/s)

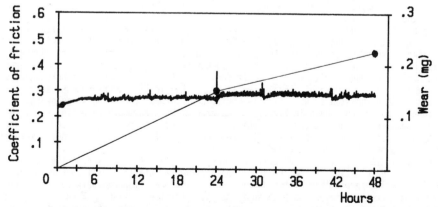

Fig.4. Friction coefficient and wear of the composite tested in normal direction. (normal load = 49 N, sliding speed = 0.5 m/s)

a) Longitudinal

b) Transverse

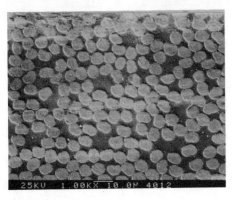

c) Normal

Fig.5. Micrographs of the wear track for different fiber orientations. (normal load = 49 N, sliding speed = 0.5 m/s)

It was not possible to measure any significant weight loss after the specimens were tested in wear condition I. It is believed that the applied normal and frictional stresses are carried by the fibers without considerable fiber breakage or thinning. In more harsh wear conditions, fibers in the transverse and longitudinal orientation must be damaged by the counterface asperity interaction until considerable wear film is formed on the surface, In longitudinal sliding, the fibers will be damaged by pushing, bending, buckling, or pulling as the micrographs of the worn surface suggest. The SEM picture implies that the fibers were thinned down by grinding and broken away from the surface owing to the counterface asperity motion. In the transverse direction, the fibers are damaged by bending but the fiber breakage will not be so severe because of fast wear film generation on the surface. Once stable wear film is formed on the counterface the fiber damage will be minimized. In the case of transverse sliding, the initial wear process seems to be important since the stable wear film is built up in about three hours according to the frictional behavior as shown in Fig.7. Deep grooves were formed both on the specimen and on the metal counterface at the same position when the specimen was tested in normal direction under the condition III. With the condition II, one or more grooves were also observed both on the counterface and the specimen. The mechanism of groove formation in the normal sliding is proposed as follows. The stiff graphite fibers plow the counterface causing large deformation at the surface. The deformed counterface will in turn damage the composite specimen generating fiber debris, which will be trapped between the sliding surfaces and create grooves on both surfaces by abrasive action. When the normally oriented fibers and the fiber debris penetrate into the relatively soft counterface and the power of the test setup is not large enough, seizure will occur as Tsukizoe and Ohmae

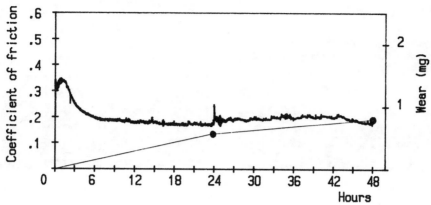

Fig.6. Friction coefficient and wear of the composite tested in longitudinal direction.
(normal load = 19.6 N, sliding speed = 2.5 m/s)

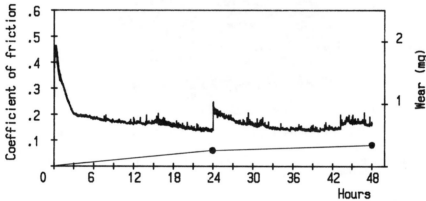

Fig.7. Friction coefficient and wear of the composite tested in transverse direction.
(normal load = 19.6 N, sliding speed = 2.5 m/s)

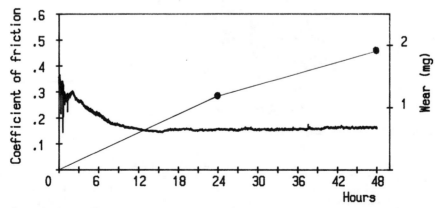

Fig.8. Friction coefficient and wear of the composite tested in normal direction.
(normal load = 19.6 N, sliding speed = 2.5 m/s)

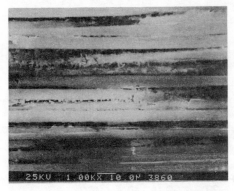

a) Longitudinal   b) Transverse

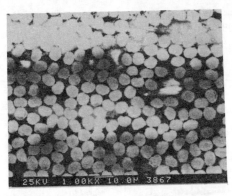

c) Normal

Fig.9.   Micrographs of the wear track for different fiber orientations. (normal load =
19.6 N, sliding Speed = 2.5 m/s)

Fig.10. Micrograph of the massive failure region.

observed. The phenomenon of massive failure region in the case of normal fiber orientation
is believed to occur owing to active interaction between the two sliding surfaces. The
friction increases just before the massive failure occurs and drops suddenly right after
the massive failure region is formed. This may be caused by the delamination type of wear
[9] or by the hot spot [10] generated by frictional heating. More experments and
theoretical investigations are required to better explain the special phenomenon.

188

CONCLUSIONS

Friction and wear behavior of unidirectionally oriented graphite fiber composite material varies depending upon sliding conditions and fiber orientations. In light sliding conditions, the maximum friction coefficient is obtained in normal direction and the minimum in longitudinal direction while the wear is negligibly small in any sliding direction when tested for 24 hours. As the wear conditions become severe, the friction coefficient is the highest in longitudinal direction under the condition II and in normal direction under the condition III, while the friction coefficient in transverse direction is the lowest under both conditions. The initial transient frictional behavior strongly depends on the fiber debris generation and asperity interaction.

The presence of wear film on the counterface or on the specimen controls the steady friction coefficient. The highest wear is obtained in the normal direction and the lowest wear in the transverse direction for the severe wear conditions. When the normally oriented fiber composite was tested, deep grooves are found on the wear track for both the composite specimen and the counterface. It must be noted that the above conclusions are valid when the high modulus graphite fiber (type I) composite is slid against stainless steel counterface.

ACKNOWLEDGEMENTS

This work is partially supported by a grant from the Korea Science and Engineering Foundation (KOSEF) and the tribology laboratory of the Korea Advanced Institute of Science and Technology (KAIST).

REFERENCES

1. Tsukizoe T. Ohmae N., "Wear Performance of Unidirectionally Oriented Carbon-fiber-reinforced Plastics," Tribol. Int., Vol.8, 1975, pp.171-175.

2. Tsukizoe T. and Ohmae N., "Wear Mechanism of Unidirectionally Oriented Fiber-reinforced Plastics," Wear of Materials, S.K. Rhee, et al., ed., ASME, 1977, pp.518-525.

3. Tsukizoe T. and Ohmae N., "Friction and Wear of Advanced Composite Materials," Fibre Sci. Tech., Vol.18, 1983, pp.265-286.

4. Sung N. and Suh N.P., "Effect of Fiber Orientation on Friction and Wear of Fiber Reinforced Polymer Composites," Wear, Vol.53, 1979, pp.129-141.

5. Giltrow J.P. and Lancaster J.K., "The Role of the Counterface in the Friction and Wear of Carbon Fiber Reinforced Thermosetting Resins," Wear, 1970, Vol.16, pp.359-374.

6. Giltrow J.P., "The Influence of Temperature on the Wear of Carbon Fiber Reinforced Resins," ASLE Transactions, 1972, Vol.16, pp.83-90.

7. Fahmy A.A. and West H.A., "The Effect of Fiber Orientation on the Wear Behavior of Graphite-Epoxy Composite Materials," Proceedings of the Sixth International Conference on Composite Materials, 1987, pp.5.311-5.323.

8. Suh N.P. and Sin H.-C., "The Genesis of Friction," Wear, 1981, Vol.69, pp.91-114.

9. Suh N.P., "An Overview of the Delamination Theory of Wear," Wear, 1977, Vol.44, pp.1-16.

10. Bonfield W. et al., "Wear Transfer Films Formed by Carbon Fiber Reinforced Epoxy Resin Sliding on Stainless Steel," Wear, 1976, Vol.37, pp.113-121.

ENVIRONMENTAL-STRESS CONSIDERATIONS IN DESIGNING WITH POLYMERS

D. M. Salisbury and K. L. DeVries
College of Engineering
University of Utah
Salt Lake City, Utah

ABSTRACT

Increasingly engineering polymers are replacing metals in many applications. Design criteria used with metals are not always applicable to polymers. Accordingly, considerable design information is needed to successfully bring about these structural products, in part due to their time, load and environmental dependencies. In some phases of the design process degradation of polymer systems must be considered, since many structural polymers are exposed to deleterious environments. Polymers are sometimes thought to be corrosion resistant and are often used to protect metals, woods, etc., from corrosion or deleterious environments, when in fact, quite the opposite may be the case. Polymers under various environments and especially under load conditions sometimes exhibit synergistically deleterious behavior. For example, the well known Kevlar fiber fails after 8 hours under a sustained load of 80% of its ultimate load in a 1% NO_x gas (a common pollutant) and air environment. The deleterious behavior of Kevlar and other advanced fiber systems under various environmental-stress conditions will be presented in this paper. The macroscopic, microscopic and molecular aspects of polymer degradation will be explored.

INTRODUCTION

Since about 1940, the polymer industry has grown at a phenomenal rate. The polymer industry will grow to an estimated $350 billion enterprise by the end of the century. According to the report *Advanced Materials by Design: New Structural Materials Technologies* from the office of Technology Assessment of the U.S. Congress, $2 billion worth of advanced structural products from the U.S. are produced annually. By the year 2000 it will grow to nearly $20 billion per year. This estimate, includes only the value of the materials and the structures not the finished products which reflects the importance of these new materials in the global competition for markets in the 1990's and beyond.[1]

Designers have found applications where polymers are almost the only viable candidate for use, but they are also finding increasing use as substitutes for metals and other materials. From the birth of the synthetic polymer age in about 1930, polymer usage has grown until today the area of polymer usage exceeds that of all metals. The reasons for this phenomenal growth are many and vary ranging from economical production methods to unique properties. This paper will be directed toward one of the less studied but important aspects of polymers. Polymers are generally considered to be corrosion resistant; they are in fact, often used as containers for corrosive substances and frequently form the basis of protective coatings for metals, wood or other materials. However, polymers are themselves often subject to chemical attack. This phenomena may limit their usefulness for some applications. If they are to be used as load bearing structural members in industrial, automotive, and aeronautic situations, it is important to have an understanding of how various industrial and automotive pollutants affect their mechanical properties. This paper will describe studies of the affects of gases such as ozone, SO_2, and NO_x, both alone and combined with sustained stress on the ultimate mechanical properties of some common and advanced polymers. These pollutants are often by-products of combustion as well as many other industrial and electrical processes.

It is well established that the properties of polymers are intricately related to their macromolecular structure. For example, polymers that are chemically very similar but with differing physical structures may have dramatically different mechanical (and other) properties. Polymers with an amorphous cross linked structure above their glass transition temperature are elastomeric with a modulus of about $4x10^6$ MPa (600 psi). In a semicrystalline or glassy amorphous state many polymer typically have moduli of 0.25 to 5 GPa (50 to 750 ksi) and strengths of 14 to 200 MPa (2 to 30 ksi). In the oriented state (fibers or films), semicrystalline polymers may have moduli as high as 7 GPa (10^6 psi) and strengths in excess of 1000 MPa (150 ksi). More recently techniques have been developed to produce "extended chain" configurations with moduli of 50 to 220 GPa (7 to $32x10^6$ psi) and strengths greater than 7 GPa (500 ksi). Since mechanical behavior is so dependent on structure it would appear to be useful to have an understanding of occurrences at the molecular level and how these are related to macroscopic behavior. Only when armed with such an insight can the designer or materials engineer hope to reliably predict long term (service) behavior from short term laboratory tests or systematically optimize performance. One might envision mechanical destruction of a polymer structure as involving any or a combination of several molecular mechanisms. The dominate mechanism might, in fact, vary from polymer to polymer or even in a given material depending on circumstances. For example, a polymer for which chain scission dominates the failure process(es) under one condition might fail through chain slippage or disentanglement under other circumstances (e.g., at high temperature). The effects of time, temperature and environment on ultimate properties will likely depend on the mechanism(s) involved. An important aspect of this paper will be to the description of experimental techniques that may be used to observe occurrences that can be directly related to the molecular phenomena associated with mechanical failure. This paper will largely concentrate on environmental-stress degradation of semicrystalline polymers in molded, extruded oriented, and extruded chain forms.

EXPERIMENTAL METHODS

In addition to mechanical loading by which the load, elongation, strain, resilience and toughness can be measured or inferred, several "molecular sensitive" experimental methods were incorporated into this investigation. These were Electron Spin Resonance spectroscopy (ESR), Fourier Transform Infrared Spectroscopy (FTIR) and measurements of changes in molecular weight (MWM). Since the reader may not be as familiar with these techniques, a very brief description will be given of each. A complete and thorough understanding of the techniques is very involved and requires an understanding of quantum mechanics and rather advanced physical organic chemistry. Fortunately, it is not necessary to have a thorough understanding of the details of spectroscopic techniques to understand this study. A simple and straightforward explanation is sufficient for our purposes and will be presented here. For more detailed developments on the subjects, the reader is referred to references.[2-5]

ESR is a type of microwave absorption spectroscopy that detects the presence of unpaired electrons. This is accomplished by measuring the microwave absorption of an assemblage of paramagnetic electrons in a magnetic field. The net absorption, due to the material in a microwave cavity in the spectrometers magnetic field, provides a measure of the number of unpaired electrons present in the cavity. When chain scission in a polymer takes place, the two electrons making up a covalent bond may be uncoupled, forming two free radicals. These free radicals may be detected and identified by ESR spectroscopy as long as they are produced in sufficient number and are stable enough to result in a detectable concentration of unpaired electrons in the ESR cavity. The spectrometer used in these studies was a Varian E-3 system operating in the X-band (ca. 10 GHz) microwave region. For high resolution spectrometers of this type, threshold sensitivity is generally about 10^{12} to 10^{13} free radicals in the cavity (25 cm long and 1 cm in diameter). To give the reader a feel for what this means in terms of the number of broken bonds in a plastic, this can be viewed in another way. Assume we have a gram of polyethylene with a repeat unit molecular weight of 14 in the cavity. This would represent 1/14 mole or $4.28x10^{22}$ carbon-carbon bonds in the back bone. In other words, since breaking each bond would produce two free radicals the spectrometer would be able, under ideal conditions, to detect one broken bond in 10 billion present. The resonance spectrum is often split into a series of peaks (hyperfine structure) due to the interaction of the unpaired electron with the magnetic moments of nearby atoms as well as with the external field. Since this hyperfine structure is characteristic of the particular surrounding nuclei, it provides a "fingerprint" of a free radical's environment. This "fingerprint", in principle, can be used to identify a particular radical. In principle then, ESR can be used to determine the number of chains broken in the failure process and the location(s) along the polymer chain at which these occur. A servo-controlled loading system has been constructed about the spectrometer magnet and the sample may be stressed while being studied by ESR. The ESR system is also equipped with a variable temperature apparatus that facilitates investigations between -150 °C and 300 °C.

As useful as it has proven to be, ESR does have a major disadvantage for studying polymeric fracture. The polymeric radicals produced by chain scission are inherently unstable entities. The unpaired electron associated with the free radical has an affinity for other "free electrons" or certain impurities. In this way the spectra due to the free radicals present decay away with time, often so rapidly that it is essentially undetectable. Radicals produced by fracture in nylon and by ozone-stress fracture of some rubbers are fairly stable with a half-life of approximately 1 hr at room temperature. Radicals in many other polymers are very unstable at room temperature with half-lives of a few seconds of less. In most plastics the radicals become stable at cryogenic temperatures and hence some previous studies have been restricted to low temperatures. With an understanding of the radical decay kinetics one might, in some cases, extrapolate from the number of radicals observed to that produced during fracture. Several studies[6-8] of decay phenomena have been undertaken and provide information helpful in this extrapolation.

From the above paragraph it would seem useful to have other tools to investigate bond rupture in terms of something more stable than free radicals. When chain scission occurs in a polymer two new chain ends result, initially with associated free radicals. These will eventually stabilize into stable end groups, that in principle should be amenable to detection by infrared (IR) spectroscopy.

Infrared (IR) spectroscopy is also a form of absorption spectroscopy. In this case photon absorption occurs at the resonance frequency of the various chemical species present along the polymer chain. Different atom and bond combinations along the polymer chain have differing resonant frequencies that under appropriate conditions can be excited by infrared radiation. In this way the plot of absorption bands versus IR frequency (the IR spectrum) is an indicator of which atoms-bond combinations are present in the polymer chain. Recently Fourier Transform Infrared (FTIR) spectrometers have become available that are not only an order of magnitude more sensitive than previous dispersive models but are greatly facilitated in the spectral analysis by their built-in computer. Simply stated, when chain scission occurs two new chain ends are formed. These new end groups will likely correspond to different natural frequencies than the other chemical species along the chain. Observation of the increase in IR absorption in the frequency band corresponding to these end groups provides a quantitative measure of the extent of bond rupture.

Chain scission also results in a change in molecular weight. Measurements of molecular weight (MWM) before and subsequent to the degradation processes therefore provide an indication of the extent of bond rupture. A variety of techniques are available for measuring polymer molecular weight.[5] The method used in this study is dilute viscosity measurement.[5,9] This method depends on the fact that the intrinsic viscosity is a function of the polymer concentration in a solvent which is directly related to the polymer's molecular weight. Measurements before and after degradation of a polymer provides a measure of the extent of chain scission.

EXPERIMENTAL PROCEDURES

Several pieces of equipment have been employed in experimental studies of chain scission associated with mechanical degradation in polymers. A Varian E-3 Electron Spin Resonance Spectrometer with a servo controlled hydraulic loading system constructed about the magnet was used to explore the possibility of any free radicals that might accompany homolytic chain scission during fracture. A Nicolet 10-DX Fourier Transform Infrared Spectrometer was used to explore the presence of new end groups resulting from molecular bond rupture. These analytical methods have been supplemented with measurement of changes in molecular weight resulting from chain rupture. This technique provided some collaborative information for the other spectroscopic techniques. Although, only ESR studies will be presented in this paper the authors felt it was necessary to briefly describe other techniques for probing the molecular fracture phenomenon.

In order to investigate the effect of environment combined with stress, specialized load frames and environmental chambers were constructed as shown schematically in Figure 1. The samples were in the form of fiber bundles. These were constructed by wrapping several windings of yarn about two 0.635 cm diameter dowels, placed about 26.35 cm apart on a board. Near the ends, the bundles were tied and bonded with epoxy, resulting in a loop on each end. Several yarns were removed from the center part effectively producing a tensile "dog-bone" specimen. The sample could then be loaded by rods through the end loops, and if enough yarn was removed, failure would occur in the central (reduced) section of the specimen. This method, for example, assured not only that there was a region of comparatively uniform tensile stress but that, during the ESR measurements, fracture occurred within the microwave cavity.

For the sustained loading-environmental tests, a tensile sample was first loaded in the load frame of Figure 1. A predetermined load was applied with the aid of an Instron Testing Machine. A screw was then tightened to transfer the machine load to the spring. This apparatus was then transferred to an environmental chamber constructed of PMMA sheet. A pre-selected concentration of air-gas mixture of ozone, SO_2, NO_x, or simply lab air was maintained in the chamber while the sample was held under a relatively constant load by the spring. After a predetermined time under the environmental-stress condition, the samples were fractured to determine the remaining strength.

EXPERIMENTAL RESULTS

Environmental-stress aging of oriented fibers and molded samples have been studied previously.[10-12] Tests were conducted with the samples stretched and maintained at various fractions of the samples ultimate strain or strength while in the selected environment. While all of the details of structure for these two different classes of material are not known it is well-established that oriented fibers and molded materials of nylon have vastly different morphologies and hence mechanical properties. Various morphological models have been proposed for oriented fibers, all of which describe the structure as a sort of network of crystalline blocks separated by less ordered amorphous regions. The molecular chains in both regions are general oriented in the direction of the drawn fiber. This type of model is consistent with X-ray, birefringence, and other experimental studies.[13-14] Molded pieces of the same polymer, on the other hand, are usually much more isotropic. The structure is semi-crystalline but the segments of varying order are more random in direction. In the most common case the material is made up of larger building elements called spherulites. The experiments for the oriented and molded materials were conducted as similar as possible, so as to facilitate a comparison of the combined effects of environment stress on these morphologically different forms of chemically similar materials.

The observed breaking stress and Young's modulus for the molded nylon under the combined degradation effects were nearly constant irrespective of how the samples were aged. For example, molded material maintained for a week at strains up to 90% of the strain at ultimate load in a NO_x environment showed almost no degradation in strength or change in modulus. This is in marked contrast to the response of oriented nylon as demonstrated in Figures 2 & 3. This figure shows the degradation in strength after exposure to a 1% NO_2 concentration for the times indicated. This appears to provide evidence of different morphological mechanisms being involved in failure of the two physically different classes of material. The above suggests it would be useful to probe for molecular evidence for these different mechanisms. Accordingly, Electron Spin Resonance (ESR) was used in an effort to gain insight into the number of primary bonds that need to be broken to accomplish the final failure. ESR is essentially a detector of free radicals present per volume. The number of free radicals is assumed proportional to the number of primary bonds broken. For example, fibers that were not degraded were fractured in the ESR cavity which produced a certain number of free radicals; in turn, fibers that were degraded and subsequently removed from the chamber and then fractured in the ESR cavity resulted in a reduction in free radicals as shown in Figure 4. This shows the remaining number of free radicals generated during loading to failure subsequent to the aging process. The fact that the number of free radicals decreases (as does strength) with degradation time can be interpreted as evidence that some of the tie chains between crystallites have been broken by the combined influence of stress and environment. Another feature of Figure 4 is that the appearance of the initial free radical concentration at various sustained loads, do not extrapolate to the initial (or, as received) sustained load as does the fracture strength in Figures 2 and 3. To understand this behavior, free radicals are produced when the fibers are monotonically loaded in the environmental chambers, thus reducing the total free radical concentration as compared against the as received sample. In addition to ESR, viscosity, and infrared spectroscopy measurements have been performed demonstrating trends similar to those observed with ESR.[10-12] The three independent methods of measurement yield essentially the same conclusions as to the molecular mechanisms involved in failure for the different physical structures.

The diffusion-reaction process of NO_x in nylon, both in a spherulitic-cast form and fiber form have previously been studied.[15,16] If nylon degraded as a result of surface rather than bulk reaction processes then one would need to be careful how large of a sample to use, such as, the comparison of molded versus fibrous nylon. Studies[15,16] indicate that the degradation process is primarily a bulk process if the nylon has been dried; whereas, the process tends toward a surface reaction process when moisture is introduced into the polymer system. The nylon in this study was dried.

This observation is contrasted with evidence obtained from samples of other materials and/or differing physical structure. The Grace polyethylene fibers we tested had a strength 40% of that of the nylon fibers. The polyethylene also experience much more plastic deformation before failure. The free radical concentration produced during fracture, in the as received fibers, was also an order of magnitude lower than that for the nylon fibers. Furthermore, the polyethylene did not experience anywhere near as large of a percentage of reduction in strength during the stress-environmental aging nor was its free radical spectra intensity reduced as much (relative to its base value) after aging. A limited number of experiments were conducted on machined polyethylene which had a spherulitic structure as the molded nylon. This material had nearly an order of magnitude lower strength than the fibers of the same material had but much larger elongations at failure. The free radical production for the molded nylon and polyethylene was at best barely detectable, they experienced hardly any reduction in strength in the same environments and comparable sustained loadings (relative to their as received strength) over the same time scale as those shown for the fibers in Figure 2; however, both the nylon and the polyethylene did experience dramatic decreases in engineering toughness associated with a significant reduction in elongation at failure.

These observations suggest that the mechanism of strength (or failure) differ for these different physical forms of chemically similar materials. Apparently, the relatively high strength and relative lack of deformation, for the nylon fibers, is due to oriented tie chains bridging the crystalline regions in the semicrystalline polymer. The experimental evidence indicates that large numbers of these must be broken in order to complete macroscopic fracture. The failure of these chains is a thermally activated process that is aided by both the stress and the aggressive environments.

At the other end of the spectrum, in the molded parts, strength is apparently largely attributed to secondary forces. Failure here involves processes such as chain slippage, pull out, etc., the energy for which is apparently not as sensitive to stress-aided environmental attack. The data suggests that there are tie chains bridging the crystalline blocks, but they are not oriented or as uniformly stressed as in the case for the fibers. They do, however, apply constraints in the molded material and as slippage occurs these tend to hold the structure together, and ultimate failure is prevented until relatively large strains have occurred. It is postulated that these are ruptured during the stress-environmental aging, resulting in the dramatic reduction in toughness observed for the aged samples. The polyethylene fibers fall somewhere between these extremes. Here, failure apparently involves both extensive slippage and chain scission. Environmental stress has a significant, but much smaller effect than that observed in oriented nylon.

In our most recent studies, we have used these same techniques to investigate failure of Kevlar. Our earlier efforts to use ESR to measure bond rupture were hampered by a large residual ESR signal. We were able to observe slight variations in this signal when the sample was stressed, but the increase was small relative to the residual signal thereby making quantitative analysis impossible.

Building upon the work of Brown, et al.,[17] we have been able to "anneal" the residual signal to an acceptable value. With samples prepared in this manner, we have been able to produce easily detectable ESR signals by several mechanical treatments. Free radical concentrations are produced by tensile loading to failure. Typical results for such tests are shown in Figure 5. As the strain is increased, the first detectable signals in this case appear at loads just before failure (as contrasted with 60% of the ultimate stress for nylon fibers). Somewhat larger ESR signals were produced by other mechanical treatments. These large spectra might be, but have not as yet been, used to analyze the fine structure of the spectra for clues as to the nature of the ruptured bonds. The largest signals were produced by scraping the fibers with a knife blade while the fibers were immersed in liquid nitrogen to retard free radical decay. Large signals were also produced by a "kinking" operation similar to that described by Dodd, et. al.,[18] in which fibers were bent and pulled about a very small radius corner. These experiments provide clear evidence that mechanical failure of Kevlar involves chain scission. This raises the concern as to whether it might experience environmental stress degradation analogous to that observed in the nylon fibers. Accordingly, experiments were conducted in which the Kevlar fiber was held under a sustained loading in a NO_x environment. After a period of time., the fibers were removed from the chamber and fractured in tension in a tensile testing machine. The results shown in Figure 6 demonstrate that little degradation in strength occurs in the absence of either the stress or the environment. However, at stresses in excess of 50% of the ultimate stress for the material, rather rapid catastrophic degradation occurs. For example, at sustained loads of 70% of the ultimate load (P_{max}) the reduction in fracture strength was about 8% after 75 hours in the chamber. At sustained loads of 80% and 90% of P_{max}, the fibers lost nearly all strength at about 4 and 9 hours respectively.

CONCLUSIONS

The most pertinent results of this study might be briefly summarized:

1. Failure of oriented nylon combined with stress and environmental agents, is dominated by the failure of tie chains. The strength of such materials is due in large part to the fact that they have not

only oriented but some extended chains and take advantage of the very high strength of covalent bonds. The environmental pollutants studied had the most deleterious effect on this type of material.

    2. Failure of molded nylon and polyethylene (with a spherulitic structure) was primarily due to chain slippage from rupture of secondary bonds. Stress and environmental effects do not have any where near as pronounced effect on these materials as they were on the oriented systems.

    3. There is a synergistic affect of stress and environmental agents on oriented nylon and Kevlar of which the combined effect was much more rapid than would be predicted by adding the individual effects.

ACKNOWLEDGEMENT

Much of this research was supported by the National Science Foundation Grant 85-07175.

REFERENCES

1   Cicen, J. R. *Forecast 89, New Directions*, <u>Advanced Materials and Processes</u>, Vol. 135, Issue 1, ASM International, Metals Park, Ohio, Jan 1989, pg 21.

2   Berson, M. and Baird, J.C., *An Introduction to Electron Paramagnetic Resonance*, W.A. Bengamin, Inc., New York, 1966.

3   Siesler, H.W., and Holland-Moritz, K., *Infrared and Raman Spectroscopy of Polymers*, Marcel Dekker, Inc., New York, 1980.

4.   Levine, I. L., *Molecular Spectroscopy*, Wiley, New York, 1975.

5.   Ezrin, M. (Editor), *Polymer Molecular Weight Methods*, American Chemical Society, Washington, D.C., 1973.

6.   Zurkov, S. N., Tomashevskii, E. E. and Zakrevskii, V. A., *Sov. Phys. Solids State*, **1962**, 3, 2075.

7.   Wilde, T. D., M.S. Thesis, University of Utah, Salt Lake City, Utah, 1974.

8.   Lloyd, B. A., Ph.D. Dissertation, University of Utah, Salt Lake City, Utah, 1973.

9.   Brandrup, J., and Immergut, E. H. (Editors), *Polymer Handbook,* Interscience Publishers, New York, 1966.

10.   Igarashi, M., and DeVries, K. L., *Polymer*, **1983**, 24, 769-782.

11.   Igarashi, M., and DeVries, K. L., *Polymer*, **1983**, 24, 1035-1041.

12.   DeVries, K. L., Murthy, S., and Igarashi, M., Environmental Stress Damage in Oriented Semi-crystalline Polymers, 84-WA/Mats-27, 1984.

13.   Zurkov, S. N., Tomashevskii, E. E. and Zakrevskii, V. A., *Sov. Phys. Solids State,* **1962**, 3, 2075.

14.   Stoeckel, T. M., Blasius, J., and Crist, B., *J. Polym. Sci.-Polym. Phys. Ed.*, **1978**, 16, 485-500.

15.   Salisbury, D.M. and DeVries, K.L., <u>Effects of Moisture and Diffusion on NO_x-Stress Degradation of Nylon 6/6 Fibers</u>, *Technical Papers, Society of Plastic Engineers*, **1987**, *33*, 491.

16.   Jellinek, H.H.G. and Chaudhuri, A.K., <u>Inhibited Degradaton of Nylon 66 in Presence of Nitrogen Dioxide, Ozone, Air, and Near-Ultraviolet Radiation</u>, *J. of Polym. Sci.* 1972, **10** (A-1), 1773-1788

17.   Brown, T. C. Sandreczki and R. J. Morgan, *Polymer*, **1984**, 25, 737.

18.   Dobb, D. J. Johnson and B. P. Saville, *Polymer*, **1981**, 22, 960.

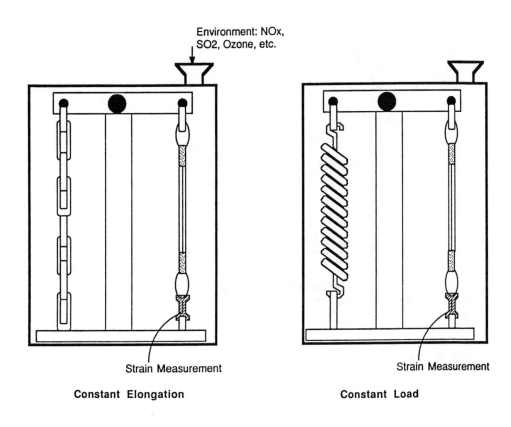

Figure 1. Loading frames for constant elongation and constant loading conditions.

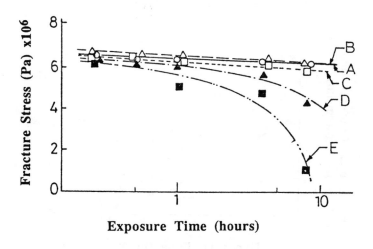

Figure 2. Degradation in strength as a function of time for nylon 6 fibers in 0.86% concentration of NO_x at 81°C. Final fracture was at room temperature.

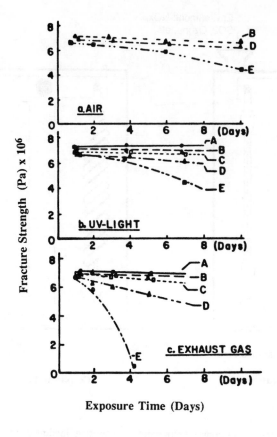

Figure 3. Degradation in strength as a funtion of time for nylon 6 fibers in various environments.

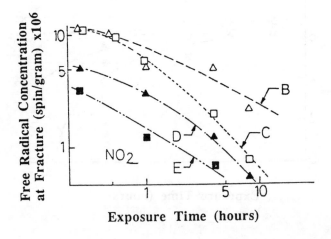

Figure 4. Free radical concentration produced during fracture after exposure to 0.86% NO_x concentration. Sustained loads of 60, 70, 80, and 90% of P_{max} correspond to curves B, C, D, and E respectively.

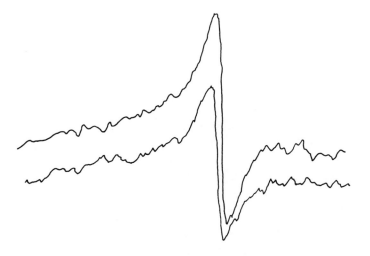

Figure 5. ESR signal of Kevlar 49 resulting from tensile loading to failure for upper curve. Lower curve is the residual signal of the same fiber.

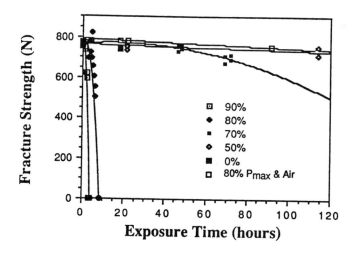

Figure 6 Fracture strength of Kevlar versus exposure time at different sustained loads and in 1% concentration of NO_x.

THE ELASTIC PROPERTIES OF WOVEN POLYMERIC FABRIC

W. E. Warren
Sandia National Laboratories
Albuquerque, New Mexico

ABSTRACT

The in-plane linear elastic constants of woven fabric are determined in terms
of the specific fabric microstructure. The fabric is assumed to be a
spatially periodic interlaced network of orthogonal yarns and the individual
yarns are modeled as extensible elastica. These results indicate that a
significant coupling of bending and stretching effects occurs during
deformation. Results of this theoretical analysis compare favorably with
measured in-plane elastic constants for Vincel yarn fabrics.

INTRODUCTION

Woven fabrics represent a class of polymeric materials which provide a number
of beneficial mechanical propeties, the most important of which include
strength, flexibility, and relatively low density. Early weaving processes
exploiting these properties developed empirically using yarns spun from
natural fibers. The development of synthetic polymeric fibers with their wide
diversity of mechanical, physical and chemical properties offered a
corresponding wide diversity of woven fabric properties together with the
possibility of engineering fabrics for specific applications. Woven polymeric
fabrics, for example, are often used as the filler in building up individual
laminates of a composite material. The desire to engineer or design fabrics
for specific applications has stimulated considerable interest in theoretical
analysis relating their effective mechanical properties to specific aspects of
the woven fabric morphology and microstructure.

The first comprehensive investigation of the relationships between various
parameters of a woven fabric was apparently the work of Peirce [1]. The
Peirce model is strictly a geometrical or kinematical model describing the
geometry of woven fabric and no consideration of forces or equilibrium is
given. A more physical model utilizing an analysis of the inextensible
elastica was developed by Olofsson [2] and Grosberg and Kedia [3]. An
excellent summary of the analysis of the mechanical properties of woven
fabrics prior to 1969 is included in the monograph by Hearle et al. [4]. More
recent summaries have been presented by Ellis [5] and Treloar [6].

This work performed at Sandia National Laboratories supported by the
U. S. Department of Energy under contract number DE-AC04-76DP00789.

The analysis of fabric structure and properties has followed primarily one of two paths: the descriptive geometrical approach based principally on the work of Peirce [1], or the mechanistic approach [7]. A number of the mechanistic approaches are based on energy methods [7,8,9] in which the strain energy due to yarn stretching is uncoupled from the strain energy due to yarn bending (crimp interchange effect). The bending energy is usually obtained from an analysis of the inextensible elastica and represented in terms of elliptic integrals which are cumbersome to interpret because of the reversal of independent and dependent variables inherent in this approach. In an effort to circumvent these complications, Leaf and Kandil [10] consider a saw-tooth model of plain-woven fabric, which also uncouples the yarn stretching energy from bending energy, and they obtain simple expressions for the effective linear elastic constants in the principal yarn directions in terms of the fabric microstructure. While these results are easily interpreted, they provide estimates of the elastic constants considerably higher than they measure experimentally. An energy method incorporating the Peirce geometry has been used by Hamed and Sadek [11] who minimize the energy using a pattern search computational program and obtain numerical results for uniaxial loading.

In this work we assume the woven fabric consists of a regular network of orthogonal interlaced yarns. We model the individual yarns as extensible elastica and thus couple stretching and bending effects at the outset. Consideration is restricted to biaxial loading in the principal yarn directions in the plane of the fabric; the case of more general loading in the plane, as suggested by the trellis model of Kilby [12], will be the subject of future analysis. The initial unloaded yarn geometry is assumed to be a sequence of alternating circular arcs of constant radius R as considered by Olofsson [13].

We first obtain the solution to the differential equation describing the non-linear deformation of an extensible elastica under the action of a normal contact force $2V$ and a mid-point force T_o. Our interest here is on the effective linear elastic constants, and for small T_o we obtain a linear displacement-force relation for each yarn which depends upon the elastic properties and intial weave geometry of that yarn. The mechanical response of the woven fabric is obtained from the interaction of two of these solutions corresponding to each yarn by enforcing equilibrium and compatibility of displacements. This provides the in-plane linear strain-stress relations for the woven fabric in the two principal yarn directions. These results indicate that a significant coupling of bending and stretching effects occurs during deformation.

The theoretical analysis has been used to estimate the in-plane Young's moduli for one group of fabrics woven from a polymeric Vincel yarn as described by Leaf and Kandil [10]. These theoretical estimates compare favorably with the experimentally measured moduli of [10].

THE EXTENSIBLE ELASTICA

The first step in obtaining expressions for the effective elastic constants for the woven fabric requires an analysis of the force-displacement relation of the individual yarns. Each yarn will be modeled as an extensible elastica, and the geometry of this model is shown in Figure 1. The extensible elastica has been considered in detail by Antman [14] and Tadjbakhsh [15], and in this analysis we make use of the intrinsic coordinates as presented for the inextensible case by Mitchell [16]. The elastica is assumed to deform in the (x,z) plane as shown, and the initial shape is taken to be an arc of a circle of radius R. The undeformed shape of the elastica is defined by the slope

$$\phi = \phi\ (s_o) = \frac{s_o}{R},\ 0 \leq s_o \leq L_o \tag{1}$$

where s_o is arc length along the undeformed curve of total length L_o, and the deformed shape is defined by the slope

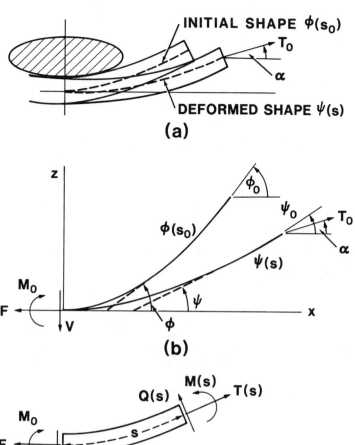

(a)

(b)

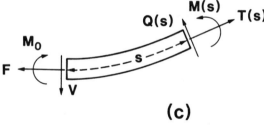

(c)

FIGURE 1. The Extensible Elastica: (a) Schematic of woven
yarn, (b) The intrinsic coordinates and applied
forces, (c) Forces acting on an element.

$$\psi = \psi \ (s), \ 0 \le s \le L \tag{2}$$

with s the arc length along the deformed curve having total length L. The
elastica is assumed to stretch linearly under the effect of the axial force
T(s) acting through the centroid of the cross-section of area A which provides
the relation

$$\frac{ds}{ds_o} = 1 + \frac{T(s)}{EA} \tag{3}$$

where E is the Young's modulus of the elastica. The shear force Q(s) and
axial force T(s) as shown in Figure 1c are given by

$$Q(s) = V \cos \psi - F \sin \psi$$
$$T(s) = F \cos \psi + V \sin \psi \tag{4}$$

203

where V and F are the forces at the symmetry point $s = 0$. In terms of the force T_o applied at the end $s = L$ we find

$$V = T_o \sin \alpha, \quad F = T_o \cos \alpha \tag{5}$$

which, with Equation (4_2), provides the important relation

$$T(s) = T_o \cos (\psi - \alpha). \tag{6}$$

The end of the elastica at $s = L$ is assumed to be an inflection point of the yarn, that is, a point of anti-symmetry of the yarn, and at this point the moment must vanish so we have $M(L) = 0$. The differential equation which describes the non-linear bending of the extensible elastica under the conditions just described is

$$EI \frac{d}{ds} \left[(1 + \frac{T}{EA}) \frac{d\psi}{ds} - \frac{d\phi}{ds_o} \right] = - \frac{dT}{d\psi} \tag{7}$$

where I is the section moment of inertia, and this equation is to be solved subject to the boundary conditions

$$\psi(0) = 0$$

$$\left[(1 + \frac{T}{EA}) \frac{d\psi}{ds} \right]_{s = L} = \frac{d\phi}{ds_o} \Bigg]_{s_o = L_o} \tag{8}$$

The second condition insures that the moment at the end $s = L$ vanishes. Since the initial shape is taken to be an arc of a circle of radius R, we have

$$\frac{d\phi}{ds_o} = \frac{1}{R}, \quad 0 \le s_o \le L_o. \tag{9}$$

A first integral of Equation (7) may be obtained, and with the boundary condition (8_2) leads to

$$R \left[1 + \kappa R^2 \gamma \cos (\psi - \alpha) \right] \frac{d\psi}{ds} =$$

$$\Big\{ 1 + 2\kappa R^2 \left[\cos (\psi_o - \alpha) - \cos (\psi - \alpha) \right]$$

$$+ \kappa^2 R^4 \gamma \left[\cos^2 (\psi_o - \alpha) - \cos^2 (\psi - \alpha) \right] \Big\}^{1/2} \tag{10}$$

where we have made use of Equation (6) and the two constants

$$\kappa = \frac{T_o}{EI}, \quad \gamma = \frac{I}{AR^2}. \tag{11}$$

The constant γ represents a measure of the relative effects of bending and stretching in the deformation of the elastica, and the case $\gamma = 0$ represents the inextensible elastica. Equation (3) provides the relation between ds and ds_o of

$$ds = \left[1 + \kappa R^2 \gamma \cos (\psi - \alpha) \right] ds_o \tag{12}$$

which is important in evaluating the deformed slope ψ_o at the end. The arc length s may be obtained from Equation (10) as an elliptic integral of Weirstrass' form [14] but this result is not particularly useful for the present application.

We now restrict interest to deformations due to small forces and assume $\kappa R^2 \ll$
1. To first order terms in κR^2, Equations (10) and (12) become

$$\frac{ds}{R} = \left\{ 1 - \kappa R^2 \left[\cos (\psi_0 - \alpha) - (1 + \gamma) \cos (\psi - \alpha) \right] \right\} d\psi \qquad (13)$$

$$\frac{ds_0}{R} = \left\{ 1 - \kappa R^2 \left[\cos (\psi_0 - \alpha) - \cos (\psi - \alpha) \right] \right\} d\psi \qquad (14)$$

Making use of the boundary condition (8_1) and the fact that the integral over ds_0 is equal to the original length $L_0 = R\phi_0$, Equation (14) provides

$$\phi_0 = \psi_0 + \kappa R^2 \left[\sin (\psi_0 - \alpha) + \sin \alpha - \psi_0 \cos (\psi_0 - \alpha) \right], \qquad (15)$$

where ψ_0 is the slope of the deformed curve at the end $s = L$ and is determined by this equation. Consistent with our first order theory, we now let

$$\psi_0 = \phi_0 - \theta, \ \theta \ll 1, \qquad (16)$$

and Equation (15) gives

$$\theta = \kappa R^2 \left[\sin (\phi_0 - \alpha) + \sin \alpha - \phi_0 \cos (\phi_0 - \alpha) \right]. \qquad (17)$$

Using the relations (Figure 1) $dx = \cos \psi ds$, $dz = \sin \psi ds$, and Equation (13) together with the conditions $x = 0$, $z = 0$ at $s = 0$, integration provides

$$x_0 = R \sin \psi_0 + \frac{1}{4} \kappa R^3 \left[2(1 + \gamma) \psi_0 \cos \alpha \right.$$
$$\left. - (1 - \gamma) \sin \alpha - (1 - \gamma) \sin (2\psi_0 - \alpha) \right]$$
$$\qquad (18)$$
$$z_0 = R(1 - \cos \psi_0) + \frac{1}{4} \kappa R^3 \left[2(1 + \gamma) \psi_0 \sin \alpha - 4 \cos (\psi_0 - \alpha) \right.$$
$$\left. + (1 - \gamma) \cos (2\psi_0 - \alpha) + (3 + \gamma) \cos \alpha \right]$$

where (x_0, z_0) is the position of the endpoint $s = L$ of the deformed elastica. The displacements u_x, u_z are given by

$$u_x = x_0 - R \sin \phi_0$$
$$u_z = z_0 - R (1 - \cos \phi_0)$$
$$\qquad (19)$$

which, with (18) and (17) gives the displacements in the form

$$u_x = \frac{1}{4} \kappa R^3 (A \cos \alpha - B \sin \alpha)$$
$$u_z = \frac{1}{4} \kappa R^3 (-B \cos \alpha + C \sin \alpha)$$
$$\qquad (20)$$

where

$$A = 2(2 + \gamma) \phi_0 + 2\phi_0 \cos 2\phi_0 - (3 - \gamma) \sin 2\phi_0$$

$$B = 4 \cos \phi_0 - (1 + \gamma) - 2\phi_0 \sin 2\phi_0 - (3 - \gamma) \cos 2\phi_0 \qquad (21)$$

$$C = 2(2 + \gamma) \phi_0 - 2\phi_0 \cos 2\phi_0 + (3 - \gamma) \sin 2\phi_0 - 8 \sin \phi_0.$$

The result (20) with (5) takes the form

$$u_x = \frac{R^3}{4EI} (AF - BV)$$

$$(22)$$

$$u_z = \frac{R^3}{4EI} (-BF + CV)$$

and we will use this result in the next section to obtain the linear elastic response of a woven fabric. We note that this result is consistent with the known existence of a Gibbs free energy $\mathcal{G}$ given by

$$\mathcal{G} = \frac{R^3}{8EI} (AF^2 - 2BFV + CV^2)$$

$$(23)$$

such that

$$u_x = \frac{\partial \mathcal{G}}{\partial F}, \quad u_z = \frac{\partial \mathcal{G}}{\partial V}$$

$$(24)$$

which follows directly from Castigliano's first theorem [17].

FABRIC ELASTIC CONSTANTS

The geometry of the woven fabric under consideration here is shown in Figure 2. With reference to Figure 2b, the usual geometrical weave parameters of pick spacing p, yarn length ℓ and crimp height h are represented in terms of the elastica parameters R and ϕ_0 by

$$p = 2R \sin \phi_0$$

$$\ell = 2R\phi_0$$

$$(25)$$

$$h = 2R (1 - \cos \phi_0).$$

To fix ideas we denote the x-direction as the warp direction and the warp yarns have mechanicl properties and an initial geometry which we identify with a subscript x. Similarly the y-direction is the weft direction and the weft yarns have mechanical properties and an initial geometry which we will identify with a subscript y.

Under the action of applied forces f_x and f_y acting in the (x,y) plane, the woven fabric will deform with each yarn behaving like the extensible elstica analysed in the previous section. Equilibrium requires the transverse contact force V to be the same for both warp and weft yarns, and the geometric compatibility of the displacements requires the displacements in the z or transverse directions to be the same. From Equations (22) for the warp or x-direction yarn

$$u_x = \frac{R_x^3}{4E_x I_x} (A_x f_x - B_x V)$$

$$(26)$$

$$u_{zx} = \frac{R_x^3}{4E_x I_x} (-B_x f_x + C_x V)$$

and for the weft or y-direction yarn

206

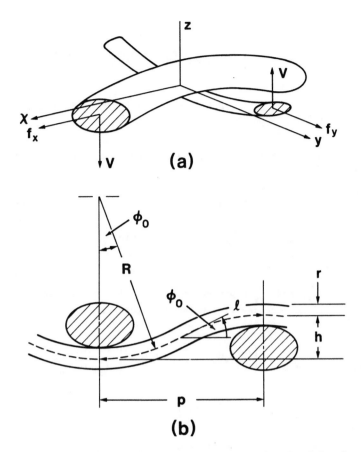

FIGURE 2. Geometry of the woven fabric: (a) Schematic of woven yarn interlace, (b) Cross-section of weave in either x (warp) or y (weft) direction.

$$u_y = \frac{R_y^3}{4E_y I_y} (A_y f_y - B_y V)$$

$$u_{zy} = - \frac{R_y^3}{4E_y I_y} (-B_y f_y + C_y V) \tag{27}$$

with

$$u_{zx} = u_{zy}. \tag{28}$$

The compatibility condition (28) determines the contact force V as

$$V = \frac{\Gamma B_x f_x + B_y f_y}{(\Gamma C_x + C_y)} \tag{29}$$

where

$$\Gamma = \left(\frac{E_y I_y}{E_x I_x} \right) \left(\frac{R_x}{R_y} \right)^3 \tag{30}$$

and the in-plane displacements u_x, u_y are

$$u_x = \frac{R_x^3}{4E_x I_x (\Gamma C_x + C_y)} \left[[\Gamma(A_x C_x - B_x^2) + A_x C_y] f_x - B_x B_y f_y \right]$$

$$(31)$$

$$u_y = \frac{R_y^3}{4E_x I_x (\Gamma C_x + C_y)} \left[[\Gamma A_y C_x + (A_y C_y - B_y^2)] f_y - \Gamma B_x B_y f_x \right]$$

The results of Equation (31) represent the displacement force relations for this woven fabric.

The effective in-plane strain-stress relations for this woven fabric may be readily obtained from the following relations established from consideration of the weave geometry shown in Figure 2 and Equations (25):

$$u_x = \epsilon_{xx} R_x \sin \phi_{ox}, \quad u_y = \epsilon_{yy} R_y \sin \phi_{oy},$$

$$f_x = 2\sigma_{xx} R_y \sin \phi_{oy}, \quad f_y = 2\sigma_{yy} R_x \sin \phi_{ox} .$$

$$(32)$$

In Equation (32), ϵ_{xx} and ϵ_{yy} represent the material strains, and σ_{xx} and σ_{yy} the effective in-plane stresses with units of force per unit length. Using the relations of (32) in Equation (31) gives the strain-stress relations

$$\epsilon_{xx} = \frac{R_x^3}{2E_x I_x (\Gamma C_x + C_y)} \left[[\Gamma(A_x C_x - B_x^2) + A_x C_y] \left(\frac{R_y \sin \phi_{oy}}{R_x \sin \phi_{ox}} \right) \sigma_{xx} \right.$$

$$\left. - B_x B_y \sigma_{yy} \right]$$

$$(33)$$

$$\epsilon_{yy} = \frac{R_y^3}{2E_y I_y (\Gamma C_x + C_y)} \left[[\Gamma A_y C_x + (A_y C_y - B_y^2)] \left(\frac{R_x \sin \phi_{ox}}{R_y \sin \phi_{oy}} \right) \sigma_{yy} \right.$$

$$\left. - \Gamma B_x B_y \sigma_{xx} \right] .$$

The effective Young's modulus $\hat{E}_x$, $\hat{E}_y$ in the x and y directions, respectively, are given by

$$\hat{E}_x = \frac{2E_x I_x (\Gamma C_x + C_y) \sin \phi_{ox}}{R_x^2 R_y [\Gamma(A_x C_x - B_x^2) + A_x C_y] \sin \phi_{oy}}$$

$$(34)$$

$$\hat{E}_y = \frac{2E_y I_y (\Gamma C_x + C_y) \sin \phi_{oy}}{R_y^2 R_x [\Gamma A_y C_x + (A_y C_y - B_y^2)] \sin \phi_{ox}}$$

We note that since the terms A_x, A_y, B_x, . . . all depend on the parameter γ_x or γ_y, the result (34) indicates that a significant coupling of bending and stretching effects occurs during fabric deformation even in the linear elastic regime.

TABLE I

Comparison of Theory and Experiment for
Young's Moduli of Woven Vincel Fabric

Fabric Number	$\hat{E}_x$ (N/cm) Measured [10]	$\hat{E}_x$ (N/cm) Theoretical	$\hat{E}_y$ (N/cm) Measured [10]	$\hat{E}_y$ (N/cm) Theoretical
A-1	9.2	8.8	25.5	13.7
A-2	9.1	8.8	14.8	8.9
A-3	2.7	9.0	13.4	10.3

RESULTS AND CONCLUSIONS

The effective fabric Young's moduli as given by Equation (34) have been used
to estimate the moduli of a group of fabrics woven from a polymeric Vincel
yarn as described by Leaf and Kandil [10]. The fabric group, group A of [10],
was woven with identical warp and weft yarns of R60/2-tex Vincel with varying
pick spacing in the warp and weft directions. A comparison of the theoretical
moduli estimates with the experimentally determined moduli as reported in [10]
for the three different fabrics of group A is shown in Table I. While the
theoretical moduli are somewhat lower than the measured values, the lower
theoretical values always correspond to the lower experimental values. This
is particularly significant when we note that for fabrics A-1, A-2, the pick
spacing in the x-direction is larger than in the y-direction while the reverse
is true for A-3. Yet for all three weaves, the smaller modulus is $\hat{E}_x$ as
predicted by the theory and verified experimentally.

In view of the uncertainties as to the Young's modulus of the Vincel yarn
itself and the degree of yarn flattening which directly effects the I, A, and
γ of each yarn, comparison between the results of the thory developed here and
experiment as shown in Table I appears to be quite good. Additional
experimental comparisons with kevlar woven fabrics are underway. This linear
theory based on the extensible elastica, which effectively couples yarn
bending and stretching effects, readily extends to large non-linear
deformations so long as the yarn force-displacement relation remains linear.
This extension to non-linear deformations is thus particularly relevant to the
large class of fabrics woven from kevlar yarns since these yarns exhibit an
essentially linear force-displacement relation up to fracture.

REFERENCES

1. Peirce, F. T., "The Geometry of Cloth Structure," J. Text. Inst., 28
   (1937) T45.

2. Olofsson, B., "A General Model of a Fabric as a Geometric-Mechanical
   Structure," J. Text. Inst., 55 (1964) T541.

3. Grosberg, P. and Kedia, S., Textile Res. J., 36 (1966) 71.

4. Hearle, J. W. S., Grosberg, P., and Backer, S., Structural Mechanics of
   Fibers, Yarns, and Fabrics, Wiley-Interscience, NY (1969).

5. Ellis, P., "Woven Fabric Geometry--Past and Present," Tex. Inst. &
   Industry, 12 (1974) 245.

6. Treloar, L. R. G., "Physics of Textiles," Physics Today, 30 (1977) 23.

7. Hearle, J. W. S. and Shanahan, W. J., "An Energy Method for Calculations in Fabric Mechanics Part I: Principles of the Method," <u>J. Text. Inst.</u>, <u>69</u> (1978) 81.

8. Shanahan, W. J. and Hearle, J. W. S., "An Energy Method for Calculations in Fabric Mechanics Part II: Examples of Application of the Method to Woven Fabrics," <u>J. Text. Inst.</u>, <u>69</u> (1978) 92.

9. de Jong, S. and Postle, R., <u>J. Text. Inst.</u>, <u>68</u> (1977) 350.

10. Leaf, G. A. V. and Kandil, K. H., "The Initial Load-Extension Behavior of Plain-Woven Fabrics," <u>J. Text. Inst.</u>, <u>71</u> (1980) 1.

11. Hamed, H. A. K. and Sadek, K. S. H., "Mechanical Properties of Woven Fabrics," <u>Proc. Inst. Mech. Engrs.</u>, <u>199</u> (1985) 67.

12. Kilby, W. F., "Planar Stress-Strain Relationships in Woven Fabrics," <u>J. Text. Inst.</u>, <u>54</u> (1963) T9.

13. Olofsson, B., <u>J. Text. Inst.</u>, <u>52</u> (1961) T272.

14. Antman, S., "General Solutions for Plane Extensible Elasticae Having Nonlinear Stress-Strain Laws," <u>Q. Appl. Math</u>, <u>26</u> (1968) 35.

15. Tadjbakhsh, I., "The Variational Theory of the Plane Motion of the Extensible Elastica," <u>Int. J. Engng. Sci.</u>, <u>4</u> (1966) 433.

16. Mitchell, T. P., "The Nonlinear Bending of Thin Rods," <u>J. Appl. Mech.</u>, <u>26</u> (1959) 40.

17. Hoff, N. J., <u>The Analysis of Structures</u>, John Wiley & Sons (1956) 377.

CHARACTERIZATION OF THE TENSILE MODULUS OF RANDOM GLASS MAT REINFORCED THERMOPLASTIC COMPOSITES

V. K. Stokes
GE Corporate Research and Development
Schenectady, New York

ABSTRACT

Random glass mat thermoplastic composites, which can be thermostamped to form complex deep-drawn parts with ribs and bosses, are excellent materials for making low-cost stiff parts with relatively high impact strength. However, the mechanical properties of these materials, as determined by standard tensile tests, are known to exhibit large scatter. In order to better characterize the tensile modulus, the local tensile moduli of long, thin, 12.7-mm-wide strips were measured along both thin edges (left and right) of the specimen at 12.7-mm intervals. While the magnitudes of the large fluctuations of the modulus along the two edges are comparable, these variations are shown to be out of phase. Direct measurements of the tensile modulus across the 12.7-mm-wide face at 12.7-mm intervals along the strips are used to show that the tensile modulus is well approximated by the arithmetic mean of the local left- and right-edge tensile moduli. This arithmetic mean modulus, which is shown to be strongly correlated to the local density, is a better practical measure of the response to tensile loads. While this mean modulus also varies along the specimen, it indicates a significantly lower level of nonhomogeneity than is indicated by the left- and right-edge moduli.

INTRODUCTION

Random glass mat reinforced thermoplastics, which come in the form of thermoplastic sheets containing nonwoven random continuous glass fibers, are relatively inexpensive and can be thermostamped at low pressures to form complex parts. These advantages make this class of materials – typified by Azdel® P100 (Azdel PM0400-101) composite sheet consisting of 40 wt% random glass mat in a polypropylene matrix – ideally suited for large, stiff, low-cost parts that have high impact strength (Tomkinson-Walles, 1988).

However, the nonuniform distribution of glass in the random mat causes the material to be both anisotropic and nonhomogeneous. As a result, elastic moduli measured by standard ASTM tests show a significant amount of scatter and the results appear to depend on the specimen size (Tomkinson-Walles, 1988).

This behavior of this class of materials raises interesting questions: How nonhomogeneous and anisotropic are the mechanical properties? How are structural parts to be designed for stiffness and strength? How does processing affect properties? How can glass flow be predicted in compression molding?

To characterize the scatter in the tensile modulus of Azdel sheet material, tensile tests were used (Stokes, 1988) to determine the tensile modulus of $12.7 \times 405 \times 3.18$-mm ($0.5 \times 16 \times 0.125$-in) strips of Azdel at 12.7-mm (0.5-in) intervals along the 405-mm length by attaching an extensometer along the right (thickness) edge, as indicated by the letter R in Figure 1. With the extensometer attached to the right edge of segment 1 (Figure 1), the specimen was pulled in tension, under strain control, at a strain rate of 10^{-2} s^{-1}, to a strain of 0.5%, and the load-strain behavior was recorded. The specimen was then unloaded and the extensometer moved down to the next 12.7-mm segment, and the test procedure repeated. In this way the elastic modulus was determined at every 12.7-mm interval along a strip. These tests, performed along 12 12.7×405-mm strips cut from a 203×405-mm (9×16-in) plaque (Stokes, 1988), produced rather unexpected results: The Young's modulus along a 12.7-mm-wide strip can vary by a factor of two over a 12.7-mm distance. The maximum and minimum values over a 150×305-mm (6×12-in) area were found to be 9.73 GPa (1,441 ksi) and a 3.14 GPa (455 ksi), respectively, giving a ratio of 3.1. The variations in the modulus appeared to have a random character with no obvious discernible pattern.

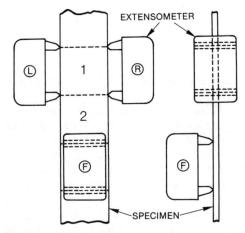

Figure 1.   Tensile test extensometer configurations for determining the left, face, and right moduli, E_L, E_F, and E_R, respectively, along a strip.

This phenomenon of the macroscale variation of the Young's modulus raises a number of interesting questions. First, if the modulus varies by a factor of two over a 12.7-mm length scale, what does an extensometer measure? What average property is measured in a flexural test? Can measured moduli from tension and bend tests be correlated? Second, does the distribution of the modulus correlate over distance? How repeatable are these data? Third, how can some "standard" properties be defined for this class of materials? How are these properties to be used in design?

As a first step toward answering these questions, a simple analytical framework was developed for interpreting the results of tests (Stokes, 1988). If the elastic modulus in a rectangular bar is assumed to vary in a one-dimensional manner $E = E(x)$ along the length (x-direction), then the effective tensile modulus E_T over a specimen length l was shown to be the *harmonic mean*

$$\frac{1}{E_T} = \frac{1}{l} \int_0^l \frac{dx}{E(x)} = \int_0^1 \frac{d\xi}{E(\xi)}, \; \xi = \frac{x}{l} \tag{1}$$

212

Since the harmonic mean of any set of numbers is smaller than their arithmetic mean, the *effective* tensile modulus will be lower than the average (arithmetic mean) modulus.

Similarly, the effective 3-point-bend-test flexural modulus, E_B, for a centrally loaded beam of length l (between $x = 0$ and $x = l$) was shown to be

$$\frac{1}{E_B} = \frac{12}{l^3} \int_0^{l/2} x^2 \left[\frac{1}{E(x)} + \frac{1}{E(l-x)} \right] dx$$

$$= \frac{3}{2} \int_0^1 \eta^2 \left[\frac{1}{E(\eta)} + \frac{1}{E(2 - \eta)} \right] d\eta, \quad \eta = \frac{2x}{l} \tag{2}$$

For a variable modulus $E = E(x)$, a comparison of Eq. (1) with Eq. (2) shows that, in general, the effective tensile and flexural moduli, E_T and E_B, respectively, are different.

Two types of tests were done (Stokes, 1988) to assess the predictions of Eqs. (1) and (2). In the first set, tensile modulus variations along the length of a specimen were determined by using 12.7-mm- and 25.4-mm-gauge-length extensometers. The 12.7-mm-gauge-length data were then used to predict the 25.4-mm-gauge-length modulus by using Eq. (1). The agreement between the measured and predicted values appeared to be "reasonably" good. In the second set of tests, the flexural moduli of selected specimens were determined for varying beam lengths after the elastic moduli had been measured at 12.7-mm intervals. The 12.7-mm tensile-modulus data were then used to predict the flexural modulus via Eq. (2). The measured flexural moduli on five specimens generally appeared to be larger than the values predicted by using the measured 12.7-mm tensile moduli.

There are at least three possible reasons for the differences between the predicted and measured values of the flexural modulus. The first, and most important, has to do with the measurement of the tensile modulus. All the tensile data in Stokes (1988) were obtained by attaching the 12.7-mm extensometer to the *thin* (3.18 mm, 0.125 in) edge of the specimen, as shown in Figure 1. The lack of a strong correlation between the local modulus and local density (Stokes, 1988) indicated that glass orientation probably has a large effect on the local modulus, so the measurement of the tensile modulus on just one thin edge may not be representative of the "mean" value over the 12.7-mm face. Since the flexural moduli were determined as an average over the 12.7-mm face, the error in the predicted flexural modulus may be understandable. Second, the 12.7-mm gauge length may not be small enough to resolve the variations in the local tensile modulus. A third possibility is through-thickness nonhomogeneity in the material, which would invalidate the derivation of Eq. (2), in which the local modulus was assumed constant across the specimen thickness.

The objective of the work discussed in this paper is to address the characterization of the tensile modulus: how should the "mean" local tensile modulus be determined if attaching the extensometer to the thin edge of a specimen does not result in a representative modulus.

CHARACTERIZATION OF THE TENSILE MODULUS

This section is concerned with developing a test procedure for characterizing the tensile modulus of Azdel P100 (Azdel PM10400-101) sheet material. First, the test procedure developed in Stokes (1988) for scanning the tensile modulus along one edge of a specimen is applied for measuring the modulus of strips along *both* edges and along the *face* of the specimen. This is schematically shown in Figure 1, in which the extensometers L, F, and R, would measure, respectively, the left-edge modulus E_L, the face modulus E_F, and the right-edge modulus E_R. These data are analyzed for defining a suitable local mean modulus along the specimen. Second, after strips have been scanned for the tensile modulus, the local average density over the

12.7×12.7-mm segments is determined by the procedure described in Stokes (1988). The local density data is then correlated with the local modulus, resulting in a new procedure for characterizing the tensile modulus of Azdel.

All the tests discussed in this report were done on 230×405-mm (9×16-in) drape-molded Azdel P100 (Azdel PM10400-101) plaques (Stokes, 1988). First, a 230×405-mm blank was cut from an Azdel sheet with the 405-mm length aligned along the machine direction – the direction along which the continuous Azdel composite sheet is laminated during the manufacturing process (Tomkinson-Walles, 1988). This blank was then heated to a nominal temperature of 215 °C (419 °F) and thermostamped at a nominal pressure of 0.1 MPa (15 psi) in a mold of the same size. During this process, called drape molding, there was no "flow" of the glass, but the plaque was subjected to the same thermal and pressure histories that the material would have to undergo during part forming. Also, it is likely that the number of voids was reduced.

Figure 2 shows a radiograph of the drape-molded plaque used in this study. Clearly, the glass distribution has a random pattern. It is this pattern that is responsible for the spatial fluctuation of the tensile modulus about some mean value.

Test Procedure

This section describes the results of different modulus scans (Stokes, 1988), along the cross-machine direction (at right angles to the machine direction) of an Azdel plaque.

Ten 12.7-mm (0.5-in) strips were marked, with a pen, on the 3.18-mm- (0.125-in-) thick drape-molded plaque along the machine direction, as shown by the dashed lines in Figure 3. Twenty-two 12.7-mm strips were then cut from the plaque along the cross-machine direction, as indicated by the solid lines in Figure 3, resulting in 22 12.7×230-mm (0.5×9-in) specimens with cross-length markers at 12.7-mm intervals, delineating 10 12.7-mm segments. The numbering system for the specimens and the 12.7-mm segments is shown in Figure 3. For the tensile tests, the specimen numbers increase from 1 at the left to 22 at the right. The 12.7-mm segments always start with the number 1 at the top and end with the number 10 at the bottom.

The mean cross-sectional area of each 12.7-mm segment on each specimen was determined by measuring the mean thickness and mean width over that 12.7×12.7-mm (0.5×0.5-in) region.

By orienting the extensometer as shown schematically in Figure 1, tensile tests were used to determine three tensile moduli – the left modulus E_L, the face modulus E_F, and the right modulus E_R, – at each 12.7-mm interval in each specimen at a strain rate of $10^{-2}\mathrm{s}^{-1}$. The following procedure was used: After clamping a 230-mm (9-in) specimen in the grips of a tensile testing machine, a 12.7-mm-gauge-length extensometer was attached to the right (thickness) edge of segment 1, as shown in Figure 1. The specimen was then pulled in tension, under strain control, at a strain rate of $10^{-2}\mathrm{s}^{-1}$, to a strain of 0.5%, during which the load-strain behavior was recorded. The specimen was then unloaded and the extensometer moved down to the next 12.7-mm segment, and the test procedure repeated for each of the 10 12.7×12.7-mm portions of the strip. The right Young's modulus E_R for each segment was then determined by using the *local* mean cross-sectional area to calculate the stress in each segment. After completing the modulus scan on the right edge of the specimen, the extensometer was attached to the left edge of the specimen and the same procedure was used to determine the left modulus E_L along the left edge. Finally, the extensometer was attached to the face of the specimen, and the same procedure was used to determine the face modulus E_F. This procedure was then repeated for the 21 remaining 12.7-mm-wide specimens. In this way, the set of three cross-machine direction Young's moduli, E_R, E_L, and E_F, were determined over every 12.7×12.7-mm region of a 127×280-mm (5×11-in) portion of the original 230×405-mm (9×16-in) plaque.

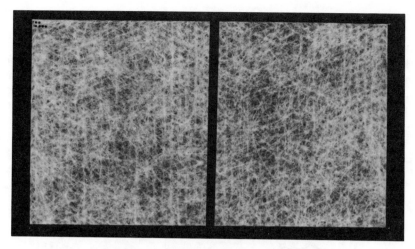

Figure 2. Radiograph showing the distribution of glass in a 230×405-mm (9×16-in) drape-molded plaque.

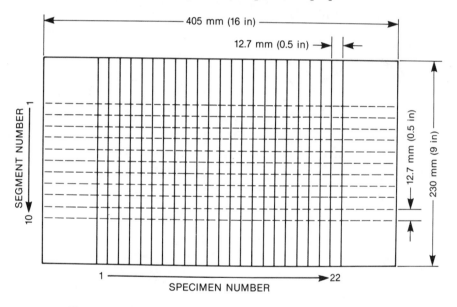

Figure 3. Layout of 22 12.7×230-mm (0.5×9-in) specimens cut from a 230×405-mm (9×16-in) plaque.

After determining the elastic moduli, the 12.7-mm-wide specimens were cut along the gauge marks into 12.7×12.7-mm coupons. The density of each coupon was then determined by measuring its mass and calculating the volume via measurement of its dimensions.

Cross-Machine Direction Tensile Modulus Data

For one plaque, the measured variations of the cross-machine-direction moduli E_L, E_R, and E_F along specimens 1 through 5, 6 through 10, 11 through 16, and 17 through 22, are shown, respectively, in Figures 4 through 7. Clearly, each of the three moduli E_L, E_R, and E_F indicate a significant level of material nonhomogeneity over distances on the order of 12.7 mm (0.5 in), just as reported in Stokes (1988). Over a 127×280-mm (5×11-in) area, the maximum and minimum values of the moduli are, respectively, 8.78 GPa (1,273 ksi) (specimen 9, segment 5) and 2.83 GPa (411 ksi) (specimen`2, segment 9) for the left modulus E_L, giving a ratio of 3.1; 9.5 GPa (1.379 ksi) (specimen 3, segment 10) and 2.89 GPa (420 ksi) (specimen 17, segment 9) for the right modulus E_R, giving a ratio of 3.3; and 10.58 GPa (1,535 ksi) (specimen 4, segment 7) and 3.37 GPa (488 ksi) (specimen 17, segment 8) for the face modulus E_F, giving a ratio of 3.1. These ratios are on the same order as those for the right modulus E_R, as reported in Stokes (1988).

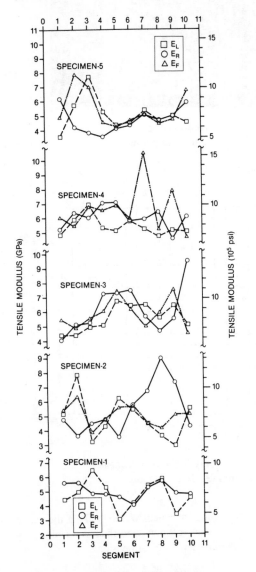

Figure 4. Variations of the cross-machine-direction Young's moduli E_L, E_R, and E_F at 12.7-mm intervals along specimens 1 through 5.

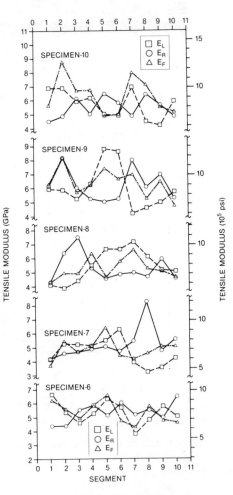

Figure 5. Variations of the cross-machine-direction Young's moduli E_L, E_R, and E_F at 12.7-mm intervals along specimens 6 through 10.

216

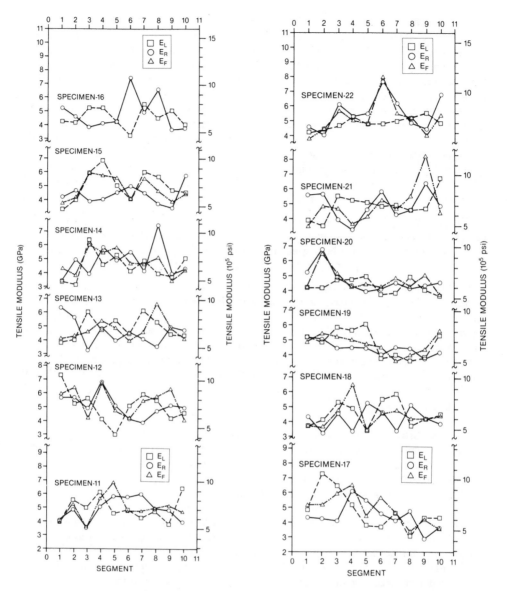

Figure 6. Variations of the cross-machine-direction Young's moduli E_L, E_R, and E_F at 12.7-mm intervals along specimens 11 through 16.

Figure 7. Variations of the cross-machine-direction Young's moduli E_L, E_R, and E_F at 12.7-mm intervals along specimens 17 through 22.

While E_L and E_R exhibit similar fluctuations along a specimen, their variations are out of phase. And, in general, in the same segment of a specimen, a high value of one is accompanied by a low value of the other. Clearly, the differences between the values of E_L and E_R in a segment are caused by differences in the local glass orientation near the left and right edges. These differences in E_L and E_R for the same segment raise an important question: for each 12.7×12.7-mm (0.5×0.5-in) segment, what constitutes a representative mean modulus?

217

The face modulus E_F is, in fact, an average modulus that might be the appropriate mean. However, because of the nonhomogeneity of the material, the extension between the anvils of the extensometer will vary along the anvils when the extensometer is attached to the face. Because the extensometer anvils are relatively stiff, they will either affect the local deformation and hence the measured strain and modulus, or the anvils will slip. In the former case the measured face modulus will be a further approximation to the modulus being measured. In the latter case, a slip in the extensometer could either result in inaccurate measurement, or, when the experiment is run in strain control, the slipping could result in the machine damaging the specimen by a sudden increase in strain as the machine tries to control the rate of extension. This is what occurred during the face-modulus tests in specimens 1 and 16, in which sudden slippage caused the machine to extend the specimens to failure. Similarly, a comparison of the data for specimen 4, segment 7 ($E_L = 5.25$, $E_R = 5.92$, $E_F = 10.58$ GPa), in which E_F is about *twice* as large as E_L and E_R, shows that the value of E_F for this segment has been grossly affected by slip at the extensometer anvils.

Figures 4 through 7 show that E_F does, in most cases, lie between E_L and E_R, thereby indicating that it represents some mean value. Given the uncertainties in the measurement of E_F, those values of E_F that do not lie between the local values of E_L and E_R are understandable. If measurement of E_F is not practicable, then how should a mean modulus be defined?

The analysis (Stokes, 1988) that resulted in Eq. (1) shows that the effective, or mean, modulus is the harmonic mean of the modulus variation when the modulus only varies (in series) along the direction of extension. However, when the mean across one segment is being considered, the material under extension is being pulled in "parallel," so that, the effective, or mean, modulus is some weighted arithmetic mean. As a first approximation, the effective modulus can be assumed to be the arithmetic mean modulus $E_A = (E_L + E_R)/2$. For purposes of comparison, contours of the cross-machine-direction moduli E_L, E_A, and E_R are shown in Figure 8. Clearly, each of the three moduli is distributed differently. The suitability of the mean tensile modulus E_A is discussed in the next section.

Comparison Between the Face and Average Moduli

For the plaque, the variations in E_F and in the arithmetic average modulus $E_A = (E_L + E_R)/2$ along specimens 1 through 5, 6 through 10, 11 through 16, and 17 through 22, are shown, respectively, in Figures 9 through 12. Clearly, E_A and E_F are close except at isolated points. This closeness is further illustrated by two additional figures: From Figure 13, which shows the cumulative distribution for $|E_F - E_A|$, it can be seen that $|E_F - E_A|$ is less than 0.5 GPa at about 58% of the points and less than 1.0 GPa at about 87% of the points. Furthermore, the values of E_F and E_A differ by less than 1.6 GPa at 95% of the points. This is important because it confirms that the average modulus is a good approximation for the local modulus. The contours of $|E_F - E_A|$ in Figure 14 show that $|E_F - E_A|$ is greater than 0.5 GPa in only a few isolated regions. The one large "spike" corresponds to the suspect value of E_F for specimen 4, segment 7.

Given the closeness between E_A and E_F, and the uncertainties in the measurement of E_F, E_A would appear to be a reasonable choice for characterizing the local mean tensile modulus.

Over a 127×280-mm (5×11-in) area, the maximum and minimum values of E_A are, respectively, 7.29 GPa (1,057 ksi) (specimen 3, segment 10) and 3.38 GPa (490 ksi) (specimen 14, segment 1), giving a ratio of 2.16. This ratio is significantly smaller than the ratios 3.1, 3.3, and 3.1 corresponding, respectively, to E_L, E_R, and E_F, thereby indicating that E_A has smaller variations over the plaque than E_L, E_R, and E_F. The relative smoothness of E_A can also be inferred from Table 1, which lists the ratio of the maximum to the minimum modulus, for each of the four moduli E_L, E_R, E_F, and E_A for each of the 22 strips. For the 22 strips, the ratios for E_L, E_R, and E_A vary between 1.31 and 2.76, 1.40 and 2.61, and 1.16 and 1.76, respectively. Clearly, the varia-

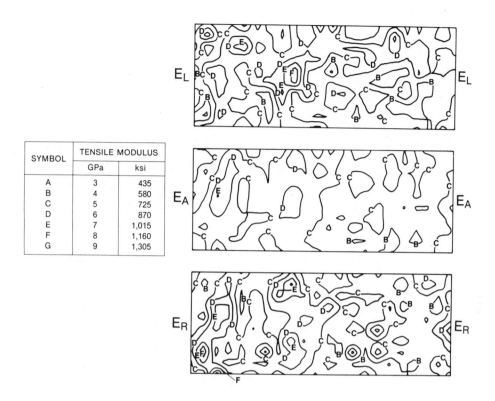

SYMBOL	TENSILE MODULUS	
	GPa	ksi
A	3	435
B	4	580
C	5	725
D	6	870
E	7	1,015
F	8	1,160
G	9	1,305

Figure 8. **Contours of the cross-machine-direction Young's moduli**
E_L, E_A, E_R **over a 127×280-mm plaque.**

tions in E_A are much smoother and exhibit much lower levels of nonhomogeneity than do the values of E_L and E_R.

Another way of assessing the relationships between E_L, E_R, and E_A is through a comparison of their harmonic means (over 10 segments), $\overline{E}_L$, $\overline{E}_R$, and $\overline{E}_A$, respectively, for each of the 22 strips, which are listed in Table 2. For the 22 specimens, $\overline{E}_R > \overline{E}_L$ for 12 specimens and $\overline{E}_L > \overline{E}_R$ for the remaining 10 specimens. Given the small sample size, these occurrences of $\overline{E}_R > \overline{E}_L$ and $\overline{E}_L > \overline{E}_R$ can be assumed to be distributed evenly. Interestingly, $\overline{E}_A$ is greater than both $\overline{E}_R$ and $\overline{E}_L$ for 10 of the 22 specimens – again, for approximately half the number of specimens. That $\overline{E}_A$ can be greater than both $\overline{E}_L$ and $\overline{E}_R$ may seem surprising at first glance, but this can be explained as follows: The harmonic mean is weighted toward the lower values in a sample. Because the arithmetic mean $E_A = (E_L + E_R)/2$ tends to reduce the amplitudes of the fluctuations in E_L and E_R, its harmonic mean, $\overline{E}_A$, is less affected by lower values, which can result in $\overline{E}_A$ being larger than $\overline{E}_L$ and $\overline{E}_R$. For the data in Table 2, the ratios of the maximum to the minimum harmonic means over the 22 specimens for $\overline{E}_L$, $\overline{E}_R$, and $\overline{E}_A$ are 1.34, 1.48, and 1.40, respectively. The *arithmetic* means of $\overline{E}_L$, $\overline{E}_R$, and $\overline{E}_A$, averaged over their values for the 22 specimens, are 4.81, 4.91, and 4.97 GPa, respectively. This shows that average modulus, $E_A = (E_L + E_R)/2$, can be expected to result in a larger harmonic mean modulus than the harmonic means for E_L and E_R. Therefore, the use of E_A will result in a higher material stiffness.

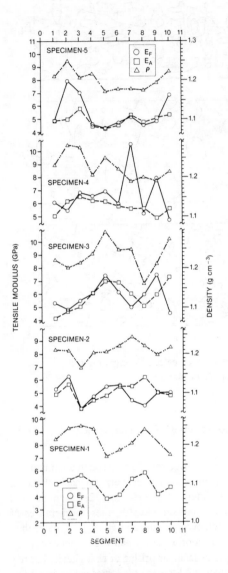

Figure 9. Variations of the cross-machine-direction Young's moduli E_F and E_A and the local density ρ at 12.7-mm intervals along specimens 1 through 5.

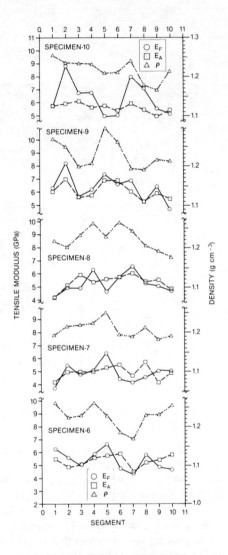

Figure 10. Variations of the cross-machine-direction Young's moduli E_F and E_A and the local density ρ at 12.7-mm intervals along specimens 6 through 10.

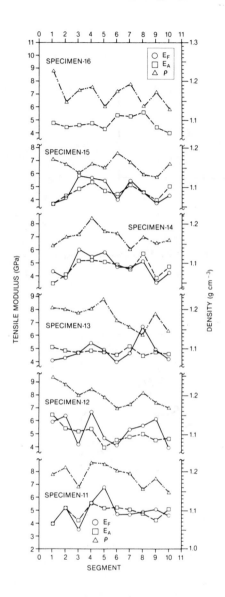

Figure 11. Variations of the cross-machine-direction Young's moduli E_F and E_A and the local density ρ at 12.7-mm intervals along specimens 11 through 16.

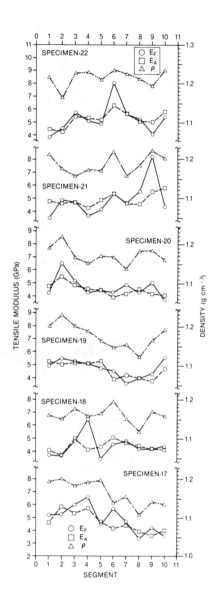

Figure 12. Variations of the cross-machine-direction Young's moduli E_F and E_A and the local density ρ at 12.7-mm intervals along specimens 17 through 22.

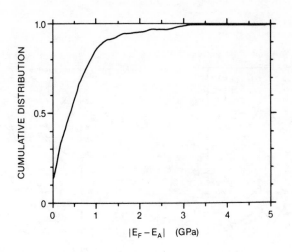

Figure 13. Cumulative distribution of $|E_F - E_A|$ for a 127×280-mm plaque.

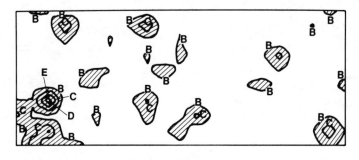

SYMBOL	$|E_F - E_A|$	
	GPa	ksi
A	0	0
B	1	145
C	2	290
D	3	435
E	4	580
F	5	725

Figure 14. Contours of $|E_F - E_A|$ over a 127×280-mm plaque.

Table 1

Variation of the Ratio of the Maximum to the Minimum Tensile Moduli
For 22 Specimens from Compression-Molded Azdel P100 Plaque

	Ratio of Maximum to Minimum Modulus										
	Specimen Number										
Modulus	1	2	3	4	5	6	7	8	9	10	11
E_L	2.15	2.76	1.56	1.47	2.20	1.75	1.97	1.85	2.12	1.66	1.71
E_R	1.40	2.61	2.39	1.57	1.75	1.50	1.97	1.77	1.63	1.43	1.70
E_F	-	1.66	1.66	2.25	1.77	1.54	1.74	1.55	1.75	1.80	1.94
E_A	1.41	1.65	1.76	1.28	1.25	1.31	1.39	1.46	1.32	1.23	1.40
	Specimen Number										
Modulus	12	13	14	15	16	17	18	19	20	21	22
E_L	2.49	1.62	1.69	1.83	1.71	2.40	1.74	1.70	1.37	1.92	1.31
E_R	1.77	1.81	2.18	1.70	2.01	2.15	1.65	1.40	1.73	1.98	1.92
E_F	1.68	1.70	1.76	1.58	-	1.96	1.92	1.57	1.76	2.36	2.12
E_A	1.63	1.16	1.69	1.46	1.41	1.62	1.40	1.41	1.40	1.36	1.48

Table 2

Harmonic Means of E_L, E_R, and E_A
Over 10 Segments for 22 Specimens

	Harmonic Mean Tensile Modulus GPa (ksi)										
	Specimen Number										
Modulus	1	2	3	4	5	6	7	8	9	10	11
E_L	4.53 (657)	4.39 (637)	5.38 (780)	5.36 (777)	4.91 (712)	5.25 (762)	4.43 (642)	5.24 (760)	5.68 (825)	5.58 (810)	4.76 (690)
E_R	4.98 (723)	4.90 (711)	5.76 (836)	5.96 (865)	4.58 (664)	5.31 (770)	5.03 (730)	5.15 (747)	6.02 (874)	5.39 (782)	4.72 (685)
E_A	4.81 (697)	4.99 (723)	5.64 (819)	5.70 (826)	4.87 (707)	5.35 (776)	4.86 (705)	5.32 (772)	6.04 (877)	5.60 (812)	4.83 (700)
	Specimen Number										
Modulus	12	13	14	15	16	17	18	19	20	21	22
E_L	4.75 (688)	4.77 (692)	4.24 (615)	4.75 (689)	4.48 (650)	4.51 (655)	4.37 (634)	4.66 (675)	4.26 (618)	4.74 (687)	4.87 (706)
E_R	4.95 (717)	4.44 (645)	4.61 (668)	4.24 (615)	4.64 (673)	4.29 (622)	4.07 (590)	4.43 (642)	4.59 (665)	4.69 (680)	5.37 (779)
E_A	4.97 (720)	4.74 (688)	4.53 (657)	4.59 (666)	4.69 (680)	4.52 (656)	4.31 (625)	4.58 (664)	4.47 (648)	4.84 (701)	5.17 (750)

Correlation of Modulus with Local Density

The local density over the 127×280-mm plaque, as measured over 12.7×12.7-mm square coupons, varies from a maximum of 1.30 g cm^{-3} (specimen 9, segment 5) to a minimum of 1.11 g cm^{-3} (specimen 17, segment 8). Contours of this density variation, together with contours of the average modulus, E_A, are shown in Figure 15. Clearly, the mean modulus E_A and the mean local density ρ are strongly correlated.

SYMBOL	DENSITY (ρ) $\text{g} \cdot \text{cm}^{-3}$
A	1.100
B	1.125
C	1.150
D	1.175
E	1.200
F	1.225
G	1.250
H	1.275
I	1.300
J	1.325
K	1.350

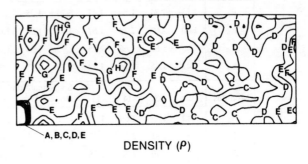

DENSITY (ρ)

SYMBOL	TENSILE MODULUS (E_A)	
	GPa	ksi
A	3.0	435
B	3.5	507
C	4.0	580
D	4.5	623
E	5.0	725
F	5.5	798
G	6.0	870
H	6.5	943
I	7.0	1,015
J	7.5	1,088
K	8.0	1,160
L	8.5	1,233
M	9.0	1,305

TENSILE MODULUS (E_A)

Figure 15. Contours of the density ρ and the mean modulus E_A over a 127×280-mm plaque.

The data for all 22 specimens are shown in Figures 16 and 17, which are plots of E_L and E_A, respectively, versus the local density ρ (the plots of E_R and E_F versus ρ are similar to that of E_L versus ρ). Here again, while the plot of E_L versus ρ shows an enormous amount of scatter, E_A appears to correlate rather well with ρ, thereby indicating the appropriateness of E_A for measuring local tensile stiffness.

CONCLUDING REMARKS

Both the left modulus, E_L, and the right modulus, E_R, exhibit large-scale fluctuations along a specimen. The ratio of the maximum to the minimum modulus over a 127×280-mm plaque area for each of these two moduli is on the order of 3. While the magnitudes of the fluctuations along specimens are comparable, the variations are out of phase. In general, in the same segment within a specimen, a high value of E_L or E_R is accompanied by a low value of the other, reflecting differences in glass concentration and orientation near the two edges. These

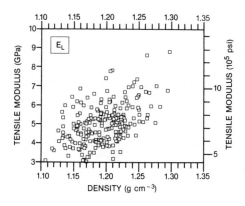

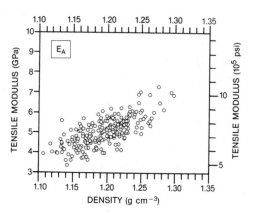

Figure 16. Variation of the cross-machine-direction left Young's modulus E_L vs the density ρ over 127×280-mm plaque.

Figure 17. Variation of the cross-machine-direction average Young's modulus E_A vs the density ρ over 127×280-mm plaque.

differences in E_L and E_R for the same 12.7×12.7-mm segment point to the need for defining a representative mean modulus. Such a mean would be some weighted arithmetic mean of E_L and E_R.

An appropriate mean modulus would be a "face modulus" that measures the average tensile response across the 12.7-mm face width. While conceptually simple and appealing, the measurement of such a face modulus presents practical difficulties. Varying extensions between the relatively stiff anvils of an extensometer, caused by the test material nonhomogeneity, will either result in the extensometer affecting the deformation, or in slip at the anvils, both of which would result in inaccurate measurements.

However, in order to measure an appropriate mean modulus, the face modulus E_F was measured subject to the uncertainties discussed above. In view of the problems involved in measuring E_F, the arithmetic mean $E_A = (E_L + E_R)/2$ was found to be a good approximation to E_F. Also, the average modulus E_A shows a much stronger correlation to the local density ρ than do E_L or E_R. For these reasons, E_A appears to be a reasonable choice for characterizing the local mean modulus. While E_A also varies along a specimen, it indicates a significantly lower level of nonhomogeneity than is indicated by measurements of E_L and E_R. Furthermore, a harmonic mean modulus based on E_A will result in a higher material stiffness than would be predicted by harmonic mean moduli based on E_L or E_R.

The results in this paper show that E_A is an appropriate measure for the local mean tensile modulus for drape-molded material (in which the glass does not flow or undergo additional orientation during processing). The next step is to establish the viability of this stiffness measure for material in which glass flow and orientation have occurred.

ACKNOWLEDGMENTS

This work was supported by GE Plastics. The help and encouragement provided by G. Tomkinson-Walles, R.P. Nimmer, T.J. Craven, J.M. Rawson, L.P. Inzinna, and K.R. Conway are gratefully acknowledged.

Special thanks are due to Julia A. Kinloch for her help and patience during the preparation of this paper.

REFERENCES

Stokes, V. K., 1988, "Random Glass Mat Reinforced Thermoplastic Composites – Part I: Phenomenology and Analysis of Tensile Modulus Variations," GE Corporate Research and Development Report No. 88CRD275, Schenectady, New York. (To appear as a two-part paper in *Polymer Composites*.)

Tomkinson-Walles, G. D., 1988, "Performance of Random Glass Mat Reinforced Thermoplastics," *Journal of Thermoplastic Composite Materials*, Vol. 1, pp. 94-106.

ISOTHERMAL RHEOLOGICAL CHARACTERIZATION OF GLASS MAT REINFORCED THERMOPLASTICS

S. M. Davis
GE Plastics, Technology Division
Pittsfield, Massachusetts

K. P. McAlea
Polymer Physics and Engineering Laboratory
GE Corporate Research and Development
Schenectady, New York

ABSTRACT

Glass mat reinforced thermoplastics (GMTs) offer a useful combination of mechanical properties and formability. In principle, these composites may be based on any thermoplastic matrix. In practice, matrix selection is limited because of its impact on the manufacturing and compression molding processes. In this work an isothermal squeezing flow technique is used to determine the apparent biaxial extensional viscosities of polycarbonate, polybutylene terephthalate, and polypropylene-based GMTs. Experimental load-deformation data is interpreted by treating the GMTs as viscous, incompressible Newtonian fluids. Two primary effects are observed: 1) the composites appear to strain harden as they are deformed; and 2) GMT apparent biaxial extensional viscosities correlate with the high rate of deformation shear viscosities of the matrices. A mechanism that explains the second result is proposed.

INTRODUCTION

GMTs are manufactured in continuous sheet form as shown in Fig. 1. This process makes conflicting demands on the thermoplastic matrix. First, the matrix must possess good melt strength to facilitate sheet extrusion. The same thermoplastic, however, must impregnate the glass mat during consolidation. Typically, these rheological requirements oppose one another. Matrices that are easily extruded in sheet form are highly viscous (and elastic), while matrices that efficiently wet-out glass mat have low shear viscosities and are inelastic. Thus, a GMT can be successfully manufactured only if the matrix has a balance of rheological characteristics. Compression molding of GMTs is accomplished in a two-step process. In the first step, a precut blank is heated in a radiant or forced convection oven. When the solidification temperature of the matrix is exceeded, elastic energy stored in the glass reinforcement is released and a lofted structure is formed. The lofted material is porous with a density determined by the glass loading, geometry, and resin viscosity. The hot lofted blank is then transferred to a matched die-set where it is squeezed to fill the tool cavity. Mold temperatures are typically 100 °C below the solidification temperature of the matrix and closing speeds in excess of 5 mm/s are common.

In the hot lofted state, a GMT has mechanical properties that are intermediate between a compliant porous solid and a viscous liquid. Upon application of a load, as in compression molding, air is expelled from the composite. When the air is completely removed the molten GMT becomes an incompressible, concentrated suspension. Literature on the rheological behavior of suspensions in viscous liquids abounds; however, much of this information is restricted to dilute suspensions, in which

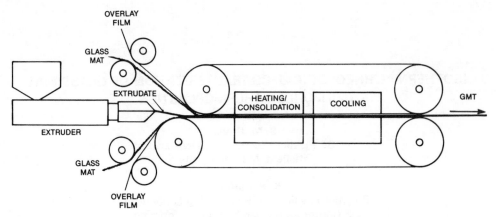

Fig. 1 GMT manufacturing process.

interparticle forces are unimportant. Theories and empirical equations that describe concentrated suspensions in viscous liquids are not as common and have been applied with a lesser degree of success. This is especially true for suspensions of rodlike particles in viscous liquids, e.g., short fiber reinforced polymers (Metzner, 1985; Ganani and Powell, 1985). GMTs present a particularly difficult challenge because the reinforcement is not in particle form. The glass mat is a non-woven textile that lends solid-like structure to the molten composite.

Despite the difficulties associated with modeling the flow behavior of such complex materials, demands for intelligent processing require development of a fundamental knowledge base. This paper describes an experimental program aimed at examining the load-deformation behavior of GMTs in squeezing flows. The purpose of this program was to determine the relationship between composite flow behavior and matrix shear viscosity. This approach was based on current rheological theory, which predicts that the bulk response of a suspension in a viscous liquid is a function of the shear viscosity of the suspending medium, regardless of deformation mode and concentration regime (Metzner, 1985).

APPARATUS

An experimental apparatus, shown in Fig. 2, was constructed to measure the flow behavior of GMTs. This instrument, called a squeeze flow rheometer, consists of two opposing heated platen assemblies mounted in an Instron model 1331 servohydraulic testing system. It is similar in concept and construction to the apparatus used by Zentner (1984) to determine the viscoelastic properties of sheet molding compound (SMC). The most important feature of this rheometer is its ability to duplicate the flow field of interest, i.e., squeezing flow. This, in fact, was the primary impetus for building the device.

The Instron servohydraulic testing system was rated at 45 kN. Pump and servovalve throughput were matched at 0.33 l/s. All GMT squeezing tests were conducted in stroke control. The stroke feedback signal was provided by an externally mounted Trans-Tek model 0217-0000 linear variable differential transducer (LVDT). This setup reduced errors caused by load frame and mounting extension compliance. The platens and all mating horizontal surfaces, including the insulation blocks, were surface-ground to insure parallelism between the opposing assemblies. A 3.85-mm diameter Teflon pin was used to center the GMT samples prior to squeezing, and to prevent unstable interlaminar shearing of the composites. The stroke command signal, which controlled the rate of platen closure, was provided by a Hewlett Packard model 6944A multiprogrammer. This instrument also logged the stroke and load feedback signals. The multiprogrammer was commanded by a Hewlett Packard series 200 computer running HP Basic 5.0/5.1. Data was processed, plotted, and stored with the aid of Hewlett Packard's DACQ/300 software.

Platen temperatures were determined by the copper heating blocks, which were controlled and monitored by Love model 32140 self tuning PID controllers. As shown in in Fig. 2, the controller feedback signals were provided by stainless steel sheathed J-type thermocouples. Each copper block

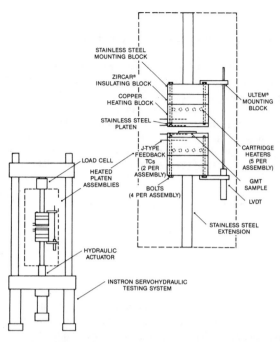

Fig. 2 Squeeze flow rheometer.

was heated by five Watlow model E4J30 cartridge heaters. Maximum power input to each block was 1 kW. The squeeze flow rheometer was routinely used to test the flow behavior of GMTs at temperatures exceeding 250 °C. Under these conditions, the largest variation in the surface temperature of the stainless steel platens, i.e., the center to edge difference, was 3 °C.

EXPERIMENTAL

GMT load-deformation behavior was measured as follows. A circular sample, 5.0 cm in diameter and 3 to 4 mm thick, was placed between the heated platen assemblies and a compressive load of approximately 10 N was applied. At this point, the servohydraulic system was operated in load control. The sample was allowed to loft, while undergoing conductive heating, for three minutes. The servohydraulic testing mode was then switched to stroke control, and the GMT sample was squeezed back to its original thickness. The purpose of the heating/lofting stage was to prevent glass-resin separation during heating and to allow the platens to recover thermally. Platen temperatures, as measured by the embedded J-type thermocouples shown in Fig. 2, dropped 4 to 6 °C during the first thirty seconds of the heating cycle. Hence, accurate control of GMT sample temperature necessitated the three-minute heating/lofting period. An additional two-minute thermal equilibration stage followed the heating/lofting period. Two minutes was selected on the basis of the classical one-dimensional slab heating result combined with experimental evidence.

The last step in the testing procedure consisted of squeezing the GMT sample to approximately 50 percent of its initial thickness. The compressive load required to achieve this deformation was recorded as a function of rate and temperature. It is emphasized that considerable effort was expended to conduct isothermal squeezing tests. The logic behind this methodology was simple. Currently, it is not possible to separate heat transfer and rheological effects in compression molding, which is a nonisothermal process. This makes it difficult to approach issues like "short shots," i.e., incompletely filled moldings, and press tonnage requirements in a rational manner. Clearly, efforts to analyze compression molding of GMTs must be preceded by isothermal rheological studies; otherwise, a complete and meaningful understanding of this process cannot be developed.

THEORY

Raw data collected during GMT squeezing experiments was of limited value, since only qualitative trends could be observed. In order to make strong conclusions regarding the influence of matrix shear viscosity on the flow behavior of GMTs, it was necessary to apply a constitutive equation and calculate apparent composite viscosities. Based on previous rheological studies of SMC (Barone and Caulk, 1986), a linear relationship between the deviatoric stress tensor and the rate of deformation tensor was assumed. In addition, inertial effects were deemed negligible and incompressibility was assumed. Thus, the GMTs were treated as viscous, incompressible Newtonian fluids.

The equations below relate compressive force (F), fluid shear viscosity (μ), closing speed (S), fluid volume (V) and gap height (h) for squeezing flow of an unconstrained circular blank of a viscous, incompressible Newtonian fluid between infinite flat parallel plates.

$$F = \frac{3\mu S V^2}{2\pi h^5} \qquad \text{[no-slip BCs]} \qquad (1)$$

and

$$F = \frac{3\mu S V}{h^2} \qquad \text{[slip BCs]} \qquad (2)$$

The differences in Eqns.(1) and (2) indicate the importance of boundary conditions (BCs) on the force required to squeeze this ideal fluid. No-slip BCs imply a shear-dominated flow, whereas slip BCs lead to a biaxial extensional flow. Equation (1) is called the Stefan equation. It is a fundamental result in lubrication theory and is derived in a number of texts on fluid mechanics (Bird, et al., 1977; Denn, 1980; Landau and Lifshitz, 1987). Equation (2) is equally basic, but appears less frequently in the literature. For this reason, it is derived in the Appendix. It is useful to rewrite Eqn. (2) as

$$F = \frac{\overline{\eta}\left(\frac{\dot{\varepsilon}}{2}\right) V}{h} \qquad (3)$$

where $\overline{\eta}$ ($= 3\mu$) is defined as the biaxial extensional viscosity and $\frac{\dot{\varepsilon}}{2}\left(=\frac{S}{h}\right)$ is the rate of biaxial extension. Equations (1) and (3) provided a starting point for interpreting data generated with the squeeze flow rheometer. The appropriateness of these results and issues related to BCs during isothermal squeezing flows of GMTs are discussed in later sections.

SAMPLE PREPARATION AND CHARACTERIZATION

A set of control materials was needed to establish a baseline for GMT flow behavior. Polycarbonate (PC) was selected as the matrix for the control group for three reasons: 1) At shear rates of 100 1/s and less, and at typical GMT compression molding temperatures, i.e., T_g + 120 °C, PC exhibits little tendency to shear thin; 2) PC is available in a wide range of molecular weights, hence matrix shear viscosity can be varied by a factor of 100 without changing chemical structure; and 3) PC exhibits reasonable thermal and oxidative stability at GMT molding temperatures. The experimental plan was to characterize the shear rheology of the various PCs, manufacture GMT samples based on these polymers, measure their load-deformation behavior, and finally, establish the role of matrix shear viscosity in the squeezing flow process. Correlations based on the control GMTs could then be tested by constructing new GMTs based on different polymers and predicting their load-deformation response. The resins selected as test matrices were polybutylene terephthalate (PBT) and polypropylene (PP).

Shear viscosity (μ) versus shear rate $(\dot{\gamma})$ data for the PCs is shown in Figs. 3 and 4. PC1, PC2, PC3, and PC4 are polycarbonates of successively increasing average molecular weights. Shear viscosity data for the PBT and PP is presented in Fig. 5. All polymers, with the exception of the PP, were supplied by GE Plastics. The PP was a 12 melt flow index grade manufactured by Aristech Corp. The viscosity data was obtained using a Rheometrics Dynamic Spectrometer (model 7700) operated in parallel plate mode. Plate diameter was 2.5 cm, and a nitrogen purge was used for all tests.

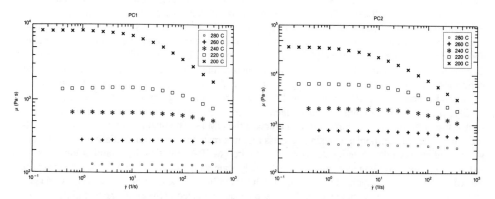

Fig. 3 Left – PC1 shear viscosity (μ) vs. shear rate ($\dot{\gamma}$).
Right – PC2 shear viscosity (μ) vs. shear rate ($\dot{\gamma}$).

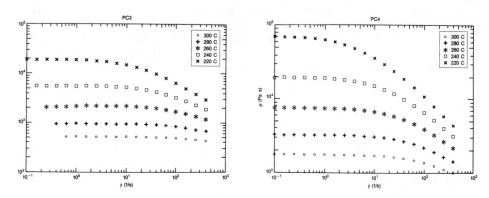

Fig. 4 Left – PC3 shear viscosity (μ) vs. shear rate ($\dot{\gamma}$).
Right – PC4 shear viscosity (μ) vs. shear rate ($\dot{\gamma}$).

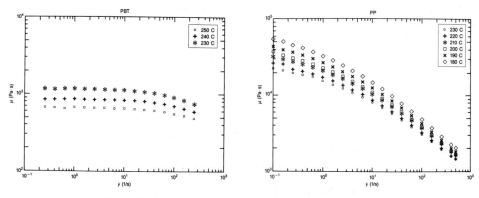

Fig. 5 Left – PBT shear viscosity (μ) vs. shear rate ($\dot{\gamma}$).
Right – PP shear viscosity (μ) vs. shear rate ($\dot{\gamma}$).

Raw data from the rheometer was recorded as complex viscosity versus rotational excitation frequency (η^* vs. ω). Justification for presenting this dynamic data as shear viscosity versus shear rate is provided by the Cox-Merz rule (Cox and Merz, 1958). This empiricism states that the magnitude of the complex viscosity of a polymer corresponds with its shear viscosity at equivalent frequencies and shear rates. Glass mat for the GMTs was supplied by PPG Industries' Fiber Glass Research Center. All mat was manufactured in a single run on one line. This was done to minimize structural variations in the mat. The mat was produced at an areal density of 0.20 kg/m^2. The glass loading level was kept constant at 20 percent by volume fraction (neglecting voids) for all GMT samples.

The circular GMT samples were machined from 9- × 13-cm plaques. Volume constraints prevented use of production equipment to manufacture the GMTs. The plaques were made by laminating sandwiches of thermoplastic resin and glass mat in a matched steel tool. Initially, the steel tool was placed between the heated platens of a Tetrahedron MTP-14 multidaylight laminating press. After the tool temperature reached the desired steady-state value, it was removed from the press and loaded with alternating layers of thermoplastic film (or pellets in the case of the PC based GMTs) and glass mat. After this step, the assembly was once again placed between the heated platens, and a nominal molding pressure of 0.2 MPa was applied for 5 min. The hot tool was then removed from the heating platens and relocated between the press cooling plates, where a nominal molding pressure of 0.2 MPa was maintained for an additional 10 min.

Press platen and tool temperatures were varied during plaque production according to matrix composition. The objective was to manufacture each GMT plaque at a temperature corresponding to a matrix zero-shear rate viscosity of 1000 Pa·s. By doing this, a similar degree of glass mat impregnation could be achieved in every experimental plaque. The processing temperatures for the PC-based GMTs were determined by fitting shear viscosity data to the WLF equation (Ferry, 1980)

$$\log a_T = \frac{-C_1(T - T_{ref})}{C_2 + (T - T_{ref})} \tag{4}$$

where a_T is the shift factor, C_1 and C_2 are constants (C_2 has units of degrees absolute), and T and T_{ref} are the temperature of interest and reference temperature, respectively. In terms of the rheological properties of a polymer

$$\log a_T = \log\left[\frac{\mu_0(T)}{\mu_0(T_{ref})}\right] \tag{5}$$

where $\mu_0(T)$ represents the zero shear rate viscosity at temperature T, and $\mu_0(T_{ref})$ represents the same quantity at temperature T_{ref}. Equating the right sides of Eqns. (4) and (5) and solving for temperature (T) gives

$$T = \frac{\left\{C_1 + \log\left[\frac{\mu_0(T)}{\mu_0(T_{ref})}\right]\right\} T_{ref} - C_2 \log\left[\frac{\mu_0(T)}{\mu_0(T_{ref})}\right]}{\left\{C_1 + \log\left[\frac{\mu_0(T)}{\mu_0(T_{ref})}\right]\right\}} \tag{6}$$

Thus, if $\mu_0(T_{ref})$, C_1, and C_2 are known, then calculation of the temperature (T) at which a zero shear rate viscosity of μ_0 occurs is elementary. For the PCs used in this work, shift factors were calculated over a broad range of temperatures using a least-squares technique. Average values of C_1 and C_2 were then determined and used with Eqn. (6), letting $\mu_0(T) = 1000$ Pa·s, to calculate the appropriate GMT plaque processing temperatures.

A similar procedure was followed for the PBT-based GMTs; however, shear viscosity data was fitted to the Arrhenius equation

$$\mu_0(T) = \mu_0(T_{ref})e^{(\Delta E/R)[(T_{ref} - T)/T_{ref}T]} \tag{7}$$

where ΔE represents the flow activation energy and R is the gas constant. All other variables in Eqn. (7) are as previously defined. The flow activation energy was also determined in a least-squares procedure and the following rearrangement of Eqn. (7) was used to calculate the PBT plaque processing temperature

$$T = \frac{T_{ref}}{\left\{1 + \frac{T_{ref}R}{\Delta E} \ln\left[\frac{\mu_0(T)}{\mu_0(T_{ref})}\right]\right\}} \tag{8}$$

A plaque processing temperature of 230 °C was selected for the PP-based samples. This temperature represented a compromise between shear viscosity requirements and concern regarding polymer degradation. Fortunately, no difficulties were encountered in producing well impregnated PP-based GMT plaques. Table 1 summarizes the conditions used to manufacture the GMTs used in this investigation.

Table 1
GMT Plaque Manufacturing Temperatures and Target Weight Fractions (W_f)

Matrix	(°C) Sample Plaque Molding Temp.	Glass Loading - Target W_f
PC1	232	0.35
PC2	250	0.35
PC3	281	0.35
PC4	320	0.35
PBT	234	0.33
PP	230	0.41

RESULTS

Early in the study, screening tests were conducted to establish bounds on squeezing rates and sample dimensions. During these preliminary tests two important observations were made: 1) the load deformation behavior of GMTs follows an unusual pattern, similar to that reported by Davis and Tucker (1988) for squeezing flows of SMC; and 2) under isothermal conditions GMTs resist gapwise shearing. A detailed discussion of the first observation will be presented with results from the control experiments. Evidence supporting the second observation is presented below.

The kinematics of GMT flows were investigated using two techniques. In the first method, multilayer black and white PBT-based GMT samples were prepared and then tested with the squeeze flow rheometer. A constant closing speed of 1 mm/s was used and platen temperatures were maintained at 230 °C. No lubricants were applied to the platens. In every case, the outer white layers flowed uniformly with the inner black core. Clearly, these materials resisted gapwise shearing. The second method of examining flow kinematics involved applying permanent ink grids to undeformed PC-based samples. The samples were then squeezed at a rate of 1 mm/s, and the grids were re-examined. Again, the outside layers slipped along the unlubricated platen surfaces. The general applicability of these results is questionable, since sample dimensions and testing conditions may have influenced the outcome. Nevertheless, within the experimental limits of this study, the dominant mode of deformation was biaxial extension.

Based on data and observations from the screening tests an experimental plan was developed. Table 2 shows the plan for the control materials. Test temperatures were selected to achieve matrix zero-shear rate viscosities of 100, 500, 1,000, 5,000 and 10,000 Pa·s, depending on the average molecular weight of the polycarbonate. These temperatures were calculated using the procedure outlined in the previous section. All tests were conducted at a constant ratio of closing speed (S) to gap height (h). This eliminated rate effects, since a steady biaxial extensional flow field was created. The single rate used for the control group tests, i.e., $S/h = 0.50$ 1/s, was chosen so that the load capacity

Table 2
Plans for Control Group Tests

Matrix Temp.		(Pa·s) μ_0	(1/s) $\dfrac{\dot{\varepsilon}}{2} = \dfrac{S}{h}$
PC1	280°C	1.0×10^2	0.50
	246°C	5.0×10^2	0.50
PC2	260°C	7.3×10^2	0.50
	249°C	1.0×10^3	0.50
PC3	281°C	1.0×10^3	0.50
	246°C	5.0×10^3	0.50
PC4	270°C	5.0×10^3	0.50
	254°C	1.0×10^4	0.50

of the testing system would not be exceeded, even under conditions that corresponded with a matrix zero-shear rate viscosity of 10,000 Pa·s. Table 3 shows the plans for the experimental materials. Test temperatures, in this case, were selected for convenience. No attempt was made to achieve specific matrix zero-shear rate viscosities. Initial tests on the experimental GMTs were conducted at an axial strain rate of –1.0 1/s. This allowed a one-to-one comparison of rheological data on the control and experimental materials. Later in the study, higher strain rate tests were conducted on the PP-based GMT.

Table 3
Plans for Experimental Group Tests

Matrix Temp.		(Pa·s) μ_0	(1/s) $\dfrac{\dot{\varepsilon}}{2} = \dfrac{S}{h}$
PBT	250°C	6.6×10^2	0.50
	240°C	8.5×10^2	0.50
	230°C	12×10^3	0.50
PP	230°C	2.3×10^3	0.50
	210°C	3.2×10^3	0.50
	190°C	4.4×10^3	0.50
	180°C	5.4×10^3	0.50

Figure 6 shows plots of gap height and compressive force as functions of time for a constant strain rate squeezing test on a PC-based GMT. Although these plots are specific to a single GMT sample under one set of conditions, the general trends exhibited are basic to all of the materials tested. The gap height-time plot in Fig. 6 is presented because it identifies the starting and stopping points in the squeezing process. In addition, the concavity in the curve shows that the rate of platen closure was not constant. As stated previously, all GMT samples were squeezed at a constant ratio of closing speed to gap height. Since slip boundary conditions prevailed, radial and tangential strain increased linearly with time. Thus, the plot of force vs. time in Fig. 6 may legitimately be referred to as a load-deformation curve. The most notable feature of this curve is that it consists of three distinct

234

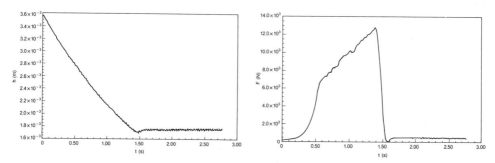

Fig. 6 Left – Gap height (*h*) vs. time (*t*) for a typical isothermal (*T* = 249 °C), constant strain rate (ε/2= 0.5 1/s) squeezing test on a PC based GMT (*V*ₓ≈ 0.20).
Right – squeezing force (*F*) vs. time (*t*) corresponding to left plot.

sections, two of which are associated with the load development process. In the first stage, the force required to deform the sample increased quickly, as air and resin were squeezed from the mat. Some of this separation, however, took place prior to squeezing. Resin was exuded in small droplets at the perimeter of the sample during the conductive heating cycle. This generated voids that contributed to the initial section of the load- deformation curve. The end of the void reduction and resin squeeze-out stage is indicated by a knee in the curve. At this point, the matrix and glass began moving together. It is during this second interval, called the composite flow stage, that the notion of a bulk suspension property, i.e., a composite viscosity, applies. The last stage of the load-deformation curve begins when squeezing ends. It consists of load decay. A lengthy relaxation period is not exhibited by this GMT; however, slight over-shoots and subsequent corrections by the closed loop servohydraulic system may have masked this effect.

All of the GMTs exhibited this three-step loading and unloading sequence. Differences between the materials manifested themselves in the length of the void reduction and resin squeeze-out stage and in the rate of load build-up. The latter result was expected, since large variations in matrix viscosity were intentionally introduced. The former result was not anticipated, but appears to have been related to resin leakage during the heating cycle. Very little leakage took place upon heating the PP-based samples. Consequently, few voids were introduced and the first stage of the load-deformation curve was brief. By comparison, resin leakage was prominent in the control GMTs, and a larger portion of the load development process was associated with void reduction. An important point regarding the composite flow stage must be mentioned. This stage did not continue indefinitely. When the axial strain exceeded –1.5 the reinforcement became compacted so thoroughly that groups of glass bundles began supporting the load. Hereafter, the force required to continue deformation rose dramatically. This obscured the bulk behavior of the suspension and rendered the load-deformation data meaningless. Composite viscosities in this study are calculated using test data with $-1.5 \leq \gamma_{zz} -1.0$, where

$$\gamma_{zz} = -\left(2\frac{S}{h}\right) t = -\dot{\varepsilon}t \tag{9}$$

Thus, the phenomenon of fiber bundle compaction is avoided. *S*, *h*, and $\dot{\varepsilon}$ in Eqn. (9) are as defined previously (see Theory); *t* represents time. This definition of axial strain reflects the assumption of steady biaxial extensional flow.

A plot of composite biaxial extensional viscosity ($\overline{\eta}$)vs. time is presented in Fig. 7. This plot corresponds to the load-deformation curve in Fig. 6. $\overline{\eta}$ is calculated using Eqn. (3). It is emphasized that $\overline{\eta}$ has meaning only during the composite flow stage. Here it increases steadily from 5 to 6.2 MPa·s. Once again, this trend is not unique. All of the control and experimental GMTs strain hardened. As a result, only an average composite biaxial extensional viscosity could be obtained for each test. Reconstructing a typical load-deformation curve based on an average $\overline{\eta}$ results in a 10 to 20

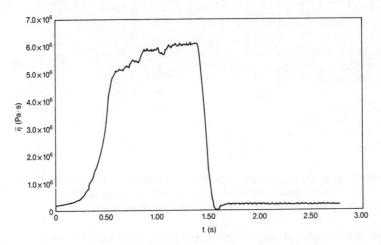

Fig. 7 Composite biaxial extensional viscosity ($\bar{\eta}$) vs. time (t) corresponding to Fig. 6.

percent overprediction of squeezing force at the onset of composite flow and a 10 to 20 percent underprediction at the end of deformation. Clearly, a linear relationship between deviatoric stress and strain rate does not adequately describe these materials. One can envision a number of causes contributing to the strain-hardening response, including resin-fiber separation and fiber bundle compaction. On the other hand, the role of surface friction cannot be dismissed without careful investigation. The assumption implicit in Eqn. (3) is that surface friction is negligible. In order to examine the importance of surface friction, PC-based GMT disks of varying aspect ratios (R/h) were squeezed under identical conditions. GMT composition and test temperature were selected to achieve a matrix zero-shear rate viscosity of 730 Pa·s. Results, shown in Fig. 8, indicated that surface friction could safely be neglected for samples having an initial aspect ratio of 8 or less. In addition, the degree of strain hardening observed in this material was not a function of sample geometry. Even the thickest disks, with an initial aspect ratio of approximately 4, exhibited a significant amount of strain hardening. Apparently, surface friction was not responsible for this unusual behavior. Since all of the GMT disks used in this investigation had aspect ratios of 6 to 8 prior to squeezing, the assumption of slip boundary conditions appears to have been justified. It can be argued that this statement is an over

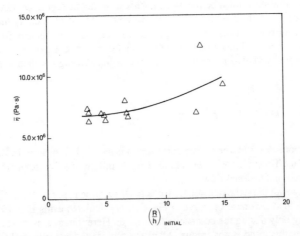

Fig. 8 Composite biaxial extensional viscosity ($\bar{\eta}$) vs. initial sample aspect ratio $\left(\frac{R}{h}\right)$ for isothermal ($T = 260$ °C), constant strain rate ($\varepsilon/2 = 0.5$ 1/s) squeezing tests for a PC2-based GMT ($V_f \approx 0.20$).

236

generalization, but experimental evidence suggests otherwise. GMT flow kinematics offer a convincing counter-argument. These materials flowed via biaxial extension because frictional resistance at the sample-platen interfaces was small, relative to the bulk or material resistance. If frictional resistance had been large, the GMT disks would have deformed by a shearing mechanism.

Table 4 summarizes the results of the control group experiments. As stated earlier, PC1, PC2, PC3 and PC4 are polycarbonates of successively higher average molecular weights. The biaxial extensional viscosities reported in Table 4 represent the averages of six tests. High and low values for each material and test condition are included as an indication of experimental scatter, which was considerable. Sources of scatter include variation in glass loading and fiber orientation. A systematic evaluation of these variables is not attempted in this study; however, it should be noted that glass loading percentages (by weight) could be maintained only within 10 percent of the targeted nominal values. This was due, almost entirely, to variability in the areal density of the nonwoven glass mat. Data regarding the starting and ending average molecular weights of the polycarbonate matrices is also presented in Table 4. Molecular weights were measured via gel permeation chromatography (GPC). The starting molecular weights indicate the condition of the matrices in their original, unprocessed form. End values reflect the state of the resin after GMT plaque preparation and squeeze-flow testing. With the exception of the highest molecular weight matrix, very little thermal/oxidative degradation of the polycarbonates occurred. Even a brief perusal of Table 4 raises serious questions regarding the relationship between GMT biaxial extensional viscosity and matrix shear viscosity. A comparison of the average biaxial viscosities of the PC2-based GMT at 249 °C and the PC3-based material at 281 °C is particularly disturbing. Here, the matrix zero-shear rate viscosities were matched, yet the biaxial extensional viscosities differ by a factor of 2.

Table 4
Control Group Results

Matrix/Temp.	(Pa·s) $<\bar{\eta}>$	(Pa·s) $\bar{\eta}_{high}/\bar{\eta}_{low}$	$<W_f>$	Start $\bar{M}_w/\bar{M}_n$	End $\bar{M}_w/\bar{M}_n$
PC1 288°C	No data. Severe resin leakage prior to testing.				
PC1 246°C	4.5×10^6	$4.9 \times 10^6 / 4.2 \times 10^6$	0.35	$30 \times 10^3 / 12 \times 10^3$	Unchanged
PC2 260°C	6.4×10^6	$7.4 \times 10^6 / 5.6 \times 10^6$	0.36	$37 \times 10^3 / 13 \times 10^3$	Unchanged
PC2 249°C	4.0×10^6	$10.3 \times 10^6 / 7.1 \times 10^6$	0.36	$37 \times 10^3 / 13 \times 10^3$	$35 \times 10^3 / 13 \times 10^3$
PC3 281°C	5.1×10^6	$6.0 \times 10^6 / 4.4 \times 10^6$	0.35	$50 \times 10^3 / 17 \times 10^3$	$45 \times 10^3 / 16 \times 10^3$
PC3 246°C	No data. Fiber bundle compaction dominated load development.				
PC4 270°C	14×10^6	$17.5 \times 10^6 / 12 \times 10^6$	0.37	$68 \times 10^3 / 23 \times 10^3$	$54 \times 10^3 / 19 \times 10^3$
PC4 254°C	18.5×10^6	$22 \times 10^6 / 15 \times 10^6$	0.37	$68 \times 10^3 / 23 \times 10^3$	$55 \times 10^3 / 20 \times 10^3$

Rather than using data from the control group tests to predict the behavior of the experimental materials, results from squeezing tests on the PBT- and PP-based materials are presented immediately in Table 5. Comparing this data with the previous results provides further reason to believe that GMT biaxial extensional viscosity is not a function of matrix zero-shear rate viscosity. Biaxial extensional viscosities of the PP-based GMT are, in certain instances, an order of magnitude less than would be expected based on the PC GMT data. These results raised an interesting possibility: perhaps matrix biaxial extensional viscosity, rather than shear viscosity, correlates with GMT deformational response? Considering observations regarding flow kinematics this seemed to be a reasonable conjecture; however, it can easily be shown that shearing was present on a local scale, despite its apparent absence on a bulk scale.

Table 5
Experimental Group Results

Matrix/Temp.	(Pa·s) $\langle \bar{\eta} \rangle$	(Pa·s) $\bar{\eta}_{high}/\bar{\eta}_{low}$	$\langle W_f \rangle$	Start $\overline{M}_w/\overline{M}_{\bar{n}}$	End $\overline{M}_w/\overline{M}_n$
PBT — 250°C	3.9×10^6	$4.4 \times 10^6/3.3 \times 10^6$	0.34	$11 \times 10^4/59 \times 10^3$	$11 \times 10^4/54 \times 10^3$
240°C	5.8×10^6	$7.0 \times 10^6/4.4 \times 10^6$	0.35	$11 \times 10^4/59 \times 10^3$	$11 \times 10^4/53 \times 10^3$
230°C	7.0×10^6	$4.0 \times 10^6/6.0 \times 10^6$	0.34	$11 \times 10^4/59 \times 10^3$	$11 \times 10^4/50 \times 10^3$
PP — 230°C	1.9×10^6	$2.1 \times 10^6/1.6 \times 10^6$	0.43	$34 \times 10^4/75 \times 10^3$	Unchanged
210°C	2.3×10^6	$2.8 \times 10^6/1.9 \times 10^6$	0.42	$34 \times 10^4/75 \times 10^3$	$34 \times 10^4/66 \times 10^3$
190°C	2.3×10^6	$2.8 \times 10^6/2.0 \times 10^6$	0.40	$34 \times 10^4/75 \times 10^3$	$32 \times 10^4/74 \times 10^3$
180°C	2.9×10^6	$3.3 \times 10^6/3.4 \times 10^6$	0.41	$34 \times 10^4/75 \times 10^3$	$31 \times 10^4/74 \times 10^3$

Consider biaxial extensional flow of a cylindrical disk relative to the coordinate system shown in Figure A1 (see Appendix). Initially, each point in the disk is described by an ordered triple (r, θ, z), where $0 \le r \le R$, $0 \le \theta \le 2\pi$, and $0 \le z \le h$. At a later time t, each point will be described by a new ordered triple $(re^{(S/h)\,t}, \theta, ze^{-(2S/h)\,t})$. This means that the instantaneous radial position of a point does not depend on its Z coordinate. Now, consider an arbitrary line segment located within the deforming disk. From the standpoint of an observer located directly above the disk, the line segment stretches but does not reorient. A glass fiber in a GMT disk undergoing biaxial extensional flow cannot stretch. Therefore, drag-induced shearing takes place at the fiber-matrix interface. Considering the enormous number of fibers present in a typical GMT and their close proximity to one another, it is reasonable to expect a great deal of localized shearing to accompany bulk extensional flow. The difficulty is in determining the characteristic rate of deformation on a local scale, which may differ significantly from the bulk rate. With this in mind, plots of composite viscosity vs. matrix shear viscosity at shear rates of 5, 50, and 500 1/s were generated. These rates were selected because they differed from the experimental rate of biaxial extension by factors of 10, 100, and 1000, respectively. Data from the PC, PBT, and PP GMT squeezing tests were included in each plot. Figure 9 shows the extreme cases. At the left is a plot of $\bar{\eta}$ vs. matrix zero shear rate viscosity. The right side shows $\bar{\eta}$ vs. matrix shear viscosity evaluated at a shear rate of 500 1/s . The conclusion is obvious. It is the shear viscosity of the matrix at high rates of deformation, i.e., $\dot{\gamma} > 100$ 1/s, that influences GMT biaxial extensional viscosity. Considering the uncertainties in glass loading level and orientation as well as other experimentally introduced errors, the correlation between $\bar{\eta}$ and $\mu|_{\dot{\gamma} = 500\,1/s}$ is surprisingly good. The investigation concluded with a brief set of rate experiments on the PP-based GMT. 5.0 cm diameter disks were squeezed at constant axial strain rates of –1.0, –2.0, –4.0 and –8.0 1/s, respectively. Results are displayed in Fig. 10. As before, each data point represents the average of six tests. This result demonstrates that the rate-dependent behavior of the PP-based GMT mimics the response of the neat matrix at high rates of deformation.

CONCLUSIONS

A squeeze flow rheometer has been built to study the relationship between matrix and viscosity and the deformational resistance of GMTs. The isothermal squeezing flow behavior of polycarbonate, polybutylene terepthalate, and polypropylene-based composites has been investigated using this apparatus. The primary findings are listed below.

- GMTs resist gapwise shearing during isothermal squeezing flow. The glass mat promotes a two-dimensional flow field with slip at the boundaries.
- At a glass loading of 20 vol%, GMT apparent biaxial extensional viscosities, calculated assuming viscous, incompressible Newtonian response, are large, e.g., 1 to 10 MPa·s. In addition, strain hardening and strain rate sensitivity are observed.

238

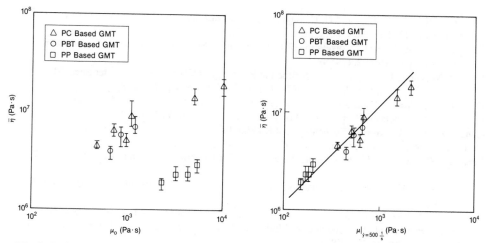

Fig. 9 Left – Composite biaxial extensional viscosity ($\bar{\eta}$) vs. matrix zero-shear rate viscosity (μ_o) $V_f \approx 0.20$.

Right – Composite biaxial extensional viscosity ($\bar{\eta}$) vs. matrix viscosity evaluated at a shear rate of 500 1/s ($\mu|_{\dot{\gamma}}$ = 500 1/s) $V_f \approx 0.20$.

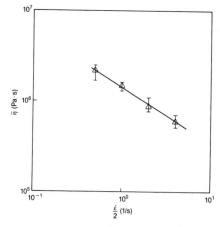

Fig. 10 Composite biaxial extensional viscosity ($\bar{\eta}$) vs. rate of biaxial extension ($\varepsilon/2$) for isothermal (T = 210 °C), constant strain rate squeezing tests on a PP-based GMT ($V_f \approx 0.20$).

The apparent biaxial extensional viscosities of GMTs correlate with the high rate of deformation shear rheology of the matrices. Apparently, the local rate of deformation during isothermal squeezing flow of GMTs is much higher than the bulk rate and is dominated by shearing rather than extension.

Translation of these results into commercially useful information will not be simple. Significant improvements in our understanding of nonisothermal flow kinematics and heat transfer during compression molding must be made. Nonetheless, the results reported in this investigation provide a starting point for matrix selection and press tonnage predictions.

REFERENCES

Barone, M.R. and D.A. Caulk, "A Model for the Flow of a Chopped Fiber Reinforced Polymer Compound in Compression Molding," *J. of Applied Mechanics 53*, pp. 361-371 (1986).

Bird, R.B., R.C. Armstrong, and O. Hassager, *Dynamics of Polymeric Liquids: Volume 1 Fluid Mechanics,* J. Wiley, New York, pp. 19-21, (1977).

Cox, W.P., and E.H. Merz, "Correlation of Dynamic and Steady Flow Viscosities," *J. of Polymer Science 28,* pp. 619-622, (1958).

Davis, S.M., and C.L. Tucker III, "Experimental Rheology of Glass Fiber/Polyester Sheet Molding Compound," *Proc., SPE 46th Annual Technical Conf.,* pp. 524-527, (1988).

Denn, M.M., "Process Fluid Mechanics," Prentice-Hall, New Jersey, pp. 255-259 (1980).

Ferry, J. D., "Viscoelastic Properties of Polymers," J. Wiley, New York, pp. 264-280, (1980).

Ganani, E. and R.L. Powell, "Suspensions of Rodlike Particles: Literature Review and Data Correlations," *J. of Composite Materials 19,* pp. 194-215 (1985).

Landau, L.D., and E.M. Lifshitz, "Fluid Mechanics," Pergammon Press, Great Britain, pp. 66-67, (1987).

Metzner, A.B., "Rheology of Suspensions in Polymeric Liquids," *J. of Rheology 29,* pp. 739-755 (1985).

Zentner, M.M., "An Exploratory Investigation of the Rheology of Sheet Molding Compound," M.S. Thesis, Department of Mechanical and Industrial Engineering, University of Illinois at Urbana-Champaign, Urbana, IL (1984).

APPENDIX

In this appendix the load deformation equation for isothermal, lubricated squeezing flow of a circular disk of a viscous, incompressible Newtonian fluid between two flat parallel plates is derived. The geometry associated with this problem is shown in Figure A1. From symmetry, the θ-direction component of the velocity vector must be zero at every point in the disk. In addition, all terms containing the operator $\frac{\partial}{\partial \theta}$ must vanish. Boundary conditions are

$$v_z = -S \quad \text{at } z = h, S > 0 \tag{A1}$$

$$v_z = 0 \quad \text{at } z = 0 \tag{A2}$$

$$v_r = 0 \quad \text{at } r = 0 \tag{A3}$$

and

$$\pi_{rr} = p_a \quad \text{at } r = R \tag{A4}$$

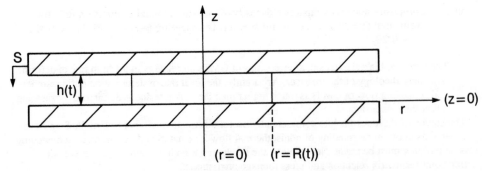

Fig. A1 Geometry for squeezing flow between parallel plates.

In Eqns. (A1) through (A4), v_z and v_r are the z direction and r direction components of the velocity vector $\underline{v}$, S is the rate of downward movement of the upper plate, π_{rr} is the rr component of the total stress tensor ($\underline{\underline{\pi}}$), and p_a represents atmospheric pressure. The notation

$$\underline{\underline{\pi}} = p\underline{\underline{\delta}} + \underline{\underline{\tau}} \tag{A5}$$

is used here, where $p\underline{\underline{\delta}}$ is the hydrostatic contribution to the total stress, and $\underline{\underline{\tau}}$ is the deviatoric or "extra" stress tensor. Compressive stresses are positive according to this sign convention.

Symmetry considerations and the assumption of incompressibility reduce the statement of conservation of mass to

$$\frac{1}{r}\frac{\partial}{\partial r}(rv_r) + \frac{\partial v_z}{\partial z} = 0 \tag{A6}$$

From Eqns. (A1) through (A3) it follows that $v_r = v_r(r)$ and $v_z = v_z(z)$. Hence,

$$\frac{1}{r}\frac{\partial}{\partial r}(rv_r) + \frac{\partial v_z}{\partial z} = c_1 \tag{A7}$$

where c_1 is a constant. Specifically, c_1 is not a function of r or z at any instant t. Integration with respect to z gives

$$v_z = -c_1 z + c_2 \tag{A8}$$

where c_2 is a constant similar to c_1. From Eqn. (A2) it is clear that c_2 is equal to zero. The remaining constraint, i.e., Eqn. (A1), requires that $c_1 = \frac{S}{h}$, so that

$$v_z = -\frac{S}{h}z \tag{A9}$$

Combining Eqn. (A3) with the left-side of Eqn. (A7) gives

$$v_r = \frac{S}{2h}r \tag{A10}$$

In order to calculate the components of the deviatoric stress tensor ($\underline{\underline{\tau}}$), it is first necessary to determine the components of the rate of strain tensor ($\underline{\underline{\dot{\gamma}}}$). For an incompressible Newtonian fluid

$$\underline{\underline{\tau}} = -\mu\underline{\underline{\dot{\gamma}}} \tag{A11}$$

where μ is the shear viscosity of the fluid, and $\underline{\underline{\dot{\gamma}}}$ is defined by

$$\underline{\underline{\dot{\gamma}}} = \left[\underline{\nabla}\,\underline{v} + (\underline{\nabla}\,\underline{v})^t\right] \tag{A12}$$

From this definition and the results given in Eqns. (A9) and (A10)

$$\dot{\gamma}_{zz} = -2\frac{S}{h} \tag{A13}$$

and

$$\dot{\gamma}_{zz} = \dot{\gamma}_{\theta\theta} = \frac{S}{h} \tag{A14}$$

It follows that

$$\tau_{zz} = 2\mu\frac{S}{h} \tag{A15}$$

and

$$\tau_{zz} = \tau_{\theta\theta} = -\mu \frac{S}{h} \tag{A16}$$

are the only non-zero components of the deviatoric stress.

Conservation of momentum for this problem reduces to

$$0 = -\left[\underline{\nabla} \cdot \underline{\underline{\pi}}\right] \tag{A17}$$

Expanding this equation and substituting the appropriate expressions for τ_{zz}, τ_{rr}, and $\tau_{\theta\theta}$ gives

$$\frac{\partial p}{\partial r} = \frac{\partial p}{\partial q} = \frac{\partial p}{\partial z} = 0 \tag{A18}$$

or

$$p = C_4 \tag{A19}$$

where C_4 is a constant. Applying Eqn. (A4) to this result yields

$$p = C_4 = p_a + \mu \frac{S}{h} \tag{A20}$$

The magnitude of the force exerted on the plates is

$$F = \left| \int_0^{2\pi} d\theta \int_0^R -\left(\pi_{zz} - p_a\right)\Big|_{z=h} r\, dr \right|$$

$$F = \pi R^2 \cdot \frac{3\mu S}{h} \tag{A21}$$

If $V = \pi R^2 h$ is the volume of the disk, then Eqn (A21) can be rewritten as

$$F = \frac{3\mu V S}{h^2} \tag{A22}$$

This is the desired result.

EFFECT OF PROCESSING PARAMETERS ON COMPRESSION MOLDED PMR-15/C3K COMPOSITES

A. Farouk and T. H. Kwon
Mechanical and Aerospace Engineering
Rutgers University
Piscataway, New Jersey

Abstract: The consolidation and curing history during the processing of composite materials affects the final properties of a part in various inter-related ways. In order to improve the quality of composites, these process-property relations must be understood in detail. Using a computer controlled compression molding and data acquisition system, the processing of PMR-15/C3k composites has been investigated. Process parameters considered were the pressure application time, pressure magnitude and crosslinking temperature. Parts were tested for inter laminar shear strength (ILSS) and measurements were made for void content and thickness. Full compaction strength ($ILSS_{fc}$) and optimum compaction strength ($ILSS_{opt}$) have been defined. The ILSS is presented as a function of void content and thickness.

1 Introduction

PMR-15 polyimide composites are beginning to find more and more usage in aerospace applications due to their higher service temperatures (around 315C) and other desirable properties. However, their processibility is still one of the important factors which restricts their widespread usage. A good survey of recent efforts at improving the processibility is presented by Serafini [1][1]. Detailed information is required concerning the processing conditions and property relations in order to be able to understand and improve the quality of these composites. As part of a project investigating the processing of these materials, experiments were performed aimed at exploring this area. Vannucci [2] and later Sims [3] et. al., have done some investigation in this area. In their study the in-process measurements were absent and the flexibility of the processing system was limited. For more detailed investigation, an advanced compression molding system is required which will have extensive data acquisition capabilities as well as flexibility. Such a system was developed specifically for this project [4]. It combined process control and monitoring functions into one unit. Using this system, the cure cycle was modified in various ways to find out the effects of process variations on the properties of the composites. The properties were determined by testing the samples thus produced. The deviations in properties were then related to the processing history. Such information permits effective modeling of the process by telling us which parts of the cure cycle affect the desired property most and in what way. This also allows more valid simplifying assumptions to be made.

---

[1]Numbers is square brackets correspond to the list of references at the end.

2 Composite Processing

In this discussion, processing of the PMR-15 composites means forming the composites from prepregs. Prepregs are graphite fiber cloths impregnated with a solution of polymer pre-cursors. The heating process controls the rates of imidization and cross-linking reactions that lead to the final form of the polymer structure. This is known as curing. The reactions which take place are documented well in literature (Serafini [1]). The pressure application helps to assure proper fiber matrix bonding, low void content and the final shape of the part. This process is called consolidation. A processing cycle includes curing and consolidation which interact with each other due to the change in physical properties of the matrix accompanying the changes in the chemical structure. A typical cure cycle for PMR-15/C3k showing the various important parameters is given in Figure 1.

During the first stage of heating, no pressure is applied. Melting and imidization occurs during this period. The temperature is raised during the second stage heating when the resin shows some melt-flow behavior in the beginning and then solidifies. Pressure is applied at the beginning of this second stage heating. This pressure application point is critical as will be shown later since the melt-flow behavior must be utilized for proper consolidation.

3 Experimental

3.1 Materials

The prepregs used were manufactured by US Polymeric, CA. in 1985 and used after a period of 3 years during which time it was stored in a freezer[2]. The matrix material is PMR-15B, while the fibers used are cross-plies of Celion 3000 Graphite fibers. Parts molded were 5in by 2in in area with 12 plies in each.

The plies were hand-cut and laid up in the mold which was sprayed with a commercial mold release. The mold was made 304L Stainless Steel. No vacuum was applied, all degassing being done at atmospheric or higher pressure. No post-curing heating was done. A computer controlled fan was used to cool down the mold and the same cooling cycle was used for all parts.

The parameter variation matrix is given in Table 1.

Table 1: Parameter Variations			
Variable and part id. #	Pressure Magnitude kPa	Pressure application, minutes from std.	Crosslinking Temperature, C
10, 11, 12	345	0	315
13, 14, 15	525	0	315
16, 17, 18	690	0	315
47, 48, 49	1375	0	315
19, 20, 21	2575	0	315
22, 23, 24	3450	0	315
25, 26, 27	1375	-10	315
47, 48, 49	1375	0	315
28, 29, 33	1375	10	315
30, 31, 32	1375	20	315
47, 48, 49	1375	0	315
34, 35, 36	1375	0	335
38, 39, 40	1375	0	345

The three important parameters of the cure cycle are: (a) the pressure magnitude, (b) pressure application time and (c) the crosslinking temperature forming three sets in the table. The standard cycle (Part #s 47, 48, 49) is the starting point whose parameters are changed one at a time. In the table, parts 47, 48, 49 are repeated in all three sets to show continuity of the parameter variations. Thus, pressure magnitude was increased from 345 kPa to 3450 kPa holding all other variables constant. The pressure application time was also varied by delaying (positive numbers)

---

[2]PMR-15 is known to undergo chemical changes during storage which may affect final properties substantially. The authors have used three other batches (including a fresh batch) of PMR-15 afterwards in similar experiments and, while the results are different, the trends observed here have been repeated. Such comparisons will be the subject matter of a future publication.

or advancing from that of the standard cycle. The 'standard cycle' as referred to in this paper is based on that suggested by the manufacturer in a private communication to the authors. The crosslinking temperature was increased in two steps from 315 C to 345 C. Each process cycle was repeated three times to reduce statistical variations from affecting the results.

3.2 Compression Molding System

Compression molding was chosen as the consolidation method for producing the samples used in this investigation. CCMS, the Computer Controlled Molding System, was used for the process and is shown in outline in Figure 2.

This system performs unattended curing and consolidation of composites based on the process cycle provided to the computer in the form of an input data file. The modifications of the cycle were made by changing this data file. The system is able to follow the reference cycle by using feedback control. Extensive measurements of temperature, pressure and mold closure rate are recorded throughout the processing period. The details of the structure and capabilities of CCMS are discussed elsewhere (Farouk [4]).

4 Testing

4.1 Inter-Laminar Shear Strength

As shown by our tests of various properties, the inter laminar shear strength shows the most significant susceptibility to the variations in the consolidation process. Therefore, this test was chosen to study effects of processing parameters on the properties. The inter laminar shear strength test, ASTM D2344-XX is frequently used to determine the inter laminar shear strength of composite materials.

Test data with Thornel-50/Epoxy 828 specimens from Hoggatt [6] have shown that as the $\frac{span}{thickness}$ ratio is increased, the failure mode changes from shear to flexure. As the mode of failure changes, the strength derived from the beam theory calculation also changes. Since the actual inter laminar shear strength is assumed not to change, the difference in measured strength is due to a difference in the mode of failure that is not being accounted for. To determine the proper $\frac{span}{thickness}$ ratio some preliminary tests were done on the specimens produced in the course of the experiments. The effect of $\frac{span}{thickness}$ ratio (Figure 3) on the inter laminar shear strength as well as the mode of failure was noted. All the specimens tested had shear failure and showed delamination. Thus, any value of $\frac{span}{thickness}$ ratio lower than 8 could be taken for this material. Compressive effects restricted the lower limit to about 4. Thus, a span length of 0.8 inches was chosen which would give a ratio of approximately 4 for the specimens.

Test specimens were prepared by cutting the samples using a diamond saw. No lubricants or coolants were used in the process. The specimens were cut out from three different areas of the sample parts and represented the edge, middle and in-between (called 'away') areas of a sample. The cutting scheme is shown in Figure 4.

4.2 Void Content

The void content was determined in the conventional way. Due to the possibility of resin density varying by itself, *i.e. not due to void content*, the ASTM (ASTM D2734-XX) method calls for resin digestion and actual weight and volume determination. This method could not be followed due to the difficulty in dissolving cured PMR-15. Instead, an approximate method was used which is described below.

The total volume of the composite is defined as

$$V = V_m + V_f + V_v \qquad (1)$$

and consequently,

$$V = \frac{W_m}{\rho_m} + \frac{W_f}{\rho_f} + V_v \qquad (2)$$

where

V_m = Volume of matrix (not including voids), cc
V_f = Volume of fibers, cc
V_v = Volume of voids, cc
W_m = Weight of matrix, gm

W_f = Weight of fibers, gm
ρ_m = Density of matrix, gm/cc
ρ_f = Density of fibers, 1.78 gm/cc (manufacturer supplied).

Again, the total volume of the composite, V, may be determined by measuring the weight loss in water, and is given by

$$V = \frac{W_{air} - W_{water}}{\rho_{water}} \qquad (3)$$

where

W_{air} = Weight of composite in air, gm
W_{water} = Weight of composite in water, gm
ρ_{water} = Density of water, gm/cc

Then, from Eq. 2, the density of the matrix is given as

$$\rho_m = \frac{W_m}{V - \left(\frac{W_f}{\rho_f} + V_v\right)} \qquad (4)$$

All quantities on the right hand side of Eq. 4 can be measured except the void content V_v itself, which is unknown. However, since all quantities are positive, the effect of not taking the void content into account is to reduce the calculated matrix density by some amount. Consequently, for a part in which the void content is smallest, the calculated resin density will be closest to the actual density. Based on the relatively small void content found in some parts (from photomicrographic evidence), the assumption is made that the highest density is indeed the actual matrix density.

Finally, the void content is given by,

$$Void\ \% = \frac{V_v}{V} \cdot 100 \qquad (5)$$

which, in terms of the measured and known quantities, becomes

$$Void\ \% = \frac{\frac{W - W_w}{\rho_w} - \frac{W - W_f}{\rho_m} - \frac{W_f}{\rho_f}}{\frac{W - W_w}{\rho_w}} \cdot 100.0 \qquad (6)$$

$$Void\ \% = \frac{\frac{W - W_w}{0.98} - \frac{W - 27.52}{1.25} - \frac{27.52}{1.78}}{\frac{W - W_w}{0.98}} \cdot 100.0 \qquad (7)$$

where the measured and manufacturer supplied values have been substituted. The fiber content remains constant since the same number of plies of the same size are being used. The resin density provided by the manufacturer is about 1.28 gm/cc. The effect of this difference on the void content measurement is very small and negligible. The manufacturer supplied density was not used since the data is an average for all batches and the effects of long term storage are not taken into account.

Finally, W and W_w are the variable quantities from which the void content is determined.

5 Results

The samples of PMR-15/C3k produced under various process conditions were tested for inter laminar shear strength. The results were then correlated to the process variables which had been varied while producing the specimens.

5.1 Pressure Magnitude Sensitivity

Utilizing the pressurizing system's present capacity of a maximum of 3.45 MPa (500psi) the specimens were produced at 345 , 525, 690, 1375, 2575, 3450 kPa (50, 75, 100, 200, 375 and 500 psi).

Shown in Figure 5, the void content decreases as pressure is increased. The strength test results are plotted in Figure 6. As pressure is increased up to about 1375kPa (200 psi), the ILSS increases. This is explained by the reduction of void content and better bonding between the laminas. The results of Vannucci [2] and later on Sims [3] with materials of the same basic type show similar trends and the data have been superimposed on the plots. With further increase in pressure, up to 3450 kPa (500 psi), reduction and apparent stabilization of the strength to a certain value is noted. A possible explanation is that as the pressure is increased beyond a certain value, the fibers become packed together and are mostly in contact with each other. The resin fills up the spaces within the network of packed fibers and its role as a binder between the layers is reduced. The part approaches full compaction thickness, d_{fc} and the strength of a fully compacted part is defined as $ILSS_{fc}$. On the other hand, there is an optimal thickness, for a part that has been compacted an optimal amount, which we define as d_{opt} and the strength of such a part as $ILSS_{opt}$. From the results we conclude that $ILSS_{fc}$ is less than $ILSS_{opt}$.

Plotting the thickness and pressure magnitude (Figure 7) it is found that the rate of decrease of the thickness decreases with increase in pressure.

5.2   Pressure Application Time Sensitivity

The pressure application time was delayed 10 and 20 minutes and advanced 10 minutes. The zero time is the time at which pressure is applied in the standard cure cycle which is that suggested by the manufacturer and is just before the second temperature ramp (220 minutes into the cycle in this case). It may be mentioned that this time is determined with an autoclave cycle in mind where such factors as the possibility of rupture of the bags has been taken into account as well as the changes in the resin chemistry. The effect on void content is shown in Figure 8. The effect is less pronounced than the effects of pressure magnitude variation. However, when the ILSS results are considered, as shown in Figure 9 the effect is more pronounced.

The ILSS falls with delay in pressure application time, since the delay affects the flow of resin and bonding of the laminas. It shows that pressure application is definitely necessary during the critical time of the cycle when the resin is more fluid. Evidence of resin viscosity minimum during this time has been found from other measurements like the melt pressure drop or dielectric measurements and is discussed in more detail elsewhere [5]. Failure to apply pressure at this time reduces strength since the polymer undergoes rapid viscosity and chemical changes and is unable to flow or bind afterwards.

However, early pressure application also reduces the strength. A possible explanation for this is that early pressure application apparently reduces the thickness from a value close to d_{opt} to one closer to the d_{fc}. The plot of pressure application time versus thickness, Figure 10, shows that thickness does reduce on early pressure application.

5.3   Crosslinking Temperature Sensitivity

In the standard cure cycle (Figure 1) the material is crosslinked at 315 C. The effect of raising the crosslinking temperature is equivalent, to some extent, to having a hot spot in the mold. It was of interest to study the effects of such hot spots since they are quite possible in practical molding situations.

The temperatures used were 315 C, 335 C and 345 C. The results are plotted in Figures 11 and 12. Void content increases with increase in temperature. ILSS decreases sharply as temperature is increased. It is interesting to note that the 'edge' ILSS is higher than the 'middle' and that the middle part appears to be more sensitive to high temperature. The mold was designed to give uniform temperature over a range of temperatures. This was verified by separate tests using thermocouple measurements inside the mold at higher temperatures. The non-uniformity was less than ±2C for up to 350C inside the mold. The effect of high temperature on resin chemistry, like higher crosslinking density or degradation of the polymer structure may be responsible for the observed variations in strength.

5.4   The Property-strength Inter-relationship

It is apparent from the results presented so far that neither void content nor thickness alone determines part quality. The inter-relationship is complicated because the same void content (or thickness) may be achieved by different cure cycles in which case the properties will not be the same. To show how strength is related to the external properties (i.e., void content and thickness), a plot of property-strength data is made. The process conditions can no longer be traced since all data are included. In the 3d-surface plot, ILSS is plotted as a function of the void content and thickness (Figure 13). It shows, for example, that the same thickness may be achieved at different void contents and the strength would be different.

Defining a void sensitivity factor as the reduction in ILSS for unit void content increase, Eq. 8,

$$K_v = \frac{ILSS \ change}{Void \ Content \ change} \tag{8}$$

it is found that ILSS has different sensitivity to voids depending on the process variable causing the void content variation, shown in Table 2. The coefficient, K_v is the slope of the regression line fitted to the ILSS vs. void content data.

Table 2: Void Sensitivity

Process Variable Varied	Void Sensitivity, K_v MPa
Pressure Magnitude	-1.8
Pressure App. point	-3.9
Crosslinking Temperature	-8.7

6   Conclusion

The objectives of this investigation were to acquire detailed information concerning the processing history and property relations. The processing history was characterized by key parameters in the process, viz., (a) the pressure magnitude, (b) pressure application time and (c) crosslinking temperature which were varied in a systematic manner and the variations correlated to the properties of the composites. The physical properties of the composite, viz., (a) void content and (b) thickness were measured along with (c) ILSS.

Pressure magnitude affects both void content and thickness as does the pressure application point. The pressure application point has lesser effect on void content than pressure magnitude, within the ranges investigated. The thickness variation for these process variables were comparable.

The ILSS tests showed that increasing pressure magnitude has a non-linear effect on the strength. The strength increases up to about 200 psi, after which it decreases and stabilizes. Pressure application point has a different effect on strength. Both early and late pressure application reduces strength.

Finally, the property-strength relations were explored using a 3D plot with strength as a function of void content and thickness. It shows that while the same void content or thickness may be achieved by different processes, the strength may not be the same. The void sensitivity factor, K_v, shows dependence on the processing history causing such voids. This illustrates the complex inter-dependence between the processing history and final properties of PMR-15 composites.

7   Acknowledgements

This study was supported by the General Electric Co., Corporate Research and Development at Schenectady who also supplied the PMR-15/C3k prepregs. The authors would like to thank Dr. R. K. Upadhyay for valuable data and technical discussions.

References

[1] Serafini, T.T., ed, High Temperature Polymer Matrix Composites, Noyes Data Corporation, NJ, 1987.

[2] Vannucci, R. D., "Effect of Processing Parameters on Autoclaved PMR Polyimide Composites", Proceedings, 9th National SAMPE Technical Conference, Oct, 1977, Vol. 9, pp 177-199, 1977.

[3] Sims, D.F., Weed D. N. and Francis, P. H. "Experimental Evaluation of Polyimide Composite Materials". ICCM/2, Proceedings of the 1978 International Conference on Composite Materials, The Metallurgical Society of AIME, NY, 1978.

[4] Farouk, A. & Kwon, T. H. "A Computer Controlled Compression Molding System", Submitted to Review of Scientific Instruments.

[5] Farouk, A. & Kwon, T. H. "A Study of the Consolidation process of composites", Proceedings of the 4th Technical Conference, American Society of Composites, October, 1989 (in press).

[6] Hoggatt, J. T. "Test methods for High Modulus Carbon Yarn and Composites", Composite Materials, testing and design, ASTM STP 460, pp 48-61, 1969.

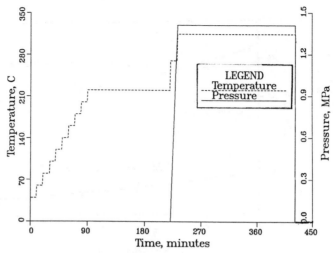

Figure 1: Typical processing cycle for PMR-15/C3k

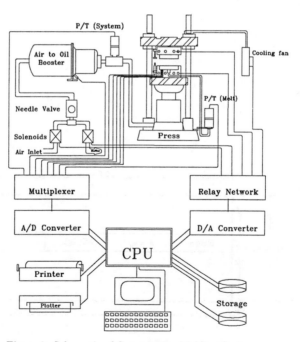

Figure 2: Schematic of Compression Molding System

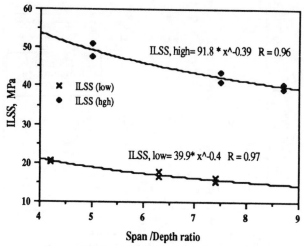

Figure 3: $\frac{Span}{thickness}$ ratio versus ILSS of PMR-15/C3k.

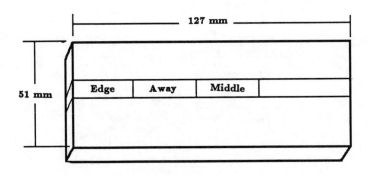

Figure 4: Cutting scheme for ILSS testing

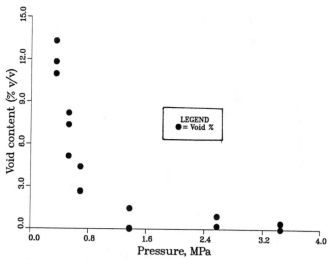

Figure 5: Void content and pressure magnitude variation.

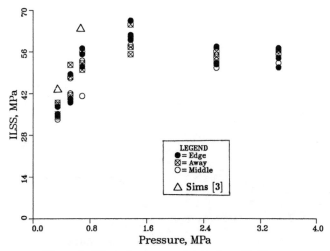

Figure 6: Effect of pressure magnitude on ILSS.

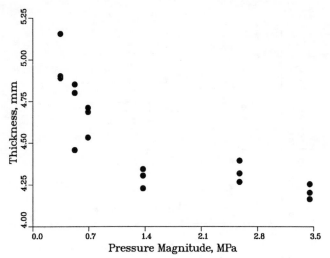

Figure 7: Part thickness (average) and pressure magnitude.

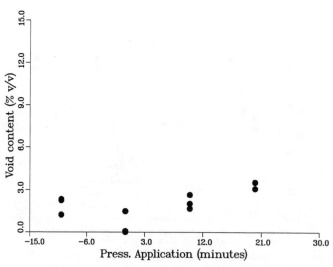

Figure 8: Void content and pressure application point.

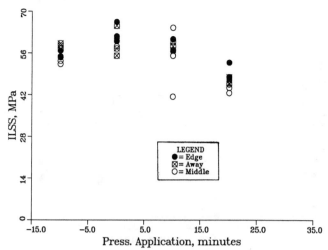

Figure 9: Effect of pressure application point on ILSS.

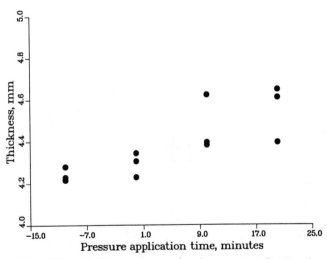

Figure 10: Part thickness (average) and pressure application time.

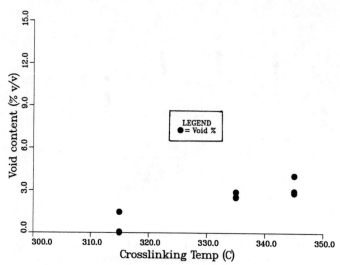

Figure 11: Void content and crosslinking temperature variation.

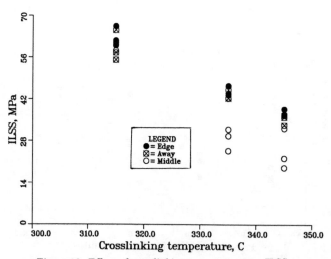

Figure 12: Effect of crosslinking temperature on ILSS.

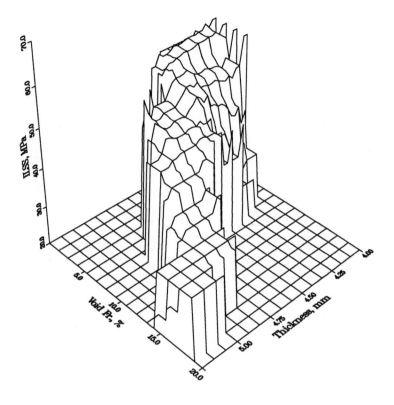

Figure 13: ILSS as a function of thickness and void content

THREE-DIMENSIONAL FINITE ELEMENT SIMULATION OF THERMOFORMING

H. F. Nied
GE Corporate Research and Development
Schenectady, New York

C. A. Taylor
GE Plastics
Pittsfield, Massachusetts

H. G. deLorenzi
GE Corporate Research and Development
Schenectady, New York

ABSTRACT

The development of a finite element program specifically designed to simulate thermoforming in complex three-dimensional geometries is described. The large strain finite element model developed for simulating thermoforming processes is based on a total Lagrangian formulation which results in a nonlinear system of equations that must be solved iteratively. The nonlinear material behavior and contact between the polymer and mold surfaces leads to additional complications in the numerical solution of the thermoforming simulation problem. In an effort to verify the the accuracy of the finite element model developed in this study, analyses are compared with measurements obtained from three-dimensional thermoformed parts.

INTRODUCTION

In basic thermoforming operations, a thermoplastic sheet is heated to a temperature well above its glass transition temperature and then quickly "inflated" into a mold cavity. The simplest of all thermoforming operations is the straight vacuum forming operation depicted in Figure 1. The vacuum forming operation shown schematically in this figure represents a quick and efficient technique for forming large thin-walled structures. However, parts formed using straight vacuum forming typically have considerable thickness variations throughout the part. A plastic part formed in the manner depicted in Figure 1, usually exhibits considerable thinning in the corners and bottom edges where the most severe drawing has occured (Figure 1c).

The objective of computer simulation for thermoforming processes is to provide accurate predictions of final part thicknesses. Without this fundamental information, it is not possible to rationally design molds and plan thermoforming processing stategies based on quantitative parameters. Once thickness distributions have been estimated using computer simulation of a given process, structural design considerations based on maximum deflection and stresses can be properly taken into account at an early phase in the engineering design process.

Recent advances in computer simulation of thermoforming have relied heavily on the finite element method [1-6]. The advantages of using a finite element formulation for analysis of thermoforming are twofold: first, the formulation is not restricted to any particular geometry and second, the highly nonlinear behavior typically associated with large strains and nonlinear polymer material behavior can be directly accommodated.

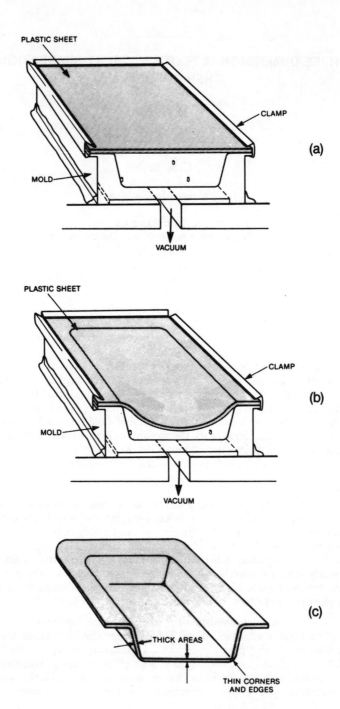

Figure 1. Schematic view of straight vacuum thermoforming.

258

In an effort to make the simulation model of thermoforming tractable, a number of assumptions have been made which greatly simplify the formulation of the problem while still incorporating the most significant physical aspects of polymer forming behavior. For example, the model should simulate the nonuniform thinning of the polymer which occurs during the free inflation stage of the forming process prior to contact with the mold. This is expecially critical if there is considerable inflation, i.e., large deformation, prior to contact with the mold surface. In addition, the model should incorporate the observation that, upon contact with the mold surface, adhesion takes place between the polymer and the mold surfaces, resulting in essentially no additional stretching of any plastic material that is in contact with the mold surface. Because most of the items that are thermoformed can be considered to be thin-walled shell structures, the hot polymer is modeled as a membrane. Thus, the bending resistance of the hot polymer is neglected and the material thickness is assumed to be small when compared with other dimensions of the structure. The mold surface in this finite element model is a rigid predefined boundary through which the plastic membrane is not permitted to penetrate. During the analysis, collision calculations are performed to determine whether contact has occurred between the plastic membrane and the rigid mold surface. When contact does occur, the membrane is permanently affixed to the mold surface at the point of contact after equilibrium has been reestablished. Because the material that contacts the mold surface is not permitted to move after contact occurs, the final thickness of the plastic in this location remains constant after contact. The consequence of this assumption is that the final thickness of the membrane is solely determined by the stretching which occurs during inflation and the sequence in which material elements come into contact with the mold surface.

The plastic itself is modeled as a "rubbery" (i.e., nonlinear, elastic, incompressible) material and therefore does not exhibit time-dependent behavior. What at first might seem to be a somewhat questionable assumption regarding the behavior of a thermoplastic material at elevated temperatures seems to be supported by strong experimental evidence [7-9]. In general, thermoplastics are considered to be viscoelastic materials, and a rigorous analysis valid for arbitrary strain rates and deformation history should incorporate a nonlinear viscoelastic constitutive relationship suitable for large biaxial deformations. However, the experimental evidence cited indicates that, at relatively high strain rates, the polymer's behavior is dominated by an elastic behavior which can be adequately modeled using constitutive relationships that have traditionally been used to model rubbery materials. Fortunately, since most industrial thermoforming processes of interest are usually conducted at relatively high strain rates in the temperature range where the polymer often exhibits this "rubberlike" behavior, the hyperelastic material models used to model rubber deformation seem to be quite reasonable. By default, such hyperelastic models appear to be the simplest reasonable constitutive relationships available for modeling thermoforming, based on our current knowledge of polymer behavior at typical thermoforming temperatures and strain rates.

FINITE ELEMENT FORMULATION

Modeling the inflation portion of the thermoforming process with the finite element method introduces many of the most difficult aspects of finite element formulation and simulation. These difficulties arise because of the large strains, nonlinear material behavior, contact between polymer and mold, and, in certain instances, membrane instability during inflation of the polymer. As would be expected for such a problem, the formulation results in a nonlinear system of equations that have to be solved in an iterative manner. The details of the finite element formulation are given in [5] and [6], and thus only a general overview of the formulation will be outlined in this paper.

Following the methodology outlined in [10], the equilibrium condition for the membrane can be derived by comparing the rate of work performed by the pressure on the membrane surface with the rate of increase in internal energy in the membrane

$$\int_a p \, \mathbf{n} \cdot \mathbf{v} \, da = \frac{\partial}{\partial t} \int_v \rho \varepsilon \, dv \tag{1}$$

where p is the pressure on the membrane, $\mathbf{n}$ is the normal to the membrane, $\mathbf{v}$ is its velocity, ρ is the material density, ε is the internal energy density, a is the current surface area of the membrane, and v its volume. The integration implied in (1) is performed element by element in a local element coordinate system which corresponds to the material coordinate system usually used in a total Lagrangian formulation [10]. This coordinate system (R,S), is fixed to the element and moves with it during the deformations. As is usually done in a finite element formulation, a polynomial interpolation function $\{N\}$ is chosen to express the unknown function inside the element in terms of the nodal values. In matrix notation this is expressed as

$$\{x\} = \{N\}\{x^e\} \tag{2}$$

where $\{x\}$ represents the coordinates of a point inside an element, and $\{x^e\}$ represents the nodal coordinates for a given element. The exact form of the interpolation function $\{N\}$ depends on the particular type of element used and the number of nodes. For example, the interpolation functions which define a linear interpolation between opposing element edges in a four noded quadrilateral element are given by

$$N_1 = 1/4\,(1 + R)(1 + S) \quad N_2 = 1/4\,(1 - R)(1 + S)$$
$$N_3 = 1/4\,(1 - R)(1 - S) \quad N_4 = 1/4\,(1 + R)(1 - S) \tag{3}$$

The equilibrium equation (1) can be rewritten in terms of quantities defined in the material coordinate system (R,S) and, after some manipulation, takes the form [4]

$$\int_A \rho\sqrt{\det\{g\}}\,\mathbf{n}\cdot\ddot{x}dRdS = \int_A 1/2H\sqrt{\det\{G\}}\,T_{IJ}\frac{\partial C_{IJ}}{\partial t}\,dRdS \tag{4}$$

where index notation has been used for repeated indices, T_{IJ} are the components of the 2nd Piola-Kirchhoff stress tensor, C_{IJ} are the components of Green's deformation tensor, H is the initial polymer thickness, $\{G\}$ and $\{g\}$ are the metric tensors, and p is the pressure acting on the surface of the membrane. It is important to note that all integrations in (4) are performed over the undeformed surface area A. In matrix notation, (4) may be reexpressed as

$$\{K\}\{x\} = \{F\} \tag{5}$$

where the unknown $\{x\}$ is the vector of all nodal coordinates in the model. The global stiffness matrix $\{K\}$ and the global load vector $\{F\}$ are assembled from the individual element stiffness matrices and load vectors as outlined in [11]. The terms which comprise the element stiffness matrix and the element load vector are given in detail in [4] to [6]. The components of stress in (4), T_{IJ}, are functions of the deformation gradients and are determined from a hyperelastic constitutive relationship.

The equilibrium condition given by (5), symbolically represents a set of nonlinear equations in $\{x\}$. By assuming that an equilibrium configuration is known for some pressure p_0, this equation can be linearized by expanding the left-hand side in a Taylor series about this position. This leads to the following equation for the determination of the increment $\{\Delta x\}$ in the nodal coordinates

$$\{K\}_0\{x\}_0 + \left[\{K\} + \left\{\frac{\partial K}{\partial x}\right\}\{x\}\right]_o \{\Delta x\} = \{F\} \tag{6}$$

The first product in (6) is commonly called the residual load vector, $\{R\}$, while the premultiplier for the unknown vector $\{\Delta x\}$ is usually called the tangent stiffness matrix K_t. Symbolically, the linearized system of equations represented by (6) can be written

$$\{K_t\}\{\Delta x\} = \{F\} - \{R\} \tag{7}$$

260

The detailed expressions for the element tangent stiffness matrix and the residual load vector are given in [4] to [6].

The unknown deformation increment can be determined from (7) for a small increase in pressure when an equilibrium configuration is known. Because the initial state without any loads is an equilibrium state, the analysis is started from the zero load condition. During each small load increment the deformations often change enough for the tangent stiffness matrix to change significantly, and a full Newton iteration scheme is thus used for each pressure increment. However, even though the pressure remains constant, the load vector still has to be recalculated during each iteration because the load includes an area integration and the shape and orientation of the membrane continuously changes.

While the membrane is inflating, collision with the mold surface must be continuously checked for all active nodes. Once contact has been detected, the nodes that have contacted are permanently fixed to the mold surface at the contact points, and equilibrium is reestablished. The degrees of freedom for the nodes that have contacted the mold are eliminated from the remaining portion of the analysis. Eliminating these degrees of freedom results in an increasingly smaller set of equations that have to be solved as the calculation proceeds and greatly speeds up the computation as more and more of the membrane contacts the mold.

MATERIAL BEHAVIOR

The assumption of nonlinear elastic behavior greatly simplifies the formulation of the problem, because the mechanics of nonlinear finite strain elasticity is a relatively well understood subject. Extensive classical treatises on this subject can be found in [12-15]. Two hyperelastic constitutive relationships have been successfully used to produce the results presented in this paper. They are the generalized Mooney-Rivlin model and Ogden models. Both of these constitutive relationships were originally developed for ideally elastic solids, such as rubber, and are often called hyperelastic models. Neither model is intrinsically better than the other for modeling purposes, but we have found the more recently developed Ogden model easier to curve fit to actual experimental data. In this paper, only expressions pertaining to the Ogden model will be explicitly examined. The formulation using a Mooney-Rivlin model results in similar expressions, derived in essentially the same manner.

For a hyperelastic material, i.e., an ideally elastic material that possesses a stress potential, the components of the 2nd Piola-Kirchhoff stress tensor $\{T\}$ can be expressed as the derivative of the strain energy function W. This expression is given by

$$T_{IJ} = \frac{\partial W}{\partial C_{IJ}} + \frac{\partial W}{\partial C_{JI}} = 2\frac{\partial W}{\partial C_{IJ}}, \ (I,J = 1,3) \tag{8}$$

where the last equality follows if W is written as a symmetric function in the components of Green's deformation tensor C_{IJ}. Materials that undergo little change in volume, such as thermoformed plastics, can usually be treated as incompressible. In terms of the principal stretches, this incompressibility condition is given by

$$\lambda_1 \lambda_2 \lambda_3 = 1 \tag{9}$$

The Ogden model is based on a strain energy function W which is expressed as a function of the principal stretches. In the Ogden model, the strain energy is written as an expansion in the principal stretches, $\lambda_1, \lambda_2, \lambda_3$, and has the form [16]

$$W = \sum_{n=1}^{r} \frac{\mu_n}{\alpha_n} \left[\lambda_1^{\alpha_n} + \lambda_2^{\alpha_n} + \lambda_3^{\alpha_n} - 3 \right] \tag{10}$$

where μ_n and α_n are experimentally determined constants. The constants, μ_n and α_n can be noninteger and negative, with the only restriction being that the total summation in (10) result in a positive strain energy function. The number of terms r in (10) is arbitrary, but most of the constants determined for various rubbers have been given for a 3-term expansion, i.e., $r = 3$.

The 2nd Piola-Kirchhoff stress tensor components for an incompressible material, using the Ogden model, are determined from (8) to (10). The resulting relationship between stress and stretch for a membrane subjected to in-plane stretching is given by

$$T_{11} = \sum_{n=1}^{r} \frac{\mu_n}{2} \left[\lambda_1^{\alpha_n - 2} (1 + D) + \lambda_2^{\alpha_n - 2} (1 - D) - 2C_{22} \lambda_3^{\alpha_n + 2} \right] \tag{11}$$

$$T_{22} = \sum_{n=1}^{r} \frac{\mu_n}{2} \left[\lambda_1^{\alpha_n - 2} (1 - D) + \lambda_2^{\alpha_n - 2} (1 + D) - 2C_{11} \lambda_3^{\alpha_n + 2} \right] \tag{12}$$

$$T_{12} = \sum_{n=1}^{r} \frac{\mu_n}{2} C_{21} \left[\frac{\left(\lambda_1^{\alpha_n - 2} - \lambda_2^{\alpha_n - 2} \right)}{B} + 2\lambda_3^{\alpha_n + 2} \right] \tag{13}$$

where

$$D = \left(C_{11} - C_{22} \right) / 2B \tag{14}$$

$$\lambda_3 = \left(C_{33} \right)^{1/2} = \left(C_{11}C_{22} - C_{12}C_{21} \right)^{-1/2} \tag{15}$$

$$\lambda_{1,2} = (A \pm B)^{1/2} \tag{16}$$

$$A = \frac{C_{11} + C_{22}}{2} \tag{17}$$

$$B = \left[1/4 \left(C_{11} - C_{22} \right)^2 + C_{12}C_{21} \right]^{1/2} \tag{18}$$

The expression for the stress components in terms of the Ogden constants simplify considerably for the special case when shear deformations are absent. For example, for equibiaxial stretching, i.e., $\lambda_1 = \lambda_2 = \lambda$, the Cauchy stress components are given by

$$\sigma_{11} = \sigma_{22} = \lambda^2 T_{11} = \lambda^2 T_{22} = \sum_{n=1}^{r} \mu_n \left[\lambda^{\alpha_n} - \lambda^{-2\alpha_n} \right] \tag{19}$$

It is highly desirable that the experimental measurements used to obtain the Ogden constants simulate the actual deformation modes observed in thermoforming. This means that extensional measurements should be performed on the polymers of interest at elevated temperatures. Unfortunately, extensional measurements are not easy to perform on plastics at high temperatures, especially for biaxial deformations. The use of bubble and tube inflation to obtain biaxial material measurements seems to the most promising approach, but temperature and strain rate control are very difficult to maintain during such an inflation experiment. In addition, it is not a simple task to measure stretch on the hot membrane surface during inflation. In [9] biaxial data is reported for polystyrene at elevated temperatures and relatively low strain rates. The data in [9] was obtained using a special extensometer to measure equibiaxial stretch at the pole of the inflated membrane. At present, the type of biaxial experiments needed to adequately characterize the behavior of hot polymers for thermoforming are not routinely performed over a wide range of temperatures and strain rates. This contrasts with uniaxial extension tests that can be more or less routinely performed on thermoplastics at elevated temperatures. By photographing the specimen while it is being stretched axially, it is possible to generate true stress vs. stretch curves for polymers at elevated temperatures [17]. Until more complete information is available on the large biaxial deformation behavior of polymers at high temperatures, constitutive relationships will have to be based solely on uniaxial data. An example of this is given in Figure 2, where true stress vs. stretch data is shown for modified Polyphenylene Oxide (PPO), NORYL®, at a single strain rate. The curves in this figure were generated from a nonlinear least squares curve fit of the experimental data for a single term Ogden model, i.e., $r = 1$. In this particular figure, it can be seen that a single term Ogden model can represent the data for simple extension reasonably well over a large strain range.

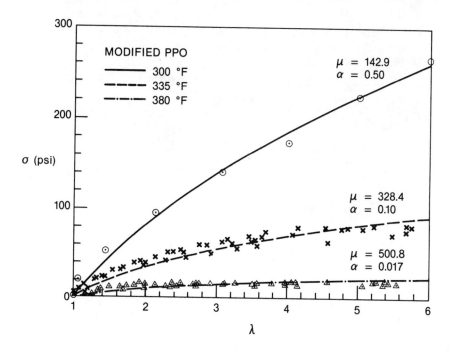

Figure 2. Stress vs. stretch data for PPO. Curve fit with single term Ogden expansion. (Displacement rate = 1.0 in./sec).

It is noteworthy to point out that in all of the isothermal finite element simulations of thermoforming that we have performed to date, different material properties only seem to strongly affect the pressure and stability characteristics of the membrane during inflation. As long as isothermal conditions are maintained, different material properties do not seem to have a strong influence on the inflated shape (except at very large stretches) and the final part thickness. This observation is not true for calculations that have been performed taking temperature gradients into account. In nonisothermal simulations, the variable "stiffness" of the nonisothermal polymer can cause dramatic thickness differences between nonisothermal and isothermal calculations.

THREE-DIMENSIONAL EXAMPLE – THERMOFORMING OF A RECTANGULAR BOX WITH AN UNDERCUT

Figure 3 shows the final thermoformed shape of a rectangular box with an undercut. The undercut in this particular geometry makes thermoforming simulation of this part particularly challenging, because elements that end up on the top surface of the undercut must rotate 180 ° in space. The box shape was formed on a single station shuttle thermoforming machine from sheets measuring 56.5 cm × 36.8 cm, with an initial thickness of 3.2 mm. The final rectangular box

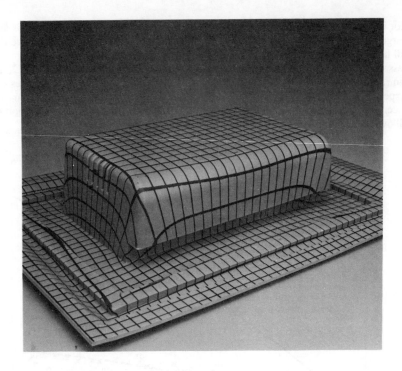

Figure 3. Thermoformed box with an undercut.

shown has a length of 24.6 cm, a width at the base of 13.75 cm, and is 7.6 cm deep. The silk-screened grid provides a good representation of the stretching that the plastic was subjected to during thermoforming. The grid consisted of 1 cm × 1 cm squares on the initial thermoformed sheet. As can be seen, considerable stretching occurs along the part edges and in the corners.

Processing simulation of the undercut box was performed only for the portion of plastic from which the part was formed, i.e., the entire sheet shown in Figure 3 was not modeled. Though this simplifies the processing simulation, it does result in some minor error in predicting the thickness of the final part. The reason for this is that when the sheet is heated to its forming temperature, it sags approximately 5 to 7 cm along the sheet centerline. This results in some initial nonuniform thinning even before vacuum is applied and, more importantly, permits a greater volume of material to be "pulled" into the mold cavity, because of the curved initial shape of the plastic sheet.

The finite element mesh used to simulate the processing of the plastic part shown in Figure 3 is depicted in Figure 4. The actual analysis took advantage of symmetry conditions and only required one-half of the mesh shown in Figure 4, i.e., calculations were performed using 3096 elements. As can be seen, the final mesh is highly refined in regions that will end up along edges and in corners. Since the final coordinates of the nodes in the deformed state are not known a priori, the refined mesh shown was obtained iteratively following successive analyses. For the mesh refinement shown, five successive calculations and mesh refinements were performed. It should also be noted that the entire mesh consists of 3-noded triangular elements. The use of linear elements, as opposed to higher-order elements, was found to greatly improve the stability of the solution, i.e., elements did not fold in upon themselves. In addition, the use of linear elements simplifies calculations during collision checks made with the mold surface, because the elements themselves do not become curved surfaces.

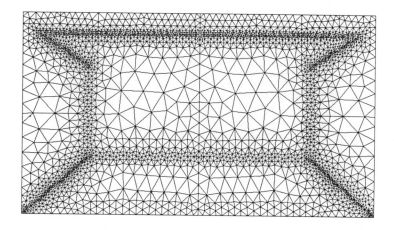

Figure 4. Refined mesh for simulating thermoforming of the box with an undercut.

Figure 5 shows three successive stages in the finite element simulation of thermoforming the undercut box. In Figure 5a, the membrane has inflated to a point where it is just contacting the top and vertical sides of the mold cavity. Figure 5b depicts a point in the inflation sequence when the membrane has almost completely filled the main cavity of the mold and is now "wrapping" around the undercut at the base of the mold. Finally, in Figure 5c, the plastic membrane has "filled" the mold cavity with all the nodes now contacting the mold surface. Careful examination of the final figure in the inflation sequence shows that even this mesh is not refined enough to completely fill all the sharp edges and corners of the mold. Indeed, a certain jaggedness is observable along the edges as elements get locked into position when the element's nodes contact intersecting mold surfaces. The solutions shown in Figure 5 were performed on a VAX 8800 and required approximately four cpu hours to run to completion. Contours of constant thickness, calculated for the undercut box are shown in Figure 6a. In the calculations, a two-term Ogden constitutive relationship was used, i.e., $r = 2$, with $\mu_1 = 142.9$ psi (0.985 MPa), $\alpha_1 = 0.50$, $\mu_2 = 1.4$ psi (9.65 kPa), $\alpha_2 = 4.0$. It was felt that these constants for the Ogden model were fairly representative of modified PPO at 300 °F (149 °C) (Figure 2). The second term was added to this calculation in an effort to stabilize the free inflation behavior of the membrane at large stretches, because it was found that without additional stiffening, numerical results became exceeding difficult, if not impossible, to obtain at large stretches. A brief review of instabilities that arise during free inflation and the nonuniqueness of the final deformed state of inflated membranes is given in [6]. Much work remains to be done in understanding the mechanics of membrane inflation and devising numerical algorithms capable of computing final deformed states following unstable behavior. The material constants used in the analysis shown in Figure 5 forced the solution to remain well behaved during process simulation. However, the actual thermoforming temperature of modified PPO can be on the order of 450 °F (232 °C) and thus the material constants used in the calculation were overly stiff. This results in an overprediction for the forming pressure, but as long as the process being modeled is isothermal, such a modification of the constitutive behavior does not seem to affect the final thickness distribution.

In an effort to compare the thicknesses predicted from the finite element simulation of the undercut box with actual thermoformed parts, parts such as the one shown in Figure 3 were vacuum formed out of ABS (CYCOLAC™), Modified PPO (NORYL®), PC-PBT blend (XENOY®), and PEI (ULTEM®). A variety of forming temperatures were used for the different plastics, ranging in magnitude from 400 °F (204 °C) for ABS to 600 °F (315 °C) for PEI. In all cases an effort was made to maintain isothermal conditions on the plastic sheet prior to ther-

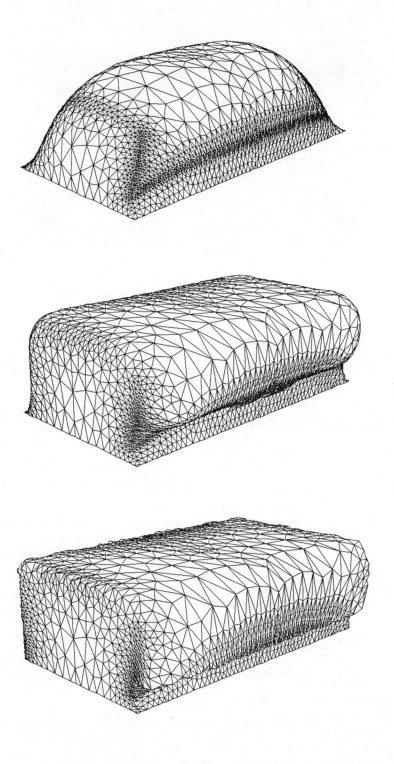

**Figure 5. Three successive stages in the thermoforming of the box with an under-
cut.**

266

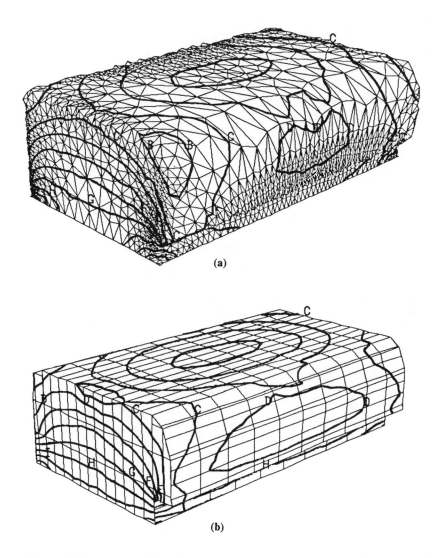

(a)

(b)

Figure 6. Comparison of predicted and measured normalized thickness distributions t/t_0. (a) results from finite element simulation; (b) Measured thickness distribution for part formed from PC-PBT alloy. (A=0.0, B=0.1, C=0.2, D=0.3, E=0.4, F=0.5, G=0.6, H=0.7, I=0.8, J=0.9, K=1.0).

moforming. Thicknesses were systematically measured on the formed parts at grid locations shown in Figure 6b. In Figure 6b a typical contour plot is shown depicting measured thicknesses, in this case for a part formed out of the PC-PBT alloy. As can be seen, the general distribution of thickness in the computed solution (Figure 6a) is very similar to the measured results (Figure 6b).

A more detailed examination of the measured thickness is shown in Figures 7a, 7b, and 7c at different slices through the undercut box. In Figure 7a the normalized thickness as a function of arc length along the plane of symmetry is shown for all four of the materials tested. Arc length S is measured from the part edge in the direction of the arrow shown in the figure inset. Figure 7b shows the thickness along the edge of the box, and Figure 7c contains thickness vs. arc length

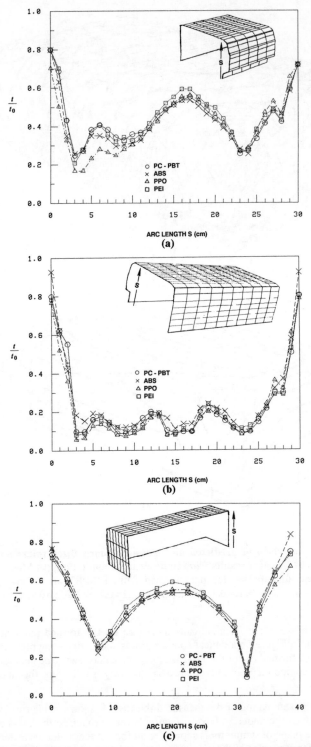

Figure 7. Measured thicknesses for four different polymers at three different cross sections.

268

measurements along the long axis of the box. The nonsymmetry of measured thicknesses in Figure 7c is due to the fact that the actual part had corrugations approximately 6 cm in arc length from the left edge of the box (see Figure 3). The corrugations, which were not modeled in the finite element calculation, resulted in a substantial thinning of the plastic in this location. The thinnest measured points in this part (excluding the corrugated zone), occur along the edge shown in Figure 7b. In the undercut and in the corners, the ratio of final thickness to initial thickness t/t_0 gets as small as 0.06, or almost 17 times thinner than the original sheet thickness. Considering that the parts were formed using different plastics at different temperatures and thus with differing degrees of initial sag prior to forming, it is remarkable that all the thickness measurements are virtually identical. The biggest difference in measured thickness among the four plastics is between PPO and the other three plastics along the centerline cut shown in Figure 7a, at a location approximately 6 cm in arc length from the front edge of the box. This point is located between the undercut and the top of the box. It is not clear why PPO was thinner than the other materials at this location, expecially since all the other PPO thickness measurements were virtually identical to the other plastics at all other points. As a result of having slightly less thickness in the region around the undercut, the parts formed out of PPO contained a smaller volume of material than the parts formed using the other plastics. It is reasonable to assume that the difference in thickness at this location and the overall difference in volume can be attributed to a difference in sag between PPO and the other plastics. That is, PPO probably had a smaller amount of sag just prior to the application of vacuum and thus the final part was formed utilizing a smaller cross-sectional area on the plastic sheet. In the long axis direction (Figure 7c), all four plastics had virtually identical thicknesses.

Comparison of the computed nondimensional thicknesses (t/t_0) with the measured thicknesses for ABS along the same slices shown in Figures 7a and 7c are given in Figures 8a and 8b. The overall behavior of the thickness prediction closely follows the experimental results. However, in both plots, the finite element prediction underpredicts the thickness of the actual part. It is obvious that the finite element calculations result in a part with a smaller predicted volume of material than the actual part. It is thought that the additional amount of material in the experimental results can be attributed to the fact that the plastic sheet sagged during heating and thus was curved prior to the application of vacuum. This meant that more plastic was introduced into the mold than if the sheet remained flat prior to forming, as in the idealized model. In addition, not all details of the actual mold were modeled in the finite element calculation. Certain small steps in the mold and the corrugated region on the side of the mold were omitted from the analytic model. The "jaggedness" in the finite element thickness predictions also demonstrates that the level of mesh refinement, especially in the corners, is not sufficient for "smooth" thickness plots. In spite of these differences, the finite element predictions and the measured thicknesses are in remarkable agreement, considering the complexity of the molding process and the large amount of deformation to which the plastic has been subjected.

CONCLUSIONS

Finite element simulation of the thermoforming process is a relatively new approach for predicting thicknesses in thermoformed parts. As shown in this paper, the method seems to yield excellent results in predicting final part thicknesses for complex 3-D moldings. It is somewhat surprising that the predictions are so close to actual measurements, since a relatively large number of simplifying assumptions were used to develop a model for an admittedly complex forming process. It appears that finite element simulation of thermoforming has the potential to have a major impact on the way new parts and processes are designed and could lead to drastic savings in material costs through design optimization. Much work, however, remains to be done in improving the analytic model. The current isothermal analysis should be extended to include nonisothermal process simulation. It is well known in thermoforming that temperature gradients can have a strong influence on final part thickness and this fact is often taken advantage of by using zone heating to assist in the fabrication of parts that cannot be formed isothermally. Finally, much more work remains to be done in measuring the biaxial stretching behavior of polymers at elevated temperatures and different strain rates. With a better understanding of how polymers

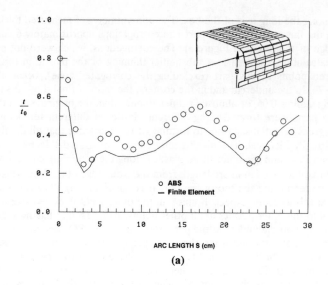

(a)

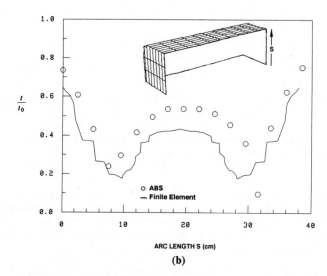

(b)

Figure 8. Comparison of predicted thicknesses with measured thicknesses in ABS undercut box at two different cross sections.

deform under biaxial stretching at elevated temperatures, more sophisticated constitutive models can be constructed. It is anticipated that these models will permit even more precise simulation of thermoforming under a wide variety of processing conditions.

ACKNOWLEDGMENT

The authors gratefully acknowledge the support provided by GE Plastic Business, Pittsfield, MA. Special thanks are due to Mr. L.P. Inzinna and Mr. K.R. Conway at GE Corporate Research and Development for performing numerous experimental measurements on polymers at high temperatures and measuring the thicknesses in the thermoformed parts.

REFERENCES

[1] Allard, R., Charrier, J.-M., Ghosh, A., Marangou, M., Ryan, M.E., Shrivastava, S. and Wu, R., "An Engineering Study of the Thermoforming Process: Experimental and Theoretical Considerations," *Journal of Polymer Engineering*, Vol. 6, Nos. 1-4, pp. 363-394, 1986.

[2] Charrier,M.-M., Shrivastava,S., Wu,R., "Free and Constrained Inflation of Elastic Membranes in Relation to Thermoforming − Axisymmetric Problems," *Jour. Strain Anal.*, Vol. 22, No. 2, pp. 115-125, 1987.

[3] Nied, H.F. and deLorenzi, H.G., "Finite Element Simulation of Thermoforming and Blow Molding," *Proceedings of the Society of Plastics Engineers 45th Annual Technical Conference*, pp. 418-420, 1987.

[4] deLorenzi, H.G. and Nied, H.F.,"Blow Molding and Thermoforming of Plastics: Finite Element Modeling," *Computers & Structures*, Vol. 26, No. 1/2, pp. 197-206, 1987.

[5] deLorenzi, H.G., Nied, H.F., and Taylor, C.A., "A Numerical/Experimental Approach to Software Development for Thermoforming Simulations," Computational Experiments, ASME PVP - Vol. 176, *Proceedings of the ASME Pressure Vessels and Piping Conference*, Honolulu, Hawaii, 1989.

[6] deLorenzi, H.G. and Nied, H.F., "Finite Element Simulation of Thermoforming and Blow Molding," to be published in *Progress in Polymer Processing*, A.I. Isayev, ed., Hanser-Verlag

[7] Schmidt, L.R. and Carley, J.F., "Biaxial Stretching of Heat-Softened Plastic Sheets Using an Inflation Technique," *Int. J. Eng. Sci.* Vol. 13, pp. 563-578, 1975.

[8] Schmidt, L.R. and Carley, J. F., "Biaxial Stretching of Heat-Softened Plastic Sheets: Experiments and Results," *Polym. Eng. Sci.*, Vol. 15, No.1, p. 51, 1975.

[9] DeVries, A.J., Bonnebat, C. and Beautemps, J., "Uni- and Biaxial Orientation of Polymer Films and Sheets," *Journal of Polymer Science: Polymer Symposium 58*, p. 109-156, 1977.

[10] Oden, J.T., *Finite Elements of Nonlinear Continua*, McGraw-Hill, New York, 1972.

[11] Zienkiewicz, O.C., *The Finite Element Method*, McGraw-Hill, London, 1977.

[12] Truesdell, C. and Toupin, R., *The Classical Field Theories*. In Handbuch der Physik (Edited by S. Flugge), Springer, Berlin, 1960.

[13] Eringen, A.C., *Nonlinear Theory of Continuous Media*, McGraw-Hill, New York, 1962.

[14] Green, A.E. and Adkins, J.E., *Large Elastic Deformations*, Oxford University Press, 1960.

[15] Green, A.E. and Zerna, W., *Theoretical Elasticity*, Oxford Press, 1954.

[16] Ogden, R. W., "Large Deformation Isotropic Elasticity - On the Correlation of Theory and Experiment for Incompressible Rubberlike Solids," *Proc. R. Soc. Lond. A.*, *326*, pp. 565-584, 1972.

[17] Nied, H. F., Stokes, V. K. and Ysseldyke, D. A., "High-Temperature Large-Strain Behavior of Polycarbonate, Polyetherimide and Poly(Butylene Terephthalate)," *Polym. Eng. Sci.*, Vol. 27, No. 1, p. 101-107, January 1987.

A MODEL FOR COMPOSITE LAMINATE BENDING UNDER ARBITRARY CURVATURE DISTRIBUTION AND EXTENSIVE INTERLAMINAR SLIDING

M. F. Talbot and A. K. Miller
Department of Materials Science and Engineering
Stanford University
Stanford, California

ABSTRACT

While thermoplastic matrix composites are of increasing interest for a variety of applications, methods of forming these materials into useful parts are still under development. One way of hastening their development and utilization is to develop models which predict the material response to processing. One such model is under development and will predict the response of a consolidated laminate which is being formed by bending to a singly-curved shape. The model accounts for the unique features of thermoplastic composites, namely the presence of continuous inextensible fibers and the need for extensive relative ply sliding during processing. At this stage in model development, it treats a laminate as a series of alternating layers of elastically hard and elastically soft material. The hard layers can slide over each other without length changes because of the soft shearable layers between them. In its present stage of development, the model predicts, for any arbitrarily complex singly-curved shape, the extent of relative ply sliding, the stresses generated, and the array of point forces that are required to produce the specified shape.

A single dimensionless parameter, the Interply Sliding Parameter or ISP, predicts the extent of relative ply sliding in these simulated laminates. The ISP depends on the relative moduli of the two materials and on the geometry. The behavior of a beam with extensive ply sliding is very different from that described by classical beam theory. First, the ply normal stresses depend not only on the local curvature and distance from the neutral axis, but also on the position along the laminate. Second, the relative ply sliding and the bending stresses occur not only in bent sections of the laminate, but also in nearby straight sections. Thus, the response of a given element of material depends not only on the local conditions, but also on the curvature distribution throughout the laminate.

NOMENCLATURE

E	elastic modulus of hard material, Pa
G	shear modulus of soft material, Pa
L	length of a segment, m
M	total number of segments in laminate
N	number of plies in laminate
q	displacement caused by relative sliding of plies, m
R^{inv}	reciprocal of radius of curvature of a segment, 1/m
t^{mat}	thickness of the matrix interlayers

t^{ply}	thickness of layers of hard material
γ	shear strain in soft layers
ε	bending strain in hard layers
Θ	angle through which a segment is bent, rad
σ	in-plane normal stress in hard layers, Pa
τ	shear stress in hard and soft layers, Pa

Subscripts
m	segment number or segment boundary number
n	ply number

INTRODUCTION

Thermoplastic polymer matrices are a major development in composite materials of the last decade or so. Thermoplastics are not cured, as are thermosetting polymers, but can be softened by heating, reshaped, and resolidified as often as desired. This feature gives thermoplastic composites potential reductions in manufacturing cost. Other advantages of thermoplastics include higher toughness, better property retention at high temperatures, better damage tolerance, and better solvent resistance than most thermosets and no need to refrigerate the prepreg to prevent premature curing. Accordingly, the interest in thermoplastic matrices for composite materials has led to a great deal of research.

However, thermoplastic composites will not realize their full potential unless cost-effective forming methods which take advantage of the matrices' thermoplasticity are developed. A number of such processes are under study, including thermoforming, or post-forming of consolidated laminates [1; 2], and die-less forming [3]. The development of these manufacturing methods will be greatly assisted, and their usefulness enhanced, by models which can accurately predict the material response during processing.

Such models must describe a very complicated material. The presence of continuous and virtually inextensible fibers causes composites to behave very differently from homogeneous materials. If a metal sheet, for example, is bent to some shape, the material on the outside of the bend will deform plastically in tension and that on the inside in compression. The metal far from the bent region will not be affected. With a fiber composite, however, such bending can cause problems. The fibers cannot stretch in tension and will buckle in compression. To avoid buckling, the plies on the inside of the bend must slide outward relative to those on the outside. This sliding must take place over large portions of the length of the composite, so that, in contrast to metals, the material far from a bent region will be significantly affected. Figure 1 illustrates this necessity. A process model for metals need not allow for such relative ply sliding, but one for thermoplastic composites must. Whether ply sliding will take place and, if so, to what extent, must be predicted for any possible forming operation and processing conditions. In addition, the model must predict the stresses that forming will generate, whether the shape formed will deviate from that desired, whether buckling or deconsolidation will occur during forming, what stresses will remain in the laminate after the forming loads are removed, and how much shape recovery those stresses will cause.

Gutowski et al [4; 5] have developed a model which predicts the stresses in a laminate being formed, the extent to which the composite plies slide over each other during forming, and whether buckling will occur. This model can describe thermoforming, a process in which the desired shape is imposed on the composite. More complex manufacturing methods, such as die-less forming and others, will however require a more extensive model. In die-less forming, a shape is imposed on one local section of the laminate in each forming operation while nearby material will be unconstrained. The method involves rather complicated loading and temperature histories, which other models cannot treat. Accordingly, a model is under development which will describe the effects of complicated histories and simultaneous curvature and force boundary conditions on a thermoplastic composite laminate during and after forming. This model should be able to predict the shape recovery of a die-formed laminate after it is removed from the mold and under no loads.

At its present stage of development, our model treats the behavior of laminates containing alternating layers of elastically hard and elastically soft material. Such material types cannot, of course, be used to simulate the behavior of history-dependent thermoplastic composites during forming, but have been used solely for ease of model development. However,

274

some of the behaviors which this version of the model has predicted are relevant to composite forming and interesting in their own right. Future versions of the model will predict residual stresses and shape recovery using history-dependent visco-plastic material properties.

MODEL DEVELOPMENT

Representation of Laminate

It is assumed that the composite to be formed consists of alternating layers of two types: hard, stiff layers representing the fiber-containing regions and soft layers representing the thin matrix material between these. A hard layer and the soft layer on top of it constitute a "ply." The total number of plies is N. The laminate is divided into M segments along its length, each of which has its own length and radius of curvature, so that, with a sufficient number of segments, an arbitrarily complex curvature distribution can be imposed. A segment of a ply is called an "element." A node is placed at each end of each element, in the middle of the hard layer. This representation is shown in Figure 2. An element is identified with two subscripts, the segment number m followed by the ply number n. A node has the same subscripts as the element to its left, and the nodes on the left edge of the laminate have zero as their first subscripts.

In this model, the laminate is assumed to be in a state of plane strain in the plane shown inFigure 2. Normal stresses parallel to the laminate are calculated only for the hard layers. The shear stress in this plane is calculated for the soft layers, and the shear stress in a hard layer is assumed equal to the average of the shear stresses in the soft layers above and beneath it. Other stress components are assumed to be irrelevant. Thus, σ_{12} is the normal stress in the hard layer of element 1 of ply 2 and τ_{54} is the shear stress in the soft layer of element 5 of ply 4. The shear stress in the hard layer of element 54 is the average of τ_{54} and τ_{44}. The neglect of normal stresses in the matrix material is justified by the very high tensile stiffness of the graphite or Kevlar fibers used in many composites. Even if the normal stress in the fibers is very high, their strain will be at most a few per cent. As the matrix is very soft, a normal strain of a few per cent will give rise to a negligible stress.

Effect of Relative Ply Sliding

When a laminate is bent, the hard layers can slide over each other because of the soft shearable layers between them. Therefore, the usual strength-of-materials analysis for bending, which assumes that plane sections which are initially perpendicular to the laminate surface remain plane and nearly perpendicular, does not apply. Plane sections in fact become stepped and slanted, as shown in Figure 3. The displacement of node mn from the position it would occupy in a homogeneous beam, as shown in the figure, is called the sliding displacement, q_{mn}. q_{mn} is the primary internal variable in the analysis.

Governing Equations

Parallel Force Equilibrium. The central equation used in the model states that the forces at each node must balance in the direction parallel to the laminate at that node. In this direction, individual elements of the laminate are subjected to both normal and shear stresses from the adjacent material in the same ply and shear stresses from the material above and below; all of these stresses must be included in the force balance. If we isolate node mn and half the element on each side thereof, and require force equilibrium parallel to the laminate, then, assuming unit thickness:

$$
\begin{aligned}
0 = &\ \tau_{mn} \tfrac{1}{2} L_m \cos\left(\tfrac{1}{2}\Theta_m\right) + \tau_{mn}\left(t^{mat} + \tfrac{1}{2}t^{ply}\right)\sin\Theta_m \\
&+ \tau_{m,n-1}\tfrac{1}{2}t^{ply}\sin\Theta_m - \tau_{m,n-1}\tfrac{1}{2}L_m\cos\left(\tfrac{1}{2}\Theta_m\right) \\
&+ \tau_{m+1,n}\tfrac{1}{2}L_{m+1}\cos\left(\tfrac{1}{2}\Theta_{m+1}\right) + \tau_{m+1,n}\left(t^{mat} + \tfrac{1}{2}t^{ply}\right)\sin\Theta_{m+1} \\
&+ \tau_{m+1,n-1}\tfrac{1}{2}t^{ply}\sin\Theta_{m+1} - \tau_{m+1,n-1}\tfrac{1}{2}L_{m+1}\cos\left(\tfrac{1}{2}\Theta_{m+1}\right)
\end{aligned}
$$

$$+ \sigma_{m+1,n} t^{ply} \cos(\Theta_{m+1}) - \sigma_{mn} t^{ply} \cos(\Theta_m) \tag{1}$$

where:

t^{ply} = the thickness of the hard fiber layers

t^{mat} = the thickness of the soft matrix layers

Θ_m = the angle between the line perpendicular to the laminate at node mn and the perpendicular at the middle of segment m

Θ_{m+1} = the angle between the perpendicular at node mn and that at the middle of segment m+1

L_m = the length of segment m

L_{m+1} = the length of segment m+1

as shown in Figure 4. The components of all stresses parallel to the laminate at the location of the node have been taken, and it is assumed that the forming loads are applied in the perpendicular direction at the nodes and have no components in the parallel direction. t^{ply} may be seen in this figure to be the area over which the normal stresses in the adjacent material act, and half of L_m is similarly the area over which the shear stresses exerted by the elements above and below the node in question act. The shear stresses exerted by the material on either side of the laminate act over the thickness of the element. τ_{mn} and $\tau_{m+1,n}$ act over an area of t^{mat}, while the average of $\tau_{m,n-1}$ and τ_{mn} and the average of $\tau_{m+1,n-1}$ and $\tau_{m+1,n}$ act on t^{ply}.

In order to solve this equation for each node in the laminate, the stresses are expressed in terms of the various q_{mn}'s. First, the stresses are related to strains and then the strains are related to the q_{mn}'s.

Stress-strain Relations. At this stage in model development, stress and strain are related by an elastic constitutive equation for each material:

$$\sigma_{mn} = E\varepsilon_{mn}$$
$$\tau_{mn} = G\gamma_{mn} \tag{2}$$

where:

E is the Young's modulus of the hard material

G is the shear modulus of the soft interlaminar material

ε_{mn} is the normal strain in the hard layer of element mn

γ_{mn} is the shear strain in the soft layer of element mn

For the plies in which the bent direction is the longitudinal fiber direction, the assumption of elastic behavior is rather good. The off-angle plies and the soft matrix interlayers, however, will behave in a much more complicated fashion. When the model is more fully developed, realistic time-dependent visco-elasto-plastic properties will be included for the matrix material and for off-angle plies.

Strain-q_{mn} Relations. The normal strain in the hard layers, ε_{mn}, has two parts. The first is the strain caused by bending without interply sliding, the strain that that element would suffer if the laminate were homogeneous and elastic. The classical bending strain is $\varepsilon = z/R$, where z is the distance from the neutral axis and R the local radius of curvature. For ease in computation, we use $R^{inv} = 1/R$, rather than R itself, in the model. z is evaluated at the middle of each hard layer, where the nodes are, giving:

$$\varepsilon_{mn}^{bend} = \left(t^{ply} + t^{mat} \right) \left(n - \frac{N+1}{2} \right) R_m^{inv} \qquad (3)$$

where:

n = ply number starting at one outer surface

R_m^{inv} = reciprocal of radius of curvature of segment m

The expression for the distance from the neutral axis assumes that the laminate contains an even number of plies, which is reasonable for most practical composite laminates.

The second part of the normal strain in the hard layers comes from the relative sliding of those layers and acts to alleviate the bending strain. Figure 5 shows the relationship between this strain term and the q_{mn}'s. The sliding strain is the change in length caused by sliding of the element divided by the initial length. Note that the "initial length" is the undeformed length of the element, not the bent length without sliding. These two lengths are not, of course, very different unless the laminate is quite thick compared to its radius of curvature. The change in length caused by sliding is $q_{mn} - q_{m-1,n}$ for element mn, as can be seen in Figure 5. q_{mn} is seen to be positive while $q_{m-1,n}$ is negative. Thus, the overall change in length for element mn is positive and the sliding counteracts the compressive strain which this element, on the inside of the bend, would otherwise suffer. The situation is opposite for element m, n+1 in the figure: the change in length caused by sliding is negative so that the bending tension is mitigated. The sliding strain may be expressed as:

$$\varepsilon_{mn}^{slide} = \frac{q_{mn} - q_{m-1,n}}{L_m} \qquad (4)$$

Combining Equation 3 and Equation 4:

$$\varepsilon_{mn} = \frac{q_{mn} - q_{m-1,n}}{L_m} + \left(t^{ply} + t^{mat} \right) \left(n - \frac{N+1}{2} \right) R_m^{inv} \qquad (5)$$

The shear strain in the soft interlayers is caused entirely by the relative sliding of the hard layers above and below, as shown in Figure 6. The shear strain at, for example, the right end of the soft layer in question is the angle between the deformed and undeformed edges. This angle is equal to $\dfrac{q_{m,n+1} - q_{mn}}{t^{mat}}$, while the corresponding angle at the left end is $\dfrac{q_{m-1,n+1} - q_{m-1,n}}{t^{mat}}$. The shear strain in the soft layer of element mn is the average of these terms:

$$\gamma = \frac{1}{2 t^{mat}} \left(q_{m-1,n+1} - q_{m-1,n} + q_{m,n+1} - q_{mn} \right) \qquad (6)$$

D. Boundary Conditions

Certain boundary conditions must be satisfied at the edges of the laminate being formed. A primary assumption is that there are no shear tractions on the upper and lower surfaces of the laminate. Also, we assume at this stage that the ends are traction-free. Therefore, the boundary conditions listed in Table 1 must hold.

Table 1 Boundary Conditions in the Forming Model

Edge	Designation	Condition
Upper	n = N	$\tau_{mN} = 0$
Lower	n = 0	$\tau_{m0} = 0$

| Left | $m = 0$ | $\sigma_{0n} = 0, \ \tau_{0n} = 0$ |
| Right | $m = M + 1$ | $\sigma_{M+1, n} = 0, \ \tau_{M+1, n} = 0$ |

Note that τ_{mn} is the shear stress in the soft layer at the top of element mn, and there are no soft layers in ply N or in the imaginary ply 0.

Solution of the Governing Equations. The expressions for strain, Equation 5 and Equation 6, are substituted into the equation for stress, and the resulting equations are in turn substituted into Equation 1, the nodal force balance. One such equation in terms of q_{mn}'s and $R^{inv}{}_m$'s is set up for each of the (M+1)N nodes in the laminate to give (M+1)N simultaneous equations. If the curvature distribution is specified, the equations can be solved for the (M+1)N unknown values of q_{mn}.

Calculation of Required Imposed Forces. In forming operations which use a mold or die, such as thermoforming or autoclaving, the model which simulates the imposition of a desired shape can adequately describe the material response. In die-less forming, however, forces or displacements will be imposed at discrete points along the laminate, or part of the laminate will be allowed to deform freely, i.e. subjected to a force of zero. Our model can calculate the point forces which must be imposed at the segment boundaries to give the desired curvature distribution. The calculation is a force balance around a segment boundary, or column of nodes, as shown in Figure 7. The equation for the force F_m located at the right edge of segment m and perpendicular to the laminate at that point is:

$$F_m = t^{ply}\left(\sum_{n=1}^{N}\sigma_{mn}\right)\sin(\Theta_m) + t^{ply}\left(\sum_{n=1}^{N}\sigma_{m+1,n}\right)\sin(\Theta_{m+1})$$

$$+ \left(t^{ply} + t^{mat}\right)\left(\sum_{n=1}^{N}\tau_{mn}\right)\cos(\Theta_m) - \left(t^{ply} + t^{mat}\right)\left(\sum_{n=1}^{N}\tau_{m+1,n}\right)\cos(\Theta_{m+1}) \quad (9)$$

It is thus possible to predict the forces necessary to obtain a desired shape.

These equations have been programmed in THINK's Lightspeed Pascal 2.0 on a Macintosh. For a relatively small laminate, e.g. 4 plies and up to 10 or 12 segments, the program runs in about ten minutes on a Macintosh II with math coprocessor, or rather longer on a Mac Plus. If a large curvature is imposed, so that geometric nonlinearites become significant, the run time increases considerably.

PREDICTIONS BY THE MODEL

Uniform Imposed Curvature.

Figures 8a-8d show the output of the model for a 4-ply laminate divided into 8 segments, each of length 0.02 meters, when a curvature of $R^{inv} = 0.1$ meters^{-1} is imposed on all segments. The material characteristics of this laminate have been chosen as:

$E = 10^{13}$ Pa
$G = 10^{6}$ Pa
$t^{ply} = 120 \ \mu m$
$t^{mat} = 5 \ \mu m$

For these material characteristics, relative ply sliding is easy, as can be seen from Figure 8a, which shows the considerable sliding displacements at each node. Figure 8b shows the stresses in the hard layers. If the beam were homogeneous, the normal stress would depend only on the local radius of curvature, which is the same across the laminate, and on the distance from the neutral axis, but not on segment number. However, the normal stresses for the laminated plate are seen to be quite different. They vary from ply to ply as might be expected, but they also peak in the middle of the beam and drop off toward the ends. This variation in stress from segment to segment is actually a manifestation of shear lag. The normal stresses in each ply must be zero at the ends of the laminate, but can be made nonzero in the central portions because of the shear stresses transmitted from adjacent material. In this case where the beam has a uniform radius of curvature, the shear stresses in each half of the laminate all point in the same direction, so that the normal stress is maximum at the center. Because of the relatively large sliding displacements allowed, the length over which the normal stresses build up from the ends is much greater than that given for homogeneous materials by conventional elastic theory [6].

Figure 8c shows that the shear stresses in the soft layers change very smoothly from one end of the laminate to the other. They go through zero at the middle of the laminate, just as do the displacements, because sliding and shear stress reverse their direction at that point. Finally, Figure 8d shows the array of point forces which, if applied perpendicular to the laminate at the segment boundaries, would give the specified curvature.

Interply Sliding Parameter

The material properties of the laminate simulated in Figure 8 were such as to permit easy relative ply sliding. It is also quite possible to choose properties and geometric factors which will not permit such sliding. A single dimensionless parameter, called the Interply Sliding Parameter or ISP, governs the ability of the laminate to relieve stresses by sliding. The ISP may be expressed as:

$$ISP = \left(\frac{E}{G}\right)\left(\frac{t^{ply}}{length}\right)\left(\frac{t^{mat}}{length}\right)$$

(9)

where length is the total length of composite laminate undergoing bending. Each of the variables in the ISP affects the ability of the plies to slide over each other so that the higher the ISP, the easier is sliding. For instance, the softer the soft interlaminar matrix in relation to the hard fibrous layers, the easier ply sliding will be. Therefore, E is in the numerator and G in the denominator. If a long section of laminate is bent, sliding over a long distance will be more difficult that over a short one. The ISP is, for this reason, inversely proportional to the square of the length. A higher t^{ply} increases the thickness of the laminate and thereby increases the driving force for ply sliding, that is, the bending stress that would exist without sliding. Finally, a higher t^{mat} means that the shear stress in the matrix layer is less for a given amount of sliding. The ISP for the laminate simulated in Figure 8 is 2.34×10^{-1}.

Figure 9 shows a graph of the extent of interply sliding versus the ISP. The extent of sliding is defined as the ratio of sliding to bending strain, $\varepsilon_{mn}^{slide}/\varepsilon_{mn}^{bend}$. (See Equation 3 and Equation 4.) For this graph, this ratio is evaluated at element 11. If the sliding strain component is very small compared to the bending, no relative ply sliding can occur, while if it is large enough that most of the bending strain is alleviated, this ratio is close to -1 and interply sliding is easy. Figure 9 includes predicted points for 4-ply, 4-segment laminates with E, G, tply, and laminate length varied over several orders of magnitude. All of the calculated behavior falls on one master curve. The ISP is analogous, for elastic materials, to the characteristic time constant reported by Tam and Gutowski for elastic plies and viscous interlayers [4].

Figures 10 and 11 show the predictions of the model for material properties and geometries which give, respectively, an intermediate and a low ISP. The ISP was decreased by increasing G to make the matrix material stiffer. In Figure 10, $G = 5 \times 10^7$ Pa and ISP = 4.68×10^{-3}, while in Figure 11, $G = 10^{10}$ Pa and ISP = 2.34×10^{-5}.

Figure 10a shows the sliding displacements for the same laminate as in Figure 8, except that G and ISP are different, as stated above. It is clear that sliding is much more difficult for this beam with its intermediate ISP, especially away from the ends. In fact, the maximum q is only about 3E-7 m, compared to 1.3E-6, four times as high, for Figure 8a. In addition, the normal stresses, shown in Figure 10b, are about three times as large as those in Figure 8b. They are uniform over most of the laminate and drop off only at the ends, where the lack of constraint has allowed some sliding. In the beam of Figure 8, where ISP was high and sliding was easy, the normal stresses varied considerably along the length of the beam. The stress distribution in Figure 10b is nearer to that of a homogeneous elastic beam, where only the local curvature and the distance from the neutral axis, but not the position along the beam, affect the stress.

Figure 10c shows the shear stress profile for this intermediate-ISP case. Shear stress is relatively low over all but the ends of the composite, in agreement with the near-zero sliding displacements in the same area. Finally, Figure 10d shows the required force distribution. It is much less smooth than that shown in Figure 8 for a high ISP; it is, in fact, close to the classic 4-point bending force distribution in which the section between the inner load points has constant curvature.

Lastly, Figure 11 shows the results for a beam with a very low ISP, in which no sliding is observed. The q_{mn}'s, in fact, are seen in Figure 11a to be only about 1/20 the magnitude of those for the beam with an intermediate ISP and to flip back and forth from segment to segment. The normal stresses in Figure 11b are about as large as those in Figure 10b, but do not drop even at the ends. In Figures 11c and 11d, the shear stresses and forces are seen to flip back and forth between segments, as do the q_{mn}'s. The reason for this odd behavior is that a laminate with no interply sliding is actually behaving like a homogeneous beam. Such a beam, to have a uniform radius of curvature along its entire length, must be subjected to a constant bending moment over that length. However, the model cannot calculate bending moments, but only point forces. Therefore, the forces flip back and forth in an attempt to approximate a constant bending moment. Figure 11e is the bending moment diagram calculated from this force distribution, and it does in fact oscillate around a constant value. The shear stress in the beam naturally follows the forces; hence, it is shown as a column graph instead of the line graphs used in Figures 8 and 10. As the shear stresses in the model depend only on the q_{mn}'s, these must flip back and forth also. Such behavior is not necessary for the beam in Figure 10 because ply sliding occurs in the end sections. These end sections are therefore not required to experience the same bending moment as the middle segments, so that a more familiar result is obtained despite the uniform curvature. The model thus tries to duplicate the classical solution for a homogeneous beam as the ISP goes to zero, though it was never intended to analyze such cases.

Nonuniform Curvature

A number of simulations were run in which nonuniform curvature was imposed on the laminates. These laminates consisted of four plies and twelve segments. The material characteristics were:

$E = 10^{12}$ Pa
$t^{ply} = 120$ μm
$t^{mat} = 5$ μm
segment length = 0.002 m
total length = 0.024 m

G was varied to change the ISP. Four of the segments in each laminate were bent to Rinv = 0.1 m^{-1} and the other eight were kept straight. The location of the four bent segments varied: if the twelve segments are numbered from left to right, then either segments 1-4, 3-6, or 5-8 were bent. Figure 12 shows the sliding displacements and normal stresses in ply 4, the outermost ply on the tension side of the beam for a case in which the ISP was low, 4.2E-05 and G was quite high at 2.48E10 Pa. In Figure 12a, we see that the displacements are very small, on the order of 2-3E-9 m. They flip back and forth and are not much affected by the location of the bend. The normal stresses in Figure 12b peak at around 2E7 Pa in the bent segments and are zero elsewhere. The magnitude of the stress is not affected by the location of the bend.

As Figure 13 shows, the situation is somewhat different at a higher ISP. This simulated beam had G = 1E8 Pa and ISP = 1E-2. The relative sliding displacements in Ply 4 are seen in Figure 13a to vary almost linearly in the bent segments in all three different cases. However, they drop almost to zero far away from the bent section. Figure 13b shows the normal stresses for the same cases. They are at most only 3/4 the peak normal stresses in Figure 12b, and if the bent segments are numbers 1-4, at the end of the beam, the peak stress is smaller still. The greater ease of sliding at the end of the beam for these material properties allows the peak stress in the bent segment to drop by about one fourth when the bending is imposed near the edge.

Figure 14 shows the displacements and normal stresses for a beam with a very high ISP of 42 and G = 2.48E4 Pa. The q_{mn}'s in Figure 14a are again linear in the bent sections, but in this case sliding is so easy that all of the sliding--twice that in Figure 13a--is propagated to the end of the beam. The normal stresses shown in Figure 14b are three orders of magnitude less than those in Figure 13b for a simulated beam with an intermediate ISP. Further, the peak stress for a beam bent at the end is only half that for a beam bent in the center. Bending at the edge is easier than bending elsewhere for process conditions which allow easy sliding; the die-less forming process is designed to take advantage of this fact by introducing a bend at the edge and propagating it to the desired location [3].

Figure 15 compares these three cases. Figure 15a compares the sliding displacements in ply 4 when the bend is in the center of the beam. The maximum sliding is of course seen when the ISP is highest. As ISP declines, less sliding can take place and progressively less of the sliding that does occur is carried out to the ends of the beam. When the ISP is so low that ply sliding is virtually impossible, the magnitude of the displacements is negligible. The flipping back and forth of q_{mn} in such a case is insignificant beside the large displacements seen at higher ISP.

Figure 15b shows the normal stresses predicted in Ply 4 of a beam bent at the center for various ISP's. The highest stresses are observed at the lowest ISP; these stresses are high only in the bent section and are close to zero in the straight sections. As the ISP rises, the peak normal stress declines and the stress spreads out into the sections which are not bent. Interply sliding thus decreases the stresses caused by bending and causes segments which are straight to carry some of the stress generated by bending in nearby segments.

CONCLUSION

A model has been developed to describe the response of laminates containing alternating layers of different types of material. At its present stage, this model can predict the extent of interply sliding, the stresses and strains, and the forces required when such a laminate is deformed to some specified singly-curved shape. A dimensionless Interply Sliding Parameter (ISP), which is a combination of material properties and geometry, controls the extent of interply sliding. The ratio of sliding to bending strain at a selected element, which measures the extent to which interply sliding relieves the stress caused by bending, is a unique function of ISP.

Interply sliding causes the response of a bent laminate to differ dramatically from that of a classical bent beam. First, the stresses are lower. Also, while in classical beam theory the bending of one region does not affect the response of a nearby element of material, such an effect is clearly present if interply sliding can occur. A straight section of a beam can carry some of the stress generated by a nearby bend, and interply sliding can be propagated through a straight section from a bent section. Thus, in a laminated composite where interply sliding can take place, the material response to bending depends not only on the local curvature, but also on the entire distribution of curvature throughout the laminate.

Future work will be directed toward two objectives. First, time-dependent material behavior will be included for the matrix material and for off-axis plies. Second, an array of point forces will be imposed, rather than a distribution of curvature, and the resultant curvature distribution will be calculated. In fact, any combination of curvatures and forces will be input and the remaining forces and curvatures will be calculated.

ACKNOWLEDGMENTS

This work was supported primarily by the Lockheed Aeronautical Systems Company under the US Air Force contract Manufacturing Science of Complex Shape Thermoplastics, contract number F33615-86-C-5008. Additional funding was provided by the Stanford Institute for Manufacturing and Automation. The assistance of Dr. Toshimitsu Tanaka, formerly of the Stanford University Department of Materials Science and now with Hewlett-Packard Corporation, with the computational methods used in programming the model is very much appreciated.

REFERENCES

[1]   Griffith, G. R., J. W. Damon, T. T. Lawson, "Manufacturing Techniques for Thermoplastic Matrix Composites," SAMPE Journal, September/October 1984, pp. 32-35.

[2]   Okine, R. K., "Analysis of Forming Parts from Advanced Thermoplastic Composite Sheet Materials," *SAMPE Journal* Vol. 25 No. 3 (1989), pp. 9-19.

[3]   Miller, Alan K., et al, "Die-Less Forming of Thermoplastic-Matrix, Continuous-Fiber Composits," *Journal of Composite Materials* in press 1989.

[4]   Tam, Albert S. and Timothy G. Gutowski, "Ply Slip During the Forming of
        Thermoplastic Composite Parts," submitted to *Journal of Composite Materials* 1988.

[5]   Soll, Wendy and Timothy G. Gutowski, "Forming Thermoplastic Composite Parts,"
        SAMPE Journal, Vol. 24 No. 3 (May/June 1988), 15-19.

[6]   Timoshenko, S. P. and J. N. Goodier, *Theory of Elasticity*, 3rd ed., McGraw-Hill 1970,
        pp. 267-268.

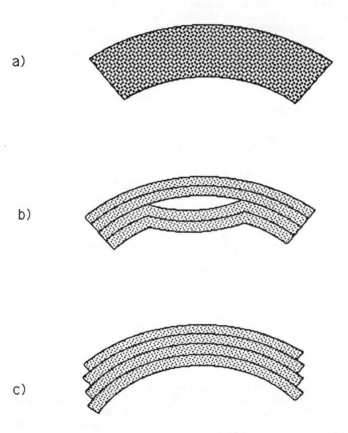

a)

b)

c)

Figure 1.   The need for interply sliding in the shaping of thermoplastic composite parts from
            consolidated sheet stock. a) A metal when bent deforms plastically in tension on the
            outside of the bend and in compression on the inside. b) The continuous fibers of a
            composite material cannot deform plastically. If a laminate is bent without permitting
            interply sliding, the plies on the inside of the bend will buckle. c) Properly
            controlled bending will, however, allow the composite plies to slide over each other
            and form a good part.

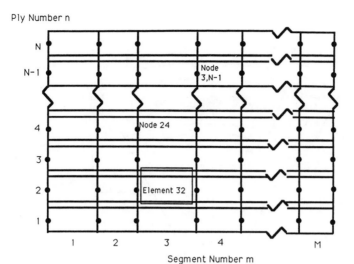

Ply Number n

Segment Number m

Figure 2. Representation of the laminate in the process model under development for thermoplastic composites

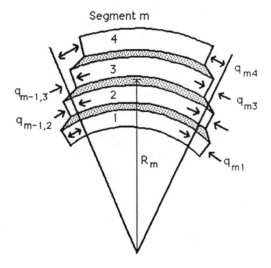

Figure 3. When segment m of the composite laminate is bent to a local radius of curvature R_m, the hard layers slide over each other as shown. The straight lines show where the edges of a homogeneous beam bent to the same curvature would lie. q_{mn} is the sliding displacement of node mn from this line.

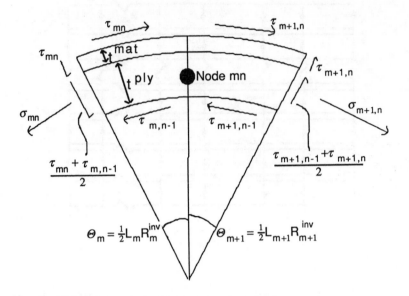

Figure 4. This diagram shows all of the stresses on node mn and the half-element on either side thereof which give rise to forces that have components in the direction parallel to the laminate at node mn. Equation 1 expresses the requirement that these force components sum to zero.

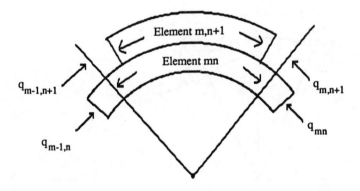

Figure 5. Illustration of the strain caused by relative ply sliding in elements mn and m, n+1. This sliding strain acts to alleviate the strain caused by bending alone.

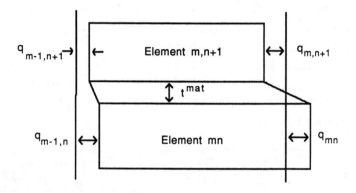

Figure 6. The shear strain in the matrix layer of element mn is caused by the relative sliding of the hard layers of elements mn and m, n+1.

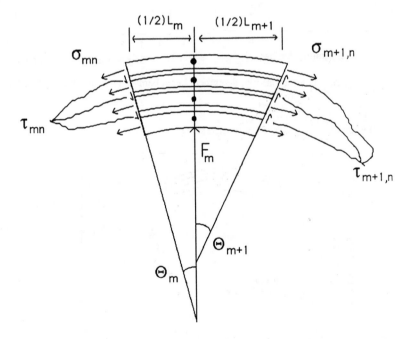

Figure 7. Force balance around nodes m1 through mN to compute the perpendicular force F_m.

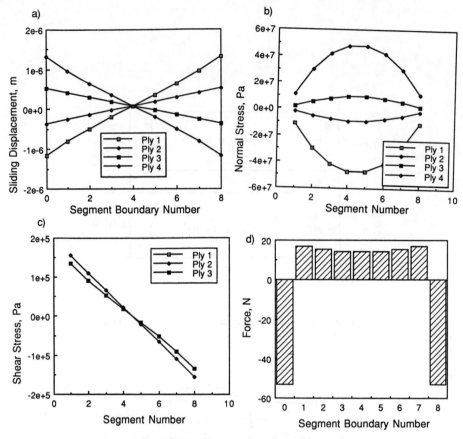

Figure 8. Simulation results for a four-ply, 8-segment beam bent to a uniform radius of curvature.'
  a) Sliding displacements
  b) Normal stresses in hard layers
  c) Shear stresses in interlayers
  d) Point forces required for specified shape.

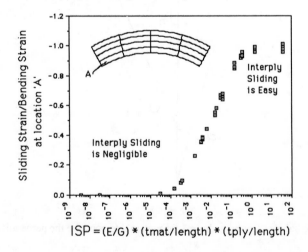

$$ISP = (E/G) * (tmat/length) * (tply/length)$$

Figure 9. The Interply Sliding Parameter (ISP) and its effect on the degree of relative ply sliding

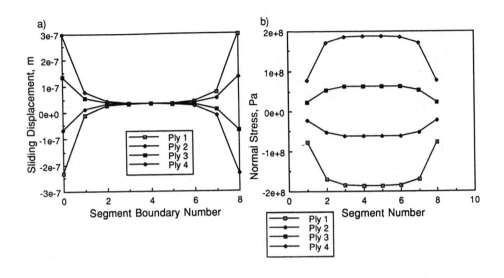

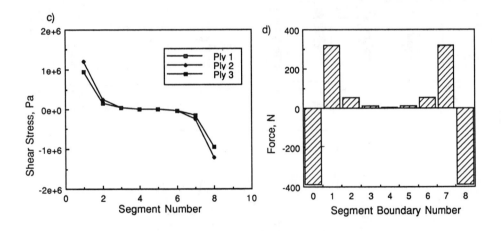

Figure 10. Simulation results for a four-ply, 8-segment beam of intermediate ISP bent to a uniform radius of curvature.
a) Sliding displacements
b) Normal stresses in hard layers
c) Shear stresses in interlayers
d) Point forces required for specified shape.

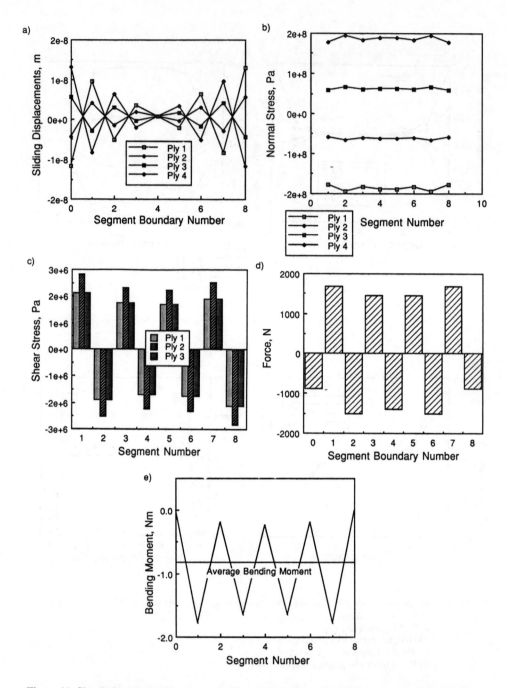

Figure 11. Simulation results for a four-ply, 8-segment beam of low ISP bent to a uniform radius of curvature.
   a) Sliding displacements
   b) Normal stresses in hard layers
   c) Shear stresses in interlayers
   d) Point forces required for specified shape
   e) Bending moment diagram showing constant average moment.

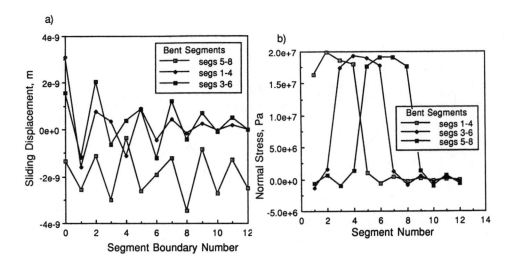

Figure 12. Simulation results for a 4-ply, 12-segment beam of low ISP bent either in segments 1-4, 3-6, or 5-8, when the segments are numbered from the left. Only the results for Ply 4, the outermost ply on the tension side, are shown.
a) Sliding displacement
b) Normal stress in hard layers.

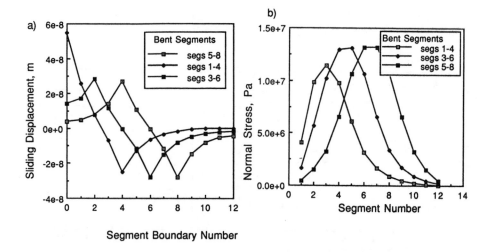

Figure 13. Simulation results for a 4-ply, 12-segment beam of intermediate ISP bent either in segments 1-4, 3-6, or 5-8, when the segments are numbered from the left. Only the results for Ply 4, the outermost ply on the tension side, are shown.
a) Sliding displacement
b) Normal stress in hard layers.

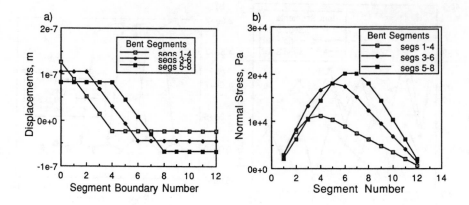

Figure 14. Simulation results for a 4-ply, 12-segment beam of high ISP bent either in segments
1-4, 3-6, or 5-8, when the segments are numbered from the left. Only the results for
Ply 4, the outermost ply on the tension side, are shown.
a) Sliding displacement
b) Normal stress in hard layers.

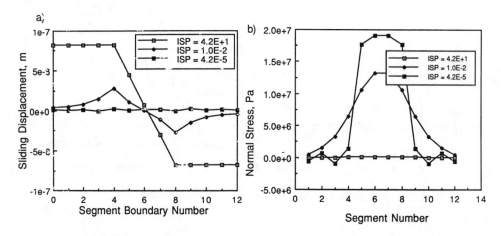

Figure 15. Comparison of the results shown in Figs. 12-14, for segments 5-8 bent, on the same
scale.
a) Sliding displacements
b)Normal stresses in hard layers.

A FINITE ELEMENT/CONTROL VOLUME APPROACH TO MOLD FILLING IN ANISOTROPIC MEDIA

M. V. Bruschke and S. G. Advani
Department of Mechanical Engineering
Center for Composite Materials
University of Delaware
Newark, Delaware

Abstract

Mold filling in anisotropic porous media is the governing phenomena in a number of composite manufacturing processes, such as resin transfer molding (RTM) and resin injection molding (RIM). In this paper we present a numerical simulation to predict the flow of a viscous fluid through a fiber network. The simulation is based on the finite element/control volume method. It can predict the movement of free surface flow front in a thin shell mold geometry of arbitrary shape and with varying thickness. The flow through the fiber network is modelled using Darcy's law. Different permeabilities may be specified in the principal directions of the preform. The simulation permits the permeabilities to vary in magnitude and direction throughout the medium. Experiments were carried out to measure the characteristic permeabilities of fiber preforms. The results of the simulation are compared with experiments performed in a flat rectangular mold using a Newtonian fluid. A variety of preforms and processing conditions were used to verify the numerical model.

Introduction

In recent years production processes involving mold filling through anisotropic porous media have become more important. Such processes include resin transfer molding (RTM) and resin injection molding (RIM). Through the design of the preform one has good control over the mechanical properties of the parts, making these processes very attractive for structural parts in high volume production. Other advantages include the ability to manufacture geometrically complex parts, and capability to produce closed parts with a core in a single step.

In general these processes may be delineated into various phases. The first phase, *preform layup* encompasses cutting of one or more pieces of the fiber mats or continuous fiber strands into specified shapes. They are then stacked and placed into the mold in desired orientations. The second phase, the *mold filling*, begins when the mold is closed and fluid is injected in the hot mold. The fluid is usually a thermoset polymeric resin, which will cure upon heating. The fluid flows around and through the fiber network until the mold is filled. Heat transfer during this phase predominantly results from conduction from the mold walls to the fluid and the flow will be non-isothermal. In the case of high speed processes, such as RIM, viscous dissipation may also play a role in the heat transfer. Once the mold is full, the phase of *curing* starts. Ideally, curing reaction should proceed after the mold is filled. The curing may affect the heat transfer substantially. The final phase, *part removal*, should start after the curing reaction is complete. Sometimes, a mold release agent is applied to the mold for easy removal of the part.

Prediction of the location of the flow front during the mold filling phase is useful in addressing several critical issues in these processes. It will allow the designer to anticipate problems such as dry spots (where no fluid impregnates the preform) and formation of voids, which can occur at merging flow fronts. The designer can optimize parameters such as pressure, flow rates, placement of gates, preform layup, etc. to achieve acceptable flow pattern before the mold is built and thus avoid potential problems.

This paper presents a numerical simulation based on a finite element/control volume method (Wang and Lee, 1989, Wang et al., 1985, Oswald and Tucker, 1989) to calculate the flow pattern (velocities and the flow front movement) for mold filling in anisotropic media. Emphasis here is on the fluid mechanics of mold filling and the influence of the anisotropic medium on the mold filling process. Presently, the simulation can model thin planar parts of otherwise arbitrary shape. The mold cavity may vary in thickness, have multiple gates and may include inserts. A key feature of the simulation is the ability to model the mold filling through a medium with non-isotropic permeability due to the presence of the fiber preforms. The simulation also captures the influence of orientations of reinforcement plies on the filling process. The effects of heat transfer are not yet included in the model. The results obtained using this computational tool are compared to experiments conducted in a simple rectangular mold with different preform lay-up, which makes up the anisotropic medium, with an insert in the mold and with a variation of thickness of the mold. In most cases, flow kinematics agree quite well. Some of the difficulties in describing the flow in the mold are discussed.

Previous Work

The flow of resin through a fiber network may be studied on two different levels, the micro- and macro-mechanical scale. The micro-mechanics deal with the complicated local flow field between the fiber bundles (Behrens, 1983, Peterson et al., 1985). An understanding of the flow on local scale should lead to equations such as Darcy's law (Darcy, 1856) for Newtonian fluids, which are applied on macro-mechanical scale to relate the average velocity to the pressure gradient. For polymeric fluids with elastic effects, the micro-mechanics are not yet fully understood. There is no quantitative analysis of this phenomenon although research is underway to study the effect of elasticity on the macroscopic flow behavior (Pilitsis and Beris, 1987).

On macroscopic scale, the flow of liquids through porous, incompressible media has been studied in great details by those interested in geological problems (Scheidegger, 1974, Lambe and Whitman, 1969). It has been observed that liquid flow through porous media can be described by Darcy's law (Darcy, 1856, Greenkorn, 1983), which demonstrated good comparison with experiments for flow through saturated soils. Practice has converged on using Darcy's law as the governing equation in composite processing for low Deborah numbers (Tadmor and Gogos, 1979) as normally the fiber network is constrained in such a way that it acts as an incompressible porous medium (Gutowsky et al., 1987, Springer 1986, Dave et al., 1987). Relatively little experimental work has been done in supporting the Darcy's law in fiber composite processes and in developing an objective method of measuring the fiber permeabilities and dependence of permeability on the fluid, local pressure, etc. (Coulter et al., 1987, Adams et al., 1986, Martin and Son, 1986).

Considerable work has been done in the area of flow through anisotropic porous media in soil mechanics (Greenkorn, 1983). An extension of Darcy's law to two or three dimensions is usually employed. The two-dimensional version of Darcy's law relates the flow velocity to the pressure gradient in the direction of flow and the pressure gradient perpendicular to the flow. This relation is somewhat more complicated than the original Darcy's law. In composites processing this theory has recently been introduced (Coulter et al., 1987). However as the porous media used in mold filling processes become more intricate and highly anisotropic, one would have to use the two and three-dimensional version to model the physics of mold filling.

An issue in the simulation of mold filling such as RTM, compression molding and injection molding is the numerical treatment of the transient free surface or the moving boundary (the boundary where the fluid is displacing air in the mold cavity). The material inside the mold is constantly changing shape as it flows. This makes it necessary to redefine the geometry of the domain in which the governing equations are to be solved after each successive time step. The governing equations for these filling models have been solved by a variety of numerical techniques (Wang and Lee, 1989, Wang et al., 1985, Oswald and Tucker, 1989, Coulter et al., 1987, Tucker, 1987, Hieber and Shen, 1980). If finite element methods are used, a new mesh must be created, every time the geometry is redefined. Mesh generation could be the most tedious part of the simulation. Finite element/control volume (FE/CV) is an attractive alternative as one does not need to re-mesh and it is possible to simulate filling in thin cavities with highly complex geometries. A key feature of the FE/CV method is a rough approximation of the domain shape combined with a thorough accounting of the mass conservation.

Theory

Most mold filling processes in anisotropic porous media deal with parts which have a shell-like geometry, with the thickness being much smaller than the other dimensions of the part. This allows us to ignore the flow in the thickness direction and model the flow as two-dimensional. As a first step, we model the resin as a Newtonian liquid. This will be a reasonable approximation provided that the Deborah number is small (Tadmor and Gogos, 1979). The fibrous preform is treated as an incompressible porous medium and a two-dimensional version of Darcy's law is used to describe the pressure flow-rate relationship in a medium with non-isotropic permeability (Greenkorn, 1983). In using Darcy's law we give up the details of the velocity profile through the thickness and use average velocities. These gap-wise averaged velocities for an isothermal Newtonian fluid can be written in matrix form as:

$$
\begin{pmatrix} \overline{u}_x \\ \overline{u}_y \end{pmatrix} = -\frac{1}{\mu} \begin{pmatrix} K_{xx} & K_{xy} \\ K_{yx} & K_{yy} \end{pmatrix} \begin{pmatrix} \partial P/\partial x \\ \partial P/\partial y \end{pmatrix} \tag{1}
$$

Here μ is the viscosity of the fluid, P is the pressure and K is the permeability tensor. The permeability is a second order tensor and measures the ease of flow. It is considered a material parameter, although there is evidence that the fluid has an effect on the permeability (Martin and Son, 1986).

In Eq.(1) we have a full symmetric permeability matrix because the pressure gradient vector may not be parallel to the velocity vector. Consider the simple case in which the mold is filled with reinforcement such that the permeabilities are homogeneous in the mold plane. This is true if mold is filled with one continuous porous medium, or if the preform has the same layup throughout the mold. For such special cases, one can always find the principal axes 1 and 2 which will diagonalize the permeability matrix. It is easier to solve Eq.(1) in this new coordinate frame (1-2) as the off-diagonal terms are zero. Applying the continuity condition in the rotated coordinate frame gives a single governing equation for the pressure distribution

$$\frac{\partial}{\partial x}\left(K_{11}\frac{\partial P}{\partial x}\right) + \frac{\partial}{\partial y}\left(K_{22}\frac{\partial P}{\partial y}\right) = 0 \tag{2}$$

where K_{11} and K_{22} are the principal permeabilities.

For this simple case, it can be shown that the flow front will always progress as an ellipse as long as it does not encounter any mold walls or other obstructions. The directions of the principal axes of the ellipse will coincide with the direction of principal permeabilities. It can be analytically shown that the ratio of the principal axes of the ellipse is equal to the square root of the ratio of the principal permeabilities. Hence, a simple experiment of injecting fluid in the center of a large flat mold will enable us to determine the ratio of the principal permeabilities. Measurement of the pressure-gradient in one of the principal directions will enable one to calculate actual values of the permeabilities.

Usually the reinforcement and hence the permeability matrix will vary throughout the mold. The reinforcement will consist of different types of mat or layups placed at different orientations in the mold. For such cases, it is not possible to find a common principal coordinate system in which the permeability matrix will diagonalize everywhere in the mold. We will now have to take into account the full permeability matrix. After applying the continuity condition to Eq.(1), the governing equation for the pressure distribution becomes

$$\frac{\partial}{\partial x}\left(K_{xx}\frac{\partial P}{\partial x}\right) + \frac{\partial}{\partial x}\left(K_{xy}\frac{\partial P}{\partial y}\right) + \frac{\partial}{\partial y}\left(K_{yx}\frac{\partial P}{\partial x}\right) + \frac{\partial}{\partial y}\left(K_{yy}\frac{\partial P}{\partial y}\right) = 0 \tag{3}$$

where the value of the permeability matrix may vary throughout the mold.

The boundary conditions on Eq.(2) and Eq.(3) are zero pressure at the free flow front and, as we assume there is no leakage through the mold walls, i.e. zero pressure-gradient normal to the mold boundary. Once the pressure distribution has been found, averaged velocities across the thickness can be calculated from Eq.(1).

A preform will normally consist of a number of layers of fiber mats stacked at different orientations. These layers may be of different materials. It would be extremely time consuming to measure the total permeability of every possible stacking sequence. Therefore, a model that can predict the total permeability of the layup given the permeabilities of the individual layers and their orientation will be useful. One way to model this effect is to find the *average gapwise permeability* by applying Darcy's law to flow through parallel layers of different permeabilities (Greenkorn, 1983). The average permeability components $(\overline{K}_{ij})$ for a layup of n layers each of thickness h^l with permeability components in the local mold coordinate frame K_{ij}^l is given as,

$$\overline{K}_{ij} = \frac{1}{H}\sum_{l=0}^{n} h^l K_{ij}^l \tag{4}$$

where H is the total thickness of the layup.

It is relatively simple to find the permeability matrix for a given preform stacking sequence. To characterize each type of preform used, only one experiment needs to be performed. Injecting fluid in the center of a flat plate mold filled homogeneously with the material will enable us to calculate the principal permeabilities and the principal directions. For any given stacking sequence we can calculate the effective average permeability using Eq.(4). The permeability matrix is now known at each spatial location in the mold. These values are used in Eq.(3) to find the pressure distribution. The velocities can be calculated from the pressure distribution using Eq.(1).

Simulation

The numerical simulation is based on the finite- element/control-volume method. The part geometry may be modeled as a thin shell in three-dimensional space, using triangular and/or quadrilateral elements of specified thickness. Different elements may have different thickness to account for variations in the gap-height of the mold. Galerkin finite element equations are used to solve the governing equation at any instant during the filling process. The mold geometry is also divided into control volumes, by associating one with every node. The control volume is bounded by the element centroids and the element mid-sides. The flow-rate between the control- volumes is calculated by multiplying the average velocities with the area connecting the two control volumes. Assuming a linear pressure profile between the nodes, and using the average element height, the equation for the flow-rate from element i to element j may be written as:

$$ q_{ij} = -\frac{S}{\mu} \left(\frac{h_i + h_j}{2} \right) \left(K_{xx} \left(\frac{P_i - P_j}{l} \right) + K_{xy} \frac{dP}{l_p} \right) \tag{5} $$

where S is the width of the connecting area and l is the distance between nodes i and j. The permeabilities K_{ij} are in the local element coordinate system, and dP/l_p is the pressure gradient perpendicular to the direction of flow. dP/l_p is obtained by calculating the pressure gradient in the direction of the surrounding nodes, and calculating the average component normal to the direction of flow.

Nodal fill factors are used to track the moving flow front. The fill factor for each node is the fraction of its control volume occupied by the fluid. Pressures are calculated at full nodes and empty nodes are ignored. Partially filled nodes are assumed to lie close to the front and the flow front boundary condition is applied there. The flow front is advanced at each time step by updating the fill factors, using the flow rates between connecting nodes (Eq.(5)), thus rigorously accounting for mass balance. With this technique, one can fill thin cavities with highly complex geometries in three-dimensional space and also account for the variation in the gap height of the part.

Experiments

A number of simple experiments were conducted to achieve a better understanding of the physics of the process and verify the results of the simulation. A flat plate mold was used consisting of two plates held apart by a spacer plate as shown in Fig. 1. The top plate was made of glass, to enable recording the experiments on a video tape. The dimensions of the mold were 46x46cm with a gap-height of 5.5mm. After positioning the preform and closing the mold, the fluid was injected at the center from a pressure tank. The pressure at the inlet gate was held constant using a pressure gauge as feedback to adjust the tank pressure. To ensure even pressure distribution in the thickness direction, a hole was cut in the preform at the inlet gate. The inlet pressure could be varied from 10^5 to $4\times10^5 \mathrm{Pa}$. The pressure at the flow-front was atmospheric at all times.

The fluid used in these experiments was SAE-30 motor-oil. It exhibits Newtonian behavior at the shear-rates present and its contrasting color made it possible to track the flow fronts through the glass top accurately. Two types of glass mats were used, unidirectional NEF240, and randomly oriented chopped strands M721. The glass fiber volume fractions were maintained at 30% for the unidirectional mat, and at 25% for the random mat. A stack of seven layers of random mat or nine layers of the unidirectional mat were used in the mold.

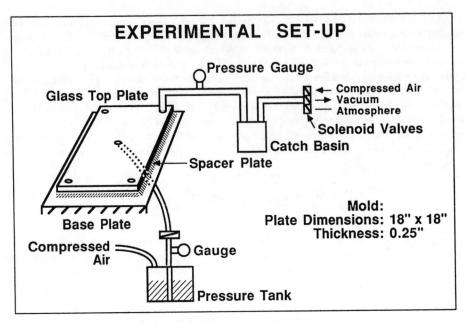

Figure 1: Experimental Set-up

A number of simple experiments with all mats oriented in the same direction were conducted to determine the ratio of principal permeabilities, K_{22}/K_{11}. After characterizing this ratio for both type of glass mats, four sets of experiments were performed.

The first set employed unidirectional fiber mats with different layups to investigate the effect of stacking plies with different orientations on the flow and compare them to the theoretical predictions. The second set of experiments used a round insert in the mold cavity with randomly oriented chopped strands to observe the influence of geometric complexity on the flow field. The third set of experiments used unidirectional mats and a discontinuous porous medium. In manufacturing parts using the RTM or RIM process, a preform layer may not be a continuous piece covering the complete mold cavity but consist of different pieces put together. In such cases, the local permeability at the interface (which we call *cutline*) will differ from the glass mat permeability and the porous medium is no longer homogeneous. To analyze the influence of non-homogeneous anisotropic medium, experiments were performed with a cutline in the preform. We used elements of different permeability at the cutline to model this effect in the simulation. The last set of experiments employed random mats in a mold with two different gap-heights. In one section of the mold the gap-height was reduced to less then half of the original gap-height. These experiments were used to investigate the effect of variation of the mold gap-height.

Results and Discussion

From the simple characterization experiments, the ratio of the principal permeabilities K_{22}/K_{11} was calculated from the ratio of the major to minor axis of the elliptical flow domain as explained earlier. For the random mat the ratio was unity and for the unidirectional mat, with the fibers oriented in the 1-direction, the value was 0.13. These experiments were performed over the complete pressure range, and no influence of pressure on the value of K_{22}/K_{11} was detected, confirming the Newtonian behavior of the fluid in this range.

To study the effect of stacking layers with different orientations, an experiment was performed using a layup of nine unidirectional mats of equal thickness with $[0, \pm 45, 90, 0, 90, \mp 45, 0]$ orientations. Using Eq.(4), the average permeability ratio was calculated to be 0.84 and the principal axes coincided with the local coordinate axes. Figure 2 shows a top view of the mold with the experimental and predicted flow-front at equal time steps. The agreement between the predicted and experimental flow front is good. However, the flow front in the experiments moved faster along the mold walls. This effect was frequently observed during the experiments. Two factors may contribute to this phenomenon. The preform did not fit the mold surface exactly which resulted in a lower fiber volume fraction near the mold walls. Secondly, the pressure difference may slightly deform the fiber ends near the walls, thus creating a flow path. No attempt was made to model this phenomenon at this time, however one can do so by assigning higher permeability values to a row of elements along the walls. It was also noted that a slight shift in the position of the preform had a significant effect on the flow near the walls. This means that the permeability near the wall is very sensitive to the positioning of the preform in the mold. Consequently, it is difficult to model this phenomenon accurately.

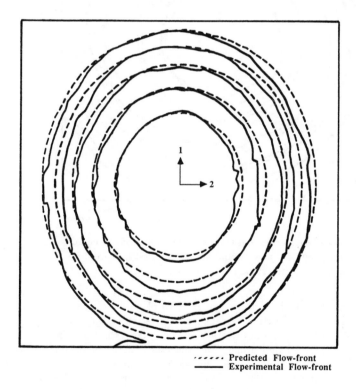

······· Predicted Flow-front
———— Experimental Flow-front

Figure 2: The experimental and predicted flow-front at equal
time steps. Lay-up of uni-directional mats with stacking
sequence $[0, \pm 45, 90, 0, 90, \mp 45, 0]$.

Figure 3 shows the comparison between the flow fronts for the mold with the round insert. For this case, random glass mats were used. The boundary conditions imposed at the insert are the same as those at the mold walls. The fronts are also spaced at equal times, except for the first one which is drawn at half the time step. Agreement between predicted and experimental values is good until the front reaches the mold wall.

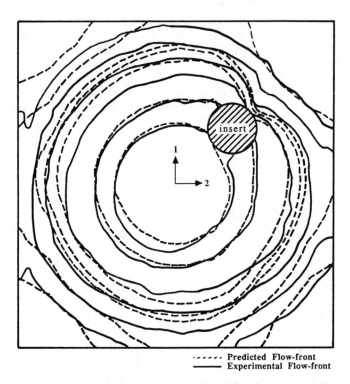

····· **Predicted Flow-front**
——— **Experimental Flow-front**

Figure 3: The experimental and predicted flow-front at equal
time steps, for flow around an insert. First flow-front is
at half the time step. Lay-up of random mat.

In the experiment with the cutline, two sections of unidirectional glass were used with the fibers aligned in the 1-direction. As Fig. 4 depicts one section of the preform covered approximately 70% of the mold and the other section covered the rest. The flow slowed down significantly when it reached the cutline. This suggests that the permeability is considerably different near the cutline. The permeability was decreased for a row of elements at the cutline by 80%. Before and after the cutline, agreement between predicted and experimental values is reasonable. However, at the cutline the flow tends to move faster in the 2-direction in the experiment. This effect seems to allude that the ratio of the principal permeabilities changes near the cutline as there is less resistance to flow in the 2-direction.

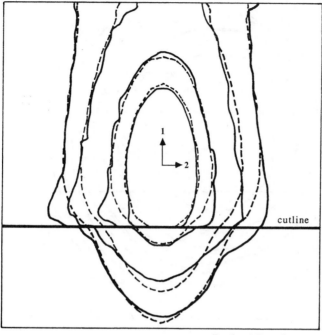

cutline

······ Predicted Flow-front
——— Experimental Flow-front

Figure 4: The experimental and predicted flow-front at equal
time steps, for flow through a cut-line. Preform consists
of two sections of uni-directional mat.

In the last set of experiments, the gap-height of the mold was decreased in one section
of the mold. This was accomplished by inserting a plate of thickness 3.4mm into the mold.
The volume fraction of fibers was 23% in the reduced thickness area and 25% in the rest of
the mold. To account for the lower fiber volume fraction the permeability was increased by
10% in the reduced thickness area. This assumes an approximately linear relationship between
permeability and volume fraction. Figure 5 shows the result for this simulation. Although both
the experiment and prediction show an increase in flow velocity in the reduced thickness area,
the experimental front progresses faster then the predicted one. We increased the permeability
by 40% in the reduced thickness area. For this case the predicted flow front matched well with
the experiments as shown in Fig. 6.

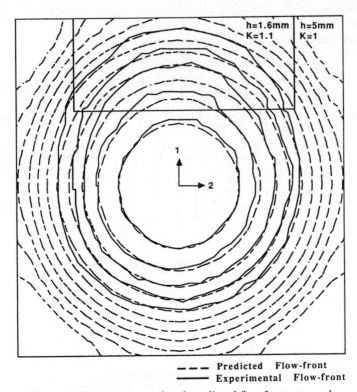

Figure 5: The experimental and predicted flow-front at equal
time steps, for flow in a mold with two different gap-
heights (h). Permeability (K) is increased by 10% in
the reduced gap-height region.

Several factors may contribute to this phenomenon. The permeability may not be a linear
function of the fiber volume fraction. However this does not account for the large change in the
permeability that we had to make in the reduced thickness section in the simulation. It is also
possible that due to the higher flow velocities, and consequently higher shear rates, the fluid
shows some shear thinning behavior in the reduced thickness region. Another factor which may
explain the discrepancy is the permeability variation through the thickness. Along the top and
bottom plate the permeability will be lower then in the centerplane. This effect is well known
in soil mechanics (Greenkorn, 1983). Due to the fact that the medium is not continuous close
to the mold wall, the fiber volume fraction will be lower. In a thin mold section, this would
have a large effect on the average permeability as compared to a thick section. To investigate
this phenomenon in more detail, it is necessary to characterize the fluid over a few decades of
shear rates, and to study the through-thickness velocity profile of the flow.

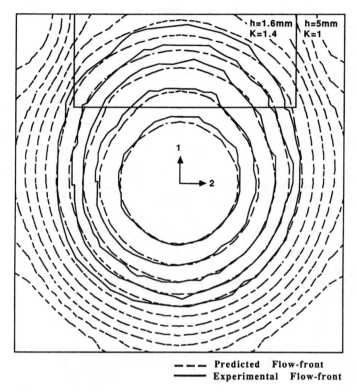

h=1.6mm h=5mm
K=1.4   K=1

1
2

– – – Predicted   Flow-front
——— Experimental Flow-front

Figure 6: The experimental and predicted flow-front at equal
time steps, for flow in a mold with two different gap-
heights (h). Permeability (K) is increased by 40% in
the reduced gap-height region.

The last two results presented are numerical and without experimental verification. The
intent is to show the versatility of the finite element/control volume approach. The first one
shows the results for a mold filled with an unidirectional mat (Fig. 7). The fluid is injected from
the two bottom corners. It demonstrates the ease with which the control volume approach can
handle this type of flow. If one used a remeshing scheme, one would need a special algorithm
that matches the two meshes when they come in contact. In the control volume approach, the
merging of the flow fronts is automatically accounted for and needs no special attention. The
finite element formulation allows one to introduce multiple gates at no extra cost.

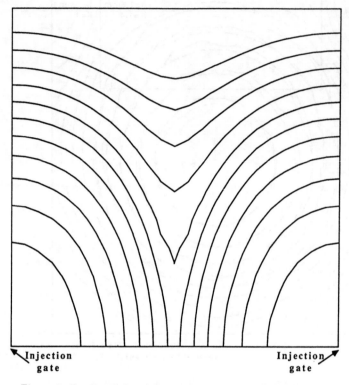

Figure 7: Predicted flow-front at equal time steps, for flow in
a mold with two injection gates.

Figure 8 shows the result for a mold filled with four unidirectional mats which all have a different orientation. As we solve the full pressure equation Eq.(3), this type of preform poses no particular problem.

It is clear from our results that although we are able to predict flow front movement in relatively complex geometries in anisotropic media, we still need further understanding of the interactions between the fluid and the porous medium. This will then permit us to predict *a priori* how the permeability of the medium is influenced by the irregularities near the mold walls, cutlines, etc. However, even within these limitations, the simulation will prove to be a valuable tool in improving our understanding of the flow in anisotropic media during mold filling.

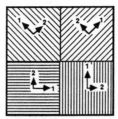

Orientation of the mats
in the four quadrants.

Figure 8: Predicted flow-front at equal time steps, for flow in
a mold with four uni-directional mats, oriented as shown
above.

Future Work

Several assumptions are made in the simulation. It would be a useful addition to be able to
model the flow of non-Newtonian fluids. This can be implemented in the model using the
theory of non-Newtonian flow through porous media [10]. At present the model only considers
isothermal flow. In reality the mold will be hot, and heat transfer and curing effects will take
place. Heat transfer and cure kinetics may be entered into the simulation by allowing for the
spatial and time varying nature of the viscosity.

Acknowledgements

The authors wish to thank Carl Johnson of the Scientific Research Lab of Ford Motor Company where the experimental part of this study was conducted. Special thanks goes to Mike Tinskey for building the experimental set-up, and assisting in the experimental work. We also gratefully acknowledge the Center for Composite Materials for the support of this project.

References

Adams, K. L., Miller, B., and Rebenfeld, L., 1986, "Forced In-Plane Flow of an Epoxy Resin in Fibrous Networks," *Polymer Engineering and Science*, Vol. 26, 20, p. 1434.

Behrens, R. A., 1983, "Transient Domain Free Surface Flows and Their Applications to Mold Filling," *Report #ccm-83-14*, Center for Composite Materials, University of Delaware, Newark, DE.

Coulter, J. P., Smith, B. F.,and Guçeri, S. I., 1987, "Expermental and Numerical Analysis of Resin Impregnation During the Manufacturing of Composite Materials," *Proceedings, 2nd Technical Conference American Society for Composites*, p. 209.

Darcy H, Les fontaines Publiques de la Ville de Dijon, 1856, Dalmont, Paris.

Dave, R., J. L. Kardos, J. L., and M. P. Dudukovic, M. P., 1987, "A Model For Resin Flow During Composite Processing: Part 1 General Mathematical Development," *Polymer Composites*, Vol. 8, 1, p. 29.

Greenkorn, R. A., 1983, Flow Phenomena in Porous Media, Marcel Dekker.

Gutowski, T. G., Tadahiko, M., and Zhong, C., 1987, "The Consolidation of Laminate Composites," *Journal of Composite Materials*, Vol. 21, 2, p. 54 .

Hieber, C. A. and Shen, S. F., 1980, *Journal of Non-Newtonian Fluid Mechanics*, Vol. 7, 1.

Lambe, T. W. and Whitman, R. V., 1969, Soil Mechanics, J. Wiley & Sons.

Martin, G. Q. and Son J. S., 1986, "Fluid Mechanics of Mold Filing for Fiber Reinforced Plastics," *Proceedings, ASM/ESD Second Conference on Advanced Composites*, p. 149.

Osswald, T. A. and Tucker, C. L., 1989, to be published in *International Polymer Processing*.

Peterson, B. K., Walton, R. B., and Gubbins, K. E., 1985, "Microscopic Study of Fluids in Pores: Computer Simulation and Mean Field Theory" *International Journal of Thermophysics*, Vol. 6, p. 585.

Pilitsis, S. and Beris, A. N., 1987, *Proceedings, 2nd Technical Conference American Society of Composite Materials*, p. 189.

Scheidegger, A. E., 1974, Physics of Flow through Porous Media, U. Toronto Press, 3rd ed..

Springer, G. S., 1986, "Modeling the Cure Process of Composites," *Proceedings, 31st International SAMPE Symposium*, p. 776,

Tadmor, Z., and Gogos, C. G., 1979, Principles of Polymer Processing, J. Wiley & Sons.

Tucker, C. L., 1987, "Compression molding of Polymers and Composites," Injection and Compression Molding Fundamentals, A. I. Isayev, ed., Marcel Dekker, New York

Wang, H. P. and Lee, H. S., 1989, Fundamentals of Computer Modeling for Polymer Processing, C. L. Tucker, Ed., Carl Hanser Verlag.

Wang, V. W., Heiber, C. A., and Wang, K. K., 1985, *SPE Technical Paper*, Vol. 31 p. 826.

HEAT TRANSFER OF SOLIDIFICATION IN ANISOTROPIC DOMAINS AND ITS APPLICATION TO THERMOPLASTIC COMPOSITES PROCESSING

S. D. Gilmore and S. I. Güçeri
Thermal Engineering and Advanced Manufacturing Group
Center for Composite Materials
Department of Mechanical Engineering
University of Delaware
Newark, Delaware

Abstract

Heat transfer analysis in thermoplastic composites manufacturing offers several challenges. The melting/solidification process results in the appearance of a moving solid–fluid interface or two-phase zone. The presence of reinforcing fibers with thermal conductivities substantially different than those of matrix materials cause the composite material to behave anisotropically. Furthermore, an analysis should be adaptable to complex geometries if it is to be useful for practical applications. In this paper, a numerical approach is presented to develop an analysis tool for the processing of thermoplastic matrix composites. A numerical grid generation technique is employed to determine the temperature distribution within the part while accounting for the heat absorption and liberation during the solidification stage. The influence of anisotropy in the domain is also accounted for by considering thermal conductivity as a second-order tensor. Sample product configurations are used to demonstrate the applicability of this technique to thermoplastic composites manufacturing processes.

NOMENCLATURE

A	surface area
$a_{11}, \ldots, a_{23}$	three-dimensional numerical grid generation coefficients
$b_{11}, \ldots, b_3$	general transformed enthalpy equation coefficients
c_p	specific heat at constant pressure
e	specific enthalpy
h	convective heat transfer coefficient
$\mathbf{K}$	thermal conductivity tensor
$k_{11}, \ldots, k_{23}$	directional thermal conductivities in fiber-oriented coordinate system
$k_{xx}, \ldots, k_{yz}$	directional thermal conductivities in cartesian coordinate system
$k_{\xi\xi}, \ldots, k_{\nu}$	transformed Enthalpy equation coefficients
L	latent heat of fusion $(L = e_f - e_s)$

$\hat{n}$	unit normal vector
P, Q, R	three-dimensional grid generation control functions
Q_b	constant term in enthalpy equation for boundary conditions
$\mathbf{q}$	heat flux vector
q	heat flow
r	general cartesian axis coordinate
T	temperature
t	time
V	control volume
x, y, z	coordinates of physical domain

Greek

β, γ	general recursive variables for computational coordinates
ρ	mass density
ξ, η, ν	coordinates of computational domain

Subscript

$1, 2, 3$	indices corresponding to computational coordinates or fiber directions
$1, 2, 3, 4, 5, 6$	indices corresponding to six boundary surfaces of domain
b	at, along, or through a boundary
f	at saturated fluid state
i	in the interior of the control volume
i, j, k	node and array indices
max	maximum allowable value
p	at constant pressure
s	at saturated solid state
w	at wall (boundary)
o	at initial conditions
∞	at ambient conditions (far away from object boundaries)

INTRODUCTION

Establishment of a sound processing science is critical to the economic manufacturing of products made from composite materials. The essential purpose of developing a science base is to increase predictability during the processing of these advanced materials. The strong coupling between the processing, resulting microstructure, and performance characteristics necessitates the development of process analysis tools which enable the designers and producers of such parts to reduce the amount of experimentation needed to generate the necessary data base for manufacturing.

Thermoplastic-matrix composites offer many advantages over thermosetting systems. These include superior fracture toughness, infinite shelf life, and easy processability. The processing of thermoplastic composite parts depends greatly upon heat diffusion. First, heat is supplied to melt the matrix material in order to improve formability; then, heat is removed from the part to solidify it. While fast cycle times require quick supply and removal of heat, material degradation can occur if safe temperatures are exceeded while attempting to reduce processing time. During the cooling phase, the rate of cooling has a very pronounced effect on the degree of crystallinity (for semi-crystalline materials) and residual stresses, which in turn significantly affect the dimensional stability and performance of the part.

The level of final crystallinity in semi-crystalline thermoplastics is primarily a function of the cooling rate, and has been investigated by several researchers, including (Erhun and Advani,

1989). Reinforcing fibers, particularly in the case of polymeric composites with carbon fibers, introduce high anisotropy which influences the heat diffusion mechanism quite significantly. The coupling of phase change and high anisotropy makes this class of problems quite challenging mathematically as well as numerically.

An example of thermoplastic composites processing is laser-assisted tape consolidation, in which laser energy is used to melt the thermoplastic matrix material (Beyeler et al., 1988). The use of unidirectional, continuous carbon fiber prepreg tapes results in an orthotropic domain for unidirectional composite laminates. A two-dimensional thermal analysis for this application has been reported (Beyeler and Güçeri, 1988) in which the melting/solidification effects were accounted for by including a heat generation term in a formulation developed for a moving coordinate system.

Many manufacturing processes involve the fabrication of parts which are both three-dimensional and irregular in shape. An analysis tool should, therefore, be capable of accounting for such geometries. The authors recently presented an application of numerical grid generation for three-dimensional solidification problems in isotropic domains (Gilmore and Güçeri, 1988). The problem was formulated in terms of temperature, and modeling of the solidification front was based on the Stefan condition where one of the mesh surfaces was coincident with the solid–fluid interface. Numerical meshes were generated in the solid region at each time step to account for the evolving domains. The primary limitation of this approach is the long computation time required for updating and regeneration of the three-dimensional grids.

The current study applies a similar numerical grid generation method to the analysis of three-dimensional, fully anisotropic domains. An enthalpy model is adopted to determine the thermal history and the position of the solid front throughout the process. This model is particularly suitable for the modeling of polymer processing since many of these materials exhibit a change of phase over an extended temperature range. The approach also allows for the generation of a single fixed mesh covering the entire domain without requiring regeneration at each time step. Several early enthalpy model formulations for phase change are cited in (Özışık, 1980). More recent investigators used an enthalpy model for the analysis of two-dimensional phase change in isotropic domains, developing both constant and variable-density formulations (Shamsundar and Sparrow, 1975; Shamsundar and Sparrow, 1976). The analysis presented here extends that work to three-dimensional, arbitrarily-shaped, fully anisotropic domains. Though this greatly increases the complexity of the problems, the solutions remain manageable due to the nature and efficiency of the enthalpy model.

PROBLEM FORMULATION

The current investigation of thermoplastic-matrix composites processing is based on a conduction model, where the entire heat transfer occurs by diffusion rather than convection from fluid motion. This is a valid assumption since the high viscosity of thermoplastic polymers and the presence of reinforcing fibers prevent fluid motion in the absence of externally imposed deformations. Furthermore, the anisotropic domain is considered to be a continuum, which leads to a macroscopic analysis of the process. Constant density in the solid, fluid, and two-phase regions is also assumed, which has a relatively minor effect on the thermal response of the system (though it has a major influence on the residual stresses).

An enthalpy model overcomes some of the limitations of temperature-only formulations for phase change by considering a single domain of fixed shape in which the enthalpy of the material becomes the dependent variable indicating its thermal state. Thus, the solid–fluid interface is considered as more of a region than a discrete surface. Since the object is considered as a single, multi-phase domain, no interpolation or grid regeneration is required and a fixed grid can be generated prior to solution of the problem. These features allow the enthalpy model to more easily accommodate complex shapes, crystallization kinetics, and extended freezing regions than the two-zone models. Other advantages of enthalpy formulations are discussed in (Shamsundar and Rooz, 1988).

The enthalpy model is based on an energy balance which, for a control volume V centered about a point in a domain, can be expressed as

$$\frac{\partial}{\partial t} \int_V \rho\, e\, dV = \int_A \mathbf{q} \cdot \hat{\mathbf{n}}\, dA \tag{1}$$

where ρ is the mass density of the phase change material, e is the specific enthalpy field distribution, t is time, and $\mathbf{q} \cdot \hat{\mathbf{n}}$ is the heat flux though the control volume surface of area A. The relationship assumes that there are no sources of energy inside V, and that there is no external work performed on the volume. This equation and a temperature–enthalpy closure relationship define the problem. Note that this method has two equations and two unknowns, whereas a temperature-only formulation has only one. Equation (1) reduces to the temperature formulation (after some simplifications) in the single phase regions and the Stefan boundary condition at the interfaces (Shamsundar and Sparrow, 1975).

For a point in the interior of a domain, the right-hand side of Equation (1) can be expressed in terms of material properties and temperature derivatives as

$$\frac{\partial}{\partial t} \int_V \rho\, e\, dV = \int_A (\mathbf{K} \cdot \nabla T) \cdot \hat{\mathbf{n}}\, dA \tag{2}$$

where $\mathbf{K}$ represents a tensor of anisotropic thermal conductivities for the material. By using the divergence theorem, this equation can be written as

$$\int_V \left[\frac{\partial}{\partial t}(\rho\, e) - \nabla \cdot (\mathbf{K} \cdot \nabla T) \right] dV = 0 \tag{3}$$

where the time derivative on the left-hand side of Equation (2) has been moved inside the integral via the Leibniz rule since the integrand is continuous throughout the control volume V and the volume does not change with time. In order for this relation to be true throughout the domain, the integrand itself must be equal to zero. Furthermore, if one assumes that the density is constant, it can be extracted from the derivative on the left-hand side to produce

$$\rho\, \frac{\partial e}{\partial t} = \nabla \cdot (\mathbf{K} \cdot \nabla T) \tag{4}$$

Temperature and enthalpy are coupled in the model via a closure formula, namely

$$c_p \equiv \left(\frac{\partial e}{\partial T} \right)_p \approx \left(\frac{de}{dT} \right)_p \tag{5}$$

the definition of specific heat at constant pressure.

Equations (4–5) are valid everywhere in the interior of the domain if the material changes phase over a range of temperatures. The solid–fluid interface must be considered separately for materials which change phase at a discrete temperature due to the discontinuity in temperature derivatives across the interface. Many thermoplastic polymers used in composite materials have an extended freezing range, so these equations are applied everywhere in the current study and temperature is assumed to vary linearly with enthalpy in the multiphase region.

The anisotropic thermal conductivity tensor is symmetric and can be expressed in a three-dimensional cartesian coordinate system as

$$\mathbf{K} = \begin{bmatrix} k_{xx} & k_{xy} & k_{xz} \\ k_{xy} & k_{yy} & k_{yz} \\ k_{xz} & k_{yz} & k_{zz} \end{bmatrix} \tag{6}$$

308

where the k_{ij} are the directional thermal conductivities in which the subscripts correspond to the (x, y, z) coordinate directions. Note that the components of this tensor may vary from point-to-point throughout the domain. In such cases, a conductivity tensor for a coordinate system oriented along the paths of the fibers must be transformed to the cartesian domain at each point.

Substitution of Equation (6) and the cartesian gradient operator into the right-hand side of Equation (4) yields

$$\rho \frac{\partial e}{\partial t} = k_{xx} \frac{\partial^2 T}{\partial x^2} + k_{yy} \frac{\partial^2 T}{\partial y^2} + k_{zz} \frac{\partial^2 T}{\partial z^2} + 2k_{xy} \frac{\partial^2 T}{\partial x \partial y} + 2k_{xz} \frac{\partial^2 T}{\partial x \partial z} + 2k_{yz} \frac{\partial^2 T}{\partial y \partial z} \qquad (7)$$

where it has been assumed that gradients in the thermal conductivity components are insignificant, which is valid for continuous-fiber composite materials. The more general case is considered by the authors in another study (Gilmore and Güçeri, 1989b). This governing equation can be simplified under certain circumstances such as isotropy (Gilmore and Güçeri, 1989a) and two-dimensionality.

Boundary conditions must be considered separately from the interior of the domain due to the environmental influences imposed. For this reason, the general governing equation for a boundary control volume is somewhat different and takes the form

$$\frac{\partial}{\partial t} \int_V \rho \, e \, dV = \int_{A_i} (\mathbf{K} \cdot \nabla T) \cdot \hat{\mathbf{n}} \, dA - \int_{A_b} \mathbf{q} \cdot \hat{\mathbf{n}} \, dA \qquad (8)$$

where A_i is the portion of the control volume surface within the phase change material and A_b is the remaining amount along the external boundary. Therefore, $A = A_i + A_b$ is the total surface area of the boundary control volume. As with the interior of the domain, $\hat{\mathbf{n}}$ is an outward unit normal vector to the control volume surface. The vector $\mathbf{q}$ represents the heat flux *into* the control volume through its external boundary A_b.

The boundary heat flux can be defined for the general case as

$$\begin{aligned} \mathbf{q} \cdot \hat{\mathbf{n}} &= -[q_w + h(T_\infty - T)] \\ &= -(q_b - hT); \qquad \text{where} \quad q_b = q_w + hT_\infty \end{aligned} \qquad (9)$$

Boundary conditions of the second and third kinds can be generated by properly assigning values to q_w, h, and T_∞, while specified temperature (first-kind) conditions require no enthalpy computations at all.

Substituting for $\mathbf{q} \cdot \hat{\mathbf{n}}$ in the general governing equation above yields

$$\frac{\partial}{\partial t} \int_V \rho \, e \, dV = \int_{A_i} (\mathbf{K} \cdot \nabla T) \cdot \hat{\mathbf{n}} \, dA + \int_{A_b} (q_b - hT) dA \qquad (10)$$

After some mathematical manipulations, including use of the divergence theorem, this expression becomes

$$\int_V \left[\frac{\partial}{\partial t} (\rho \, e) - \nabla \cdot (\mathbf{K} \cdot \nabla T) \right] dV - \int_{A_b} [(q_b - hT) - (\mathbf{K} \cdot \nabla T) \cdot \hat{\mathbf{n}}] dA = 0 \qquad (11)$$

If one assumes that all of the integrands are uniform throughout the control volume and its surfaces, then

$$\left[\frac{\partial}{\partial t} (\rho \, e) - \nabla \cdot (\mathbf{K} \cdot \nabla T) \right] \int_V dV - [(q_b - hT) - (\mathbf{K} \cdot \nabla T) \cdot \hat{\mathbf{n}}_b] \int_{A_b} dA = 0 \qquad (12)$$

This is an approximation for which the accuracy is directly related to the control volume size and gradients within it. After performing the integrations and rearranging the above equation, the following expression results

$$\rho \frac{\partial e}{\partial t} = \nabla \cdot (\mathbf{K} \cdot \nabla T) + \frac{A_b}{V} [(q_b - hT) - (\mathbf{K} \cdot \nabla T) \cdot \hat{\mathbf{n}}_b] \qquad (13)$$

The unit normal vector at the external boundary $\hat{n}_b$, control volume V, and boundary surface area A_b may be different at every point on the external boundary of the domain. In the general case they must be computed numerically given the physical coordinates of the domain. For this reason, their derivations for the cartesian coordinate system are omitted here. Notice that this equation is very similar to the result for a point in the interior of the domain, given by Equation (4), with the addition of an extra term on the right-hand side.

SOLUTION TECHNIQUE

A numerical grid generation approach is taken to solve the governing equations in three-dimensional domains of irregular geometry. This method was first brought into general use in the mid 1970's (Thompson et al., 1974; Thompson et al., 1982), and has been employed in a wide range of problems ranging from fluid mechanics and other areas of the thermal sciences, to solid mechanics. While most of the initial studies were built upon two-dimensional configurations, the method was soon extended into the third dimension (Shieh, 1982; Thomas, 1982). The technique was later generalized for two- and three-dimensional regions and refined by development of expressions for locally optimum successive over-relaxation (SOR) parameters (Thompson, 1987).

Numerical grid generation has recently been applied to a number of problems involving composites manufacturing and related areas (Farraye and Güçeri, 1985; Sottos and Güçeri, 1986; Walsh, 1987; Coulter and Güçeri, 1987; Coulter and Güçeri, 1988; Beyeler and Güçeri, 1988; Subbiah et al., 1989; Gilmore and Güçeri, 1989a; Gilmore and Güçeri, 1989b). Additional information on this technique can be found in (Güçeri, 1989).

The numerical grid generation itself is essentially a mapping operation where a problem defined over an irregular physical domain is mapped onto a computational domain of regular geometry where traditional finite difference methods can be applied to solve the governing equations. The relationship between the physical and computational coordinates can be given by partial differential equations, as well algebraic relations. In the current study, elliptic-type relations are used due to their inherent smoothness and ability to accommodate boundary discontinuities. The general form of this coordinate mapping is expressed as

$$\nabla^2 \xi = P(\xi, \eta, \nu), \qquad \nabla^2 \eta = Q(\xi, \eta, \nu), \qquad \nabla^2 \nu = R(\xi, \eta, \nu) \tag{14}$$

where the functions P, Q, and R are grid control functions. These control functions can be used to concentrate grid lines near selected points or areas as desired, or to increase the internal grid conformance with the curvatures of the bounding surfaces of the physical domain (Shieh, 1982; Thomas, 1982; Thompson, 1987). With the control functions set equal to zero for the current study, a Laplacian mapping occurs in which the grid points are distributed smoothly with no particular concentration.

The objective of the numerical grid generation process is to solve for the physical coordinates. For this purpose, one must interchange the dependent and independent variables to provide the inverse transformation. This produces a system of quasilinear elliptic equations which can be written in general form as

$$a_{11}\frac{\partial^2 r}{\partial \xi^2} + a_{22}\frac{\partial^2 r}{\partial \eta^2} + a_{33}\frac{\partial^2 r}{\partial \nu^2} + 2\left(a_{12}\frac{\partial^2 r}{\partial \xi \partial \eta} + a_{13}\frac{\partial^2 r}{\partial \xi \partial \nu} + a_{23}\frac{\partial^2 r}{\partial \eta \partial \nu}\right) = 0 \tag{15}$$

where r represents each of the physical coordinates x, y, and z. The coefficients a_{ij} are given by the following derivative combinations of the computational coordinates:

$$a_{11} = \nabla \xi \cdot \nabla \xi = \left(\frac{\partial \xi}{\partial x}\right)^2 + \left(\frac{\partial \xi}{\partial y}\right)^2 + \left(\frac{\partial \xi}{\partial z}\right)^2$$

$$a_{22} = \nabla \eta \cdot \nabla \eta = \left(\frac{\partial \eta}{\partial x}\right)^2 + \left(\frac{\partial \eta}{\partial y}\right)^2 + \left(\frac{\partial \eta}{\partial z}\right)^2$$

$$a_{33} = \nabla \nu \cdot \nabla \nu = \left(\frac{\partial \nu}{\partial x}\right)^2 + \left(\frac{\partial \nu}{\partial y}\right)^2 + \left(\frac{\partial \nu}{\partial z}\right)^2$$

$$a_{12} = \nabla \xi \cdot \nabla \eta = \frac{\partial \xi}{\partial x}\frac{\partial \eta}{\partial x} + \frac{\partial \xi}{\partial y}\frac{\partial \eta}{\partial y} + \frac{\partial \xi}{\partial z}\frac{\partial \eta}{\partial z} \qquad (16)$$

$$a_{13} = \nabla \xi \cdot \nabla \nu = \frac{\partial \xi}{\partial x}\frac{\partial \nu}{\partial x} + \frac{\partial \xi}{\partial y}\frac{\partial \nu}{\partial y} + \frac{\partial \xi}{\partial z}\frac{\partial \nu}{\partial z}$$

$$a_{23} = \nabla \eta \cdot \nabla \nu = \frac{\partial \eta}{\partial x}\frac{\partial \nu}{\partial x} + \frac{\partial \eta}{\partial y}\frac{\partial \nu}{\partial y} + \frac{\partial \eta}{\partial z}\frac{\partial \nu}{\partial z}$$

in which the partial derivatives of the transformation can be derived from the chain rule. Dirichlet boundary conditions define a unique solution for the mapping and come from the specification of the physical boundary locations of the mesh. The derivatives can be expressed by central finite differences for the Laplacian case or with one-sided differences to enhance stability when concentration is being employed through the P, Q, R Poisson mapping. When this is done, the set of resulting simultaneous equations is solved explicitly using successive over-relaxation (Thompson, 1987). Note that the a_{ij} coefficients are determined separately for each node in the domain.

Transformation of Enthalpy Equation

Following grid generation, the governing equations and boundary conditions must be mapped onto the computational domain. The derivations are lengthy and can be found elsewhere (Gilmore, 1990). The resulting expression in the interior of the computational domain is

$$\rho \frac{\partial e}{\partial t} = k_{\xi\xi}\frac{\partial^2 T}{\partial \xi^2} + k_{\eta\eta}\frac{\partial^2 T}{\partial \eta^2} + k_{\nu\nu}\frac{\partial^2 T}{\partial \nu^2}$$
$$+ 2\left(k_{\xi\eta}\frac{\partial^2 T}{\partial \xi \partial \eta} + k_{\xi\nu}\frac{\partial^2 T}{\partial \xi \partial \nu} + k_{\eta\nu}\frac{\partial^2 T}{\partial \eta \partial \nu}\right) + k_\xi \frac{\partial T}{\partial \xi} + k_\eta \frac{\partial T}{\partial \eta} + k_\nu \frac{\partial T}{\partial \nu} \qquad (17)$$

where the coefficients $k_{\beta\gamma}$ are given by the recursion formulas (β and γ being the recursion indices)

$$k_{\beta\gamma} = k_{xx}\frac{\partial \beta}{\partial x}\frac{\partial \gamma}{\partial x} + k_{yy}\frac{\partial \beta}{\partial y}\frac{\partial \gamma}{\partial y} + k_{zz}\frac{\partial \beta}{\partial z}\frac{\partial \gamma}{\partial z} + k_{xy}\left(\frac{\partial \beta}{\partial x}\frac{\partial \gamma}{\partial y} + \frac{\partial \beta}{\partial y}\frac{\partial \gamma}{\partial x}\right)$$
$$+ k_{xz}\left(\frac{\partial \beta}{\partial x}\frac{\partial \gamma}{\partial z} + \frac{\partial \beta}{\partial z}\frac{\partial \gamma}{\partial x}\right) + k_{yz}\left(\frac{\partial \beta}{\partial y}\frac{\partial \gamma}{\partial z} + \frac{\partial \beta}{\partial z}\frac{\partial \gamma}{\partial y}\right) \qquad (18)$$

$$k_\beta = k_{xx}\frac{\partial^2 \beta}{\partial x^2} + k_{yy}\frac{\partial^2 \beta}{\partial y^2} + k_{zz}\frac{\partial^2 \beta}{\partial z^2} + 2k_{xy}\frac{\partial^2 \beta}{\partial x \partial y} + 2k_{xz}\frac{\partial^2 \beta}{\partial x \partial z} + 2k_{yz}\frac{\partial^2 \beta}{\partial y \partial z}$$

$$(19)$$

and the recursion indices β, γ assume the values ξ, η, ν in all permutations. Note that these coefficients may vary from point-to-point throughout the domain (usually due to fiber orientation in composite materials). They may also change with time if temperature-dependent thermal conductivities are considered.

Next, the equations for the boundary conditions must be mapped onto the computational domain. This operation involves several mathematical steps, and the result can be expressed as

$$\rho \frac{\partial e}{\partial t} = b_{11}\frac{\partial^2 T}{\partial \xi^2} + b_{22}\frac{\partial^2 T}{\partial \eta^2} + b_{33}\frac{\partial^2 T}{\partial \nu^2} + 2\left(b_{12}\frac{\partial^2 T}{\partial \xi \partial \eta} + b_{13}\frac{\partial^2 T}{\partial \xi \partial \nu} + b_{23}\frac{\partial^2 T}{\partial \eta \partial \nu}\right)$$
$$+ \ b_1\frac{\partial T}{\partial \xi} + b_2\frac{\partial T}{\partial \eta} + b_3\frac{\partial T}{\partial \nu} + b_0 T + Q_b \tag{20}$$

where b_{ij} are the same for every case and are defined as

$$
\begin{aligned}
b_{11} &= k_{\xi\xi} & b_{12} &= k_{\xi\eta} \\
b_{22} &= k_{\eta\eta} & b_{13} &= k_{\xi\nu} \\
b_{33} &= k_{\nu\nu} & b_{23} &= k_{\eta\nu}
\end{aligned}
\tag{21}
$$

The coefficients b_i and the constant term Q_b are different for each case and are defined in Table 1. Note that the value "—" for *Face(s)* in the first row of the table indicates an added entry for the interior of the domain, which is a special case of Equation (20). Also, the subscripts $1, 2, \ldots, 6$ refer to the six faces of the computational domain denoted as follows,

$$
\begin{array}{llll}
\text{Face 1:} & \xi = 0 & ; & 0 \le \eta \le \eta_{\max}; & 0 \le \nu \le \nu_{\max} \\
\text{Face 2:} & \xi = \xi_{\max} & ; & 0 \le \eta \le \eta_{\max}; & 0 \le \nu \le \nu_{\max} \\
\text{Face 3:} & 0 \le \xi \le \xi_{\max}; & \eta = 0 & ; & 0 \le \nu \le \nu_{\max} \\
\text{Face 4:} & 0 \le \xi \le \xi_{\max}; & \eta = \eta_{\max} & ; & 0 \le \nu \le \nu_{\max} \\
\text{Face 5:} & 0 \le \xi \le \xi_{\max}; & 0 \le \eta \le \eta_{\max}; & \nu = 0 \\
\text{Face 6:} & 0 \le \xi \le \xi_{\max}; & 0 \le \eta \le \eta_{\max}; & \nu = \nu_{\max}
\end{array}
\tag{22}
$$

and the $\cap$ symbols indicate cases for the intersections of the specified faces.

The governing enthalpy equations, (17) and (20), are solved explicitly using central finite differences in the interior and one-sided differences at the boundaries, evaluated in the computational domain. Temperatures and enthalpies are coupled via Equation (5) by assuming that the material changes phase through a range of temperatures from saturated fluid (T_f) to saturated solid (T_s). The enthalpy difference between the saturated fluid and saturated solid states is equal to the latent heat of fusion of the material (L).

RESULTS

The model is demonstrated by considering solidification of an angle bracket for which the geometry is shown by the mesh in Figure 1. The part fits exactly within a cube 0.127 meter (5 in.) on each side, and the mesh has $75 \times 20 \times 40$ nodes along the bend, thickness, and depth directions, respectively. In the first of two cases, it is assumed that conductive, continuous, unidirectional reinforcing fibers in the composite material follow the curvature of the part and are parallel to the flat front and back surfaces while being orthogonal to the ends. The fiber-oriented thermal conductivity tensor for the material is

$$
\mathbf{K} = \begin{bmatrix} k_{11} & k_{12} & k_{13} \\ & k_{22} & k_{23} \\ & & k_{33} \end{bmatrix} = \begin{bmatrix} 2.000 & 0.000 & 0.000 \\ & 0.318 & 0.000 \\ & & 0.318 \end{bmatrix} \text{ W/m} \cdot {}^\circ\text{C} \tag{23}
$$

which values are typical of a carbon fiber thermoplastic composite. This tensor is transformed into the cartesian system at each node in the mesh based on the orientation of the reinforcing fibers. Thus, while orthotropic in the fiber-oriented coordinate system, this tensor becomes

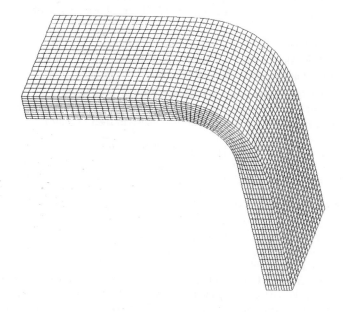

Figure 1: Physical mesh for angle bracket.

fully anisotropic when applied in the cartesian reference frame. Properties of the material system assume the constant values

$$
\begin{aligned}
c_p &= 1300 \text{ J/kg} \cdot {}^\circ\text{C} \\
\rho &= 1480 \text{ kg/m}^3 \\
L &= 460\,000 \text{ J/kg} \\
T_s &= 330\,{}^\circ\text{C} \\
T_f &= 345\,{}^\circ\text{C}
\end{aligned}
\tag{24}
$$

All of the boundaries with the exception of the back face are subjected to a Robin boundary condition with a heat transfer coefficient and ambient temperature of

$$
\begin{aligned}
h &= 5000 \text{ W/m}^2 \cdot {}^\circ\text{C} \\
T_\infty &= 80\,{}^\circ\text{C}
\end{aligned}
\tag{25}
$$

The back face is assumed to be insulated, adding additional complexity to the problem and demonstrating the versatility of the method in managing a variety of boundary conditions. These conditions are similar to those encountered when a part is cooled under pressure in a mold.

The model predicts that complete solidification of the part requires 13.8 minutes (830 seconds) after cooling under the above conditions from an initial uniform temperature of $T_\circ = 380\,{}^\circ\text{C}$.

Figure 2(a) shows the solid and fluid regions near the mid-surfaces of the part after 4 minutes of processing. That is, the part has been "sliced" open for viewing through its thickness and depth, thus exposing its inside. The corresponding temperature distribution is shown in Figure 3(a). At the instant of complete solidification the temperature distribution is as indicated in Figure 4(a). Throughout the process, symmetry about the bend of the part is maintained as expected. In addition, isotherms terminate orthogonal to the back surface, another plane of symmetry.

This first simulation requires slightly over 8.6 minutes of CPU time on an IBM 3090 with a vector facility. More than five minutes of that time are used in the vector processor, thus demonstrating that the method is highly amenable to vectorization.

In the second case, the reinforcing fibers are rotated at an angle of 45° from the orientation in the first case, about local axes through the thickness of the part. The rotated thermal conductivity tensor in the fiber-oriented coordinate system then becomes

$$\mathbf{K} = \begin{bmatrix} k_{11} & k_{12} & k_{13} \\ & k_{22} & k_{23} \\ & & k_{33} \end{bmatrix} = \begin{bmatrix} 1.159 & 0.000 & -0.841 \\ & 0.318 & 0.000 \\ & & 1.159 \end{bmatrix} \text{W/m} \cdot {}^{\circ}\text{C} \tag{26}$$

All other material properties, boundary conditions, and initial conditions are as in the first case.

The different orientation of the reinforcing fibers causes heat to be liberated slightly slower in this second case, with complete solidification occurring after 14.1 minutes (848 seconds), just 18 seconds longer than the time of the first case. This effect is primarily due to the fact that some of the fibers in the second case terminate on the insulated surface (back face) of the part, rather than the surfaces subjected to convective heat transfer. This insulated surface prohibits the highly-conductive fibers from transporting heat out of the part at that location.

Figures 2(b) and 3(b) show the solid–fluid and temperature distributions after 4 minutes for comparison with the first case. Notice that the non-symmetric orientation of the reinforcing fibers in the second case is evident in the uneven temperature distribution along the curve of the part. In both cases, one can see a definite skin–core phenomenon which may lead to undesirable shrinkage and thermal stresses being entrapped in the finished part.

CONCLUSIONS

An enthalpy formulation for phase change has been developed and shown to be very efficient for modeling phase change processes in irregular, three-dimensional domains, even when such domains have the additional complexity of being anisotropic. The method has been demonstrated with an irregular geometry subject to boundary conditions of the second and third kinds, and for two different fiber orientation states of a typical thermoplastic composite material. Computational requirements of the model for the demonstrated cases have been found to be moderate even for relatively large numbers of nodes. Furthermore, the method is capable of accounting for temperature-dependent thermal properties.

The effect of orientation of the reinforcing fibers has been shown to be very significant for the temperature distribution in a typical carbon fiber, thermoplastic-matrix composite material. In particular, skin–core effects may have a pronounced effect on the resultant product quality as thermal and shrinkage stresses become entrapped in the part. A stress analysis based on this thermal history information is required to determine the actual extent of those stresses.

ACKNOWLEDGEMENTS

This work was supported by the U.S. Army Research Office through the University Research Initiative Program (ARO-URI), grant # 24616-MS-UIF.

Face(s)	b_1	b_2	b_3	b_0	Q_b
—	k_ξ	k_η	k_ν	0	0
1	$k_\xi + k_{\xi\xi}$	k_η	$k_\nu + k_{\xi\nu}$	$-h_1\sqrt{a_{11}}$	$q_{b1}\sqrt{a_{11}}$
2	$k_\xi - k_{\xi\xi}$	k_η	$k_\nu - k_{\xi\nu}$	$-h_2\sqrt{a_{11}}$	$q_{b2}\sqrt{a_{11}}$
3	$k_\xi + k_{\xi\eta}$	$k_\eta + k_{\eta\eta}$	$k_\nu + k_{\eta\nu}$	$-h_3\sqrt{a_{22}}$	$q_{b3}\sqrt{a_{22}}$
4	$k_\xi - k_{\xi\eta}$	$k_\eta - k_{\eta\eta}$	$k_\nu - k_{\eta\nu}$	$-h_4\sqrt{a_{22}}$	$q_{b4}\sqrt{a_{22}}$
5	$k_\xi + k_{\xi\nu}$	$k_\eta + k_{\eta\nu}$	$k_\nu + k_{\nu\nu}$	$-h_5\sqrt{a_{33}}$	$q_{b5}\sqrt{a_{33}}$
6	$k_\xi - k_{\xi\nu}$	$k_\eta - k_{\eta\nu}$	$k_\nu - k_{\nu\nu}$	$-h_6\sqrt{a_{33}}$	$q_{b6}\sqrt{a_{33}}$
1∩3	$k_\xi + k_{\xi\xi} + k_{\xi\eta}$	$k_\eta + k_{\xi\eta} + k_{\eta\eta}$	$k_\nu + k_{\xi\nu} + k_{\eta\nu}$	$-h_1\sqrt{a_{11}} - h_3\sqrt{a_{22}}$	$q_{b1}\sqrt{a_{11}} + q_{b3}\sqrt{a_{22}}$
2∩3	$k_\xi - k_{\xi\xi} + k_{\xi\eta}$	$k_\eta - k_{\xi\eta} + k_{\eta\eta}$	$k_\nu - k_{\xi\nu} + k_{\eta\nu}$	$-h_2\sqrt{a_{11}} - h_3\sqrt{a_{22}}$	$q_{b2}\sqrt{a_{11}} + q_{b3}\sqrt{a_{22}}$
1∩4	$k_\xi + k_{\xi\xi} - k_{\xi\eta}$	$k_\eta + k_{\xi\eta} - k_{\eta\eta}$	$k_\nu + k_{\xi\nu} - k_{\eta\nu}$	$-h_1\sqrt{a_{11}} - h_4\sqrt{a_{22}}$	$q_{b1}\sqrt{a_{11}} + q_{b4}\sqrt{a_{22}}$
2∩4	$k_\xi - k_{\xi\xi} - k_{\xi\eta}$	$k_\eta - k_{\xi\eta} - k_{\eta\eta}$	$k_\nu - k_{\xi\nu} - k_{\eta\nu}$	$-h_2\sqrt{a_{11}} - h_4\sqrt{a_{22}}$	$q_{b2}\sqrt{a_{11}} + q_{b4}\sqrt{a_{22}}$
1∩5	$k_\xi + k_{\xi\xi} + k_{\xi\nu}$	$k_\eta + k_{\xi\eta} + k_{\eta\nu}$	$k_\nu + k_{\xi\nu} + k_{\nu\nu}$	$-h_1\sqrt{a_{11}} - h_5\sqrt{a_{33}}$	$q_{b1}\sqrt{a_{11}} + q_{b5}\sqrt{a_{33}}$
2∩5	$k_\xi - k_{\xi\xi} + k_{\xi\nu}$	$k_\eta - k_{\xi\eta} + k_{\eta\nu}$	$k_\nu - k_{\xi\nu} + k_{\nu\nu}$	$-h_2\sqrt{a_{11}} - h_5\sqrt{a_{33}}$	$q_{b2}\sqrt{a_{11}} + q_{b5}\sqrt{a_{33}}$
1∩6	$k_\xi + k_{\xi\xi} - k_{\xi\nu}$	$k_\eta + k_{\xi\eta} - k_{\eta\nu}$	$k_\nu + k_{\xi\nu} - k_{\nu\nu}$	$-h_1\sqrt{a_{11}} - h_6\sqrt{a_{33}}$	$q_{b1}\sqrt{a_{11}} + q_{b6}\sqrt{a_{33}}$
2∩6	$k_\xi - k_{\xi\xi} - k_{\xi\nu}$	$k_\eta - k_{\xi\eta} - k_{\eta\nu}$	$k_\nu - k_{\xi\nu} - k_{\nu\nu}$	$-h_2\sqrt{a_{11}} - h_6\sqrt{a_{33}}$	$q_{b2}\sqrt{a_{11}} + q_{b6}\sqrt{a_{33}}$
3∩5	$k_\xi + k_{\xi\eta} + k_{\xi\nu}$	$k_\eta + k_{\eta\eta} + k_{\eta\nu}$	$k_\nu + k_{\eta\nu} + k_{\nu\nu}$	$-h_3\sqrt{a_{22}} - h_5\sqrt{a_{33}}$	$q_{b3}\sqrt{a_{22}} + q_{b5}\sqrt{a_{33}}$
4∩5	$k_\xi - k_{\xi\eta} + k_{\xi\nu}$	$k_\eta - k_{\eta\eta} + k_{\eta\nu}$	$k_\nu - k_{\eta\nu} + k_{\nu\nu}$	$-h_4\sqrt{a_{22}} - h_5\sqrt{a_{33}}$	$q_{b4}\sqrt{a_{22}} + q_{b5}\sqrt{a_{33}}$
3∩6	$k_\xi + k_{\xi\eta} - k_{\xi\nu}$	$k_\eta + k_{\eta\eta} - k_{\eta\nu}$	$k_\nu + k_{\eta\nu} - k_{\nu\nu}$	$-h_3\sqrt{a_{22}} - h_6\sqrt{a_{33}}$	$q_{b3}\sqrt{a_{22}} + q_{b6}\sqrt{a_{33}}$
4∩6	$k_\xi - k_{\xi\eta} - k_{\xi\nu}$	$k_\eta - k_{\eta\eta} - k_{\eta\nu}$	$k_\nu - k_{\eta\nu} - k_{\nu\nu}$	$-h_4\sqrt{a_{22}} - h_6\sqrt{a_{33}}$	$q_{b4}\sqrt{a_{22}} + q_{b6}\sqrt{a_{33}}$
1∩3∩5	$k_\xi + k_{\xi\xi} + k_{\xi\eta} + k_{\xi\nu}$	$k_\eta + k_{\xi\eta} + k_{\eta\eta} + k_{\eta\nu}$	$k_\nu + k_{\xi\nu} + k_{\eta\nu} + k_{\nu\nu}$	$-h_1\sqrt{a_{11}} - h_3\sqrt{a_{22}} - h_5\sqrt{a_{33}}$	$q_{b1}\sqrt{a_{11}} + q_{b3}\sqrt{a_{22}} + q_{b5}\sqrt{a_{33}}$
2∩3∩5	$k_\xi - k_{\xi\xi} + k_{\xi\eta} + k_{\xi\nu}$	$k_\eta - k_{\xi\eta} + k_{\eta\eta} + k_{\eta\nu}$	$k_\nu - k_{\xi\nu} + k_{\eta\nu} + k_{\nu\nu}$	$-h_2\sqrt{a_{11}} - h_3\sqrt{a_{22}} - h_5\sqrt{a_{33}}$	$q_{b2}\sqrt{a_{11}} + q_{b3}\sqrt{a_{22}} + q_{b5}\sqrt{a_{33}}$
1∩4∩5	$k_\xi + k_{\xi\xi} - k_{\xi\eta} + k_{\xi\nu}$	$k_\eta + k_{\xi\eta} - k_{\eta\eta} + k_{\eta\nu}$	$k_\nu + k_{\xi\nu} - k_{\eta\nu} + k_{\nu\nu}$	$-h_1\sqrt{a_{11}} - h_4\sqrt{a_{22}} - h_5\sqrt{a_{33}}$	$q_{b1}\sqrt{a_{11}} + q_{b4}\sqrt{a_{22}} + q_{b5}\sqrt{a_{33}}$
1∩3∩6	$k_\xi + k_{\xi\xi} + k_{\xi\eta} - k_{\xi\nu}$	$k_\eta + k_{\xi\eta} + k_{\eta\eta} - k_{\eta\nu}$	$k_\nu + k_{\xi\nu} + k_{\eta\nu} - k_{\nu\nu}$	$-h_1\sqrt{a_{11}} - h_3\sqrt{a_{22}} - h_6\sqrt{a_{33}}$	$q_{b1}\sqrt{a_{11}} + q_{b3}\sqrt{a_{22}} + q_{b6}\sqrt{a_{33}}$
2∩4∩5	$k_\xi - k_{\xi\xi} - k_{\xi\eta} + k_{\xi\nu}$	$k_\eta - k_{\xi\eta} - k_{\eta\eta} + k_{\eta\nu}$	$k_\nu - k_{\xi\nu} - k_{\eta\nu} + k_{\nu\nu}$	$-h_2\sqrt{a_{11}} - h_4\sqrt{a_{22}} - h_5\sqrt{a_{33}}$	$q_{b2}\sqrt{a_{11}} + q_{b4}\sqrt{a_{22}} + q_{b5}\sqrt{a_{33}}$
2∩3∩6	$k_\xi - k_{\xi\xi} + k_{\xi\eta} - k_{\xi\nu}$	$k_\eta - k_{\xi\eta} + k_{\eta\eta} - k_{\eta\nu}$	$k_\nu - k_{\xi\nu} + k_{\eta\nu} - k_{\nu\nu}$	$-h_2\sqrt{a_{11}} - h_3\sqrt{a_{22}} - h_6\sqrt{a_{33}}$	$q_{b2}\sqrt{a_{11}} + q_{b3}\sqrt{a_{22}} + q_{b6}\sqrt{a_{33}}$
1∩4∩6	$k_\xi + k_{\xi\xi} - k_{\xi\eta} - k_{\xi\nu}$	$k_\eta + k_{\xi\eta} - k_{\eta\eta} - k_{\eta\nu}$	$k_\nu + k_{\xi\nu} - k_{\eta\nu} - k_{\nu\nu}$	$-h_1\sqrt{a_{11}} - h_4\sqrt{a_{22}} - h_6\sqrt{a_{33}}$	$q_{b1}\sqrt{a_{11}} + q_{b4}\sqrt{a_{22}} + q_{b6}\sqrt{a_{33}}$
2∩4∩6	$k_\xi - k_{\xi\xi} - k_{\xi\eta} - k_{\xi\nu}$	$k_\eta - k_{\xi\eta} - k_{\eta\eta} - k_{\eta\nu}$	$k_\nu - k_{\xi\nu} - k_{\eta\nu} - k_{\nu\nu}$	$-h_2\sqrt{a_{11}} - h_4\sqrt{a_{22}} - h_6\sqrt{a_{33}}$	$q_{b2}\sqrt{a_{11}} + q_{b4}\sqrt{a_{22}} + q_{b6}\sqrt{a_{33}}$

Table 1: Temperature derivative coefficients for general boundary condition equation (20).

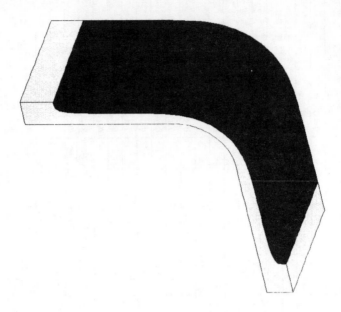

(a) Aligned fibers.

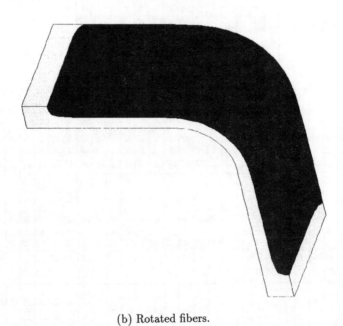

(b) Rotated fibers.

Figure 2: Solid and fluid regions near mid-surfaces after 240 seconds (dark=fluid, light=solid).

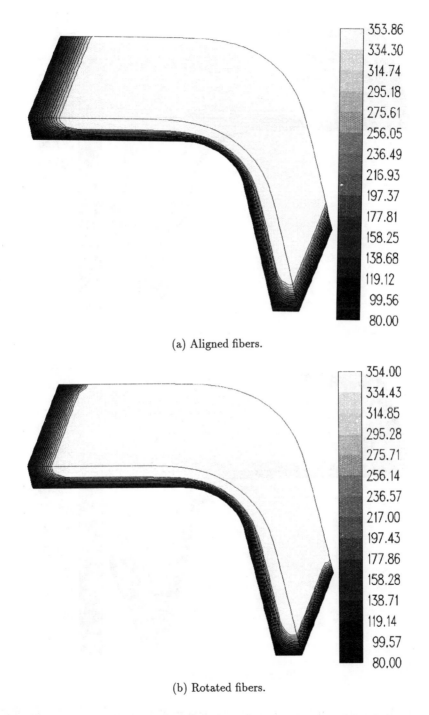

(a) Aligned fibers.

(b) Rotated fibers.

Figure 3: Temperature distribution near mid-surfaces after 240 seconds (units are °C).

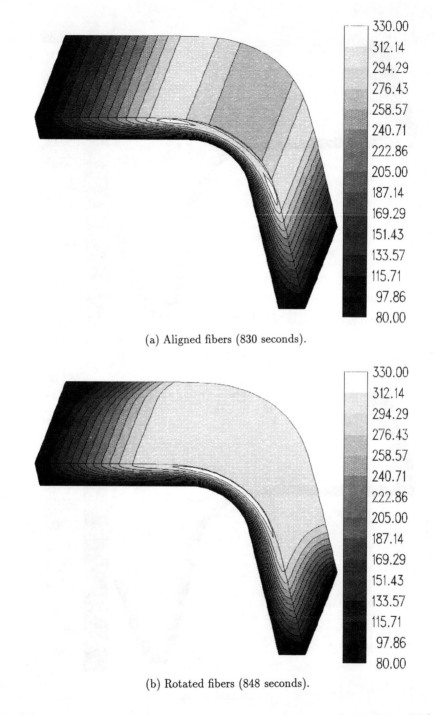

(a) Aligned fibers (830 seconds).

(b) Rotated fibers (848 seconds).

Figure 4: Temperature distribution near mid-surfaces at instant of complete solidification (units are °C).

REFERENCES

Beyeler, E. and Güçeri, S., 1988, "Thermal Analysis of Laser-Assisted Thermoplastic-Matrix Composite Tape Consolidation," *Journal of Heat Transfer*, Vol. 110.

Beyeler, E., Phillips, W., and Güçeri, S., 1988, "Experimental Investigation of Laser-Assisted Thermoplastic Tape Consolidation," *Thermoplastic Composite Materials*, Vol. 1.

Coulter, J. and Güçeri, S., 1987, "Laminar and Turbulent Natural Convection Within Irregularly Shaped Enclosures," *Numerical Heat Transfer*, Vol. 12, pp. 211–227.

Coulter, J. and Güçeri, S., 1988, "Resin Impregnation During the Manufacturing of Composite Materials Subject to Prescribed Injection Rate," *Reinforced Plastics and Composites*.

Erhun, M. and Advani, S., 1989, "A Numerical Study of Heat Flow During Crystallization," In *Symposium on Fluid Mechanics in Materials Processing*, San Francisco, ASME Winter Annual Meeting.

Farraye, E. and Güçeri, S., 1985, "Computational Investigation of Transient Heat Conduction in Composite Structures," Computer-Aided Engineering Report 4/85, Department of Mechanical Engineering, University of Delaware, Newark, Delaware 19716.

Gilmore, S., 1990, *Processing Science of Thermoplastic-Matrix Composite Materials*, PhD thesis, Department of Mechanical Engineering, University of Delaware, Newark, Delaware 19716, in progress.

Gilmore, S. and Güçeri, S., 1988, "Three-Dimensional Solidification, A Numerical Approach," *Numerical Heat Transfer*, Vol. 14, pp. 165–186.

Gilmore, S. and Güçeri, S., 1989a, "Solidification in Three-Dimensional Irregular Domains Via an Enthalpy Model," *Numerical Heat Transfer*, submitted for publication.

Gilmore, S. and Güçeri, S., 1989b, "Thermal Analysis of Thermoplastic Composites Processing," *Journal of Thermoplastic Composite Materials*, submitted for publication.

Güçeri, S., 1989, "Finite Difference Solution of Field Problems," In Tucker, C., editor, *Fundamentals of Computer Modeling for Polymer Processing*, Hanser Publishing Company, Inc., in print.

Özışık, M., 1980, *Heat Conduction*, Wiley.

Shamsundar, N. and Rooz, E., 1988, "Numerical Methods for Moving Boundary Problems," In Minkowycz, W., Sparrow, E., Schneider, G., and Pletcher, R., editors, *Handbook of Numerical Heat Transfer*, pp. 747–765, Wiley.

Shamsundar, N. and Sparrow, E., 1975, "Analysis of Multi-Dimensional Conduction Phase Change via the Enthalpy Model," *Journal of Heat Transfer*, Vol. 97, pp. 333–340.

Shamsundar, N. and Sparrow, E., 1976, "Effect of Density Change on Multi-dimensional Conduction Phase Change," *Journal of Heat Transfer*, Vol. 98, pp. 550–557.

Shieh, C., 1982, "Three-Dimensional Grid Generation Using Poisson Equations," In Thompson, J., editor, *Numerical Grid Generation*, pp. 687–694, Elsevier Science Publishing Company, Inc., New York.

Sottos, N. and Güçeri, S., 1986, "Residual and Transient Thermal Stresses in Laminated Orthotropic Composites," In Häuser, J. and Taylor, C., editors, *Numerical Grid Generation in Computational Fluid Dynamics*, Pineridge Press Limited, Swansea SA3 4BQ, U.K.

Subbiah, S., Trafford, D., and Güçeri, S., 1989, "Non-Isothermal Flow of Polymers into Two-dimensional, Thin Cavity Molds: A Numerical Grid Generation Approach," *International Journal of Heat and Mass Transfer*, Vol. 32, pp. 415–434.

Thomas, P., 1982, "Numerical Generation of Composite Three Dimensional Grids by Quasi-linear Elliptic Systems," In Thompson, J., editor, *Numerical Grid Generation*, pp. 667–686, Elsevier Science Publishing Company, Inc., New York.

Thompson, J., 1987, "A Composite Grid Generation Code For General 3-D Regions," In *AIAA 25th Aerospace Sciences Meeting*, AIAA, AIAA-87-0275.

Thompson, J., Thames, F., and Mastin, C., 1974, "Automatic Numerical Generation of Body-Fitted Curvilinear Coordinate System for Field Containing Any Number of Arbitrary Two-Dimensional Bodies," *Journal of Computational Physics*, Vol. 15, pp. 299–319.

Thompson, J., Warsi, Z., and Mastin, C., 1982, "Boundary-Fitted Coordinate Systems for Numerical Solution of Partial Differential Equations — A Review," *Journal of Computational Physics*, Vol. 47, pp. 1–108.

Walsh, R., 1987, *Thermal and Residual Stresses in Three-Dimensional Laminated Composites*, PhD thesis, Department of Mechanical Engineering, University of Delaware, Newark, Delaware 19716.

PRODUCTION OF MICROCELLULAR PLASTIC PARTS

V. Kumar
Department of Mechanical Engineering
University of Washington
Seattle, Washington

N. P. Suh
Department of Mechanical Engineering and Laboratory for Manufacturing and
Productivity
Massachusetts Institute of Technology
Cambridge, Massachusetts

ABSTRACT

Microcellular thermoplastics refer to plastic foams with cells on the order of 10 microns in diameter. The small cell size is achieved by nucleating the cells near the glass transition temperature where the cell growth rates are relatively small and can be controlled. The microstructure, however, is fragile and conventional means to impart deformation to attain a given geometry are not suitable.

In this paper we describe a process to produce plastic parts with a given geometry and microcellular structure. To meet this objective, deformation has to be integrated in the foaming process in a way that satisfies the requirements for the part geometry as well as the microstructure. The basic idea was to uncouple the cell nucleation and growth processes from deformation. The final process draws upon the physics of cell nucleation and the kinetics of cell growth to meet the design objectives. The process is demonstrated by making a microcullular polystyrene part.

1.0 INTRODUCTION

Microcellular plastics refer to thermoplastic foams with cell diameters on the order of 10μm. This idea was originally conceived as a means to reduce the amount of plastic used in mass-produced items. The rationale was that if a sufficient number of voids smaller than the critical flaw size pre-existing in polymers can be produced, then the amount of plastic used could be reduced without compromising the mechanical properties. Such a process has been developed for amorphous polymers using a thermodynamic instability phenomenon to achieve the cell nucleation (Martini, Waldman, and Suh, 1982; Martini, Suh, and Waldman, 1984).

The basic process involves saturating the polymer with an inert gas below the glass transition temperature and at a high pressure. When the pressure is removed, a supersaturated specimen is produced, since the excess gas is unable to escape the glassy polymer matrix. As the specimen is now heated above the glass transition temperature, a very large number of bubbles spontaneously nucleate. The bubbles cannot expand very rapidly due to the high viscosity of the polymer near the glass-transition temperature. It is therefore possible to obtain cell diameters on the order of 10 μm. Figure 1 shows an example of microcellular foam produced by this method.

Introducing a very large number (say 100 million per cm^3 or more) of very small bubbles leads to some interesting properties. For example, microcellular polystyrene has been found to have several-fold increase in impact strength (see Fig. 2), while maintaining a tensile strength in proportion to foam density (Waldman, 1982). These properties make it possible to foam thin-walled (say 1 to 2 mm thick) parts which, if foamed by conventional means, could suffer an excessive loss of strength. Density reductions of 50% or more have been achieved (Kumar and Suh, 1988), demonstrating the possibility of considerable savings in material costs.

Martini et al. (1982, 1984) originally produced microcellular structure in polystyrene sheets. When such a foamed sheet is deformed to obtain a desired shape, the cells become grossly distorted and are sometimes destroyed in regions of large strain gradients. To circumvent this problem, we may form the part first (say by injection molding) and then foam it using the microcellular process. This procedure results in gross distortions in part shape (see Fig. 3) due to the relaxation of the residual stresses introduced in the part-forming operation. We therefore need a process by which the desired geometry can be obtained while preserving the microcellular structure. In this paper, we describe such a process.

2.0   EXPERIMENTAL

A number of experiments were conducted in order to understand the essential physics of the phenomena involved in the microcellular process, and to establish the interrelationship among the key process variables and the process design parameters. The strategy for the experimental investigation was based on the axiomatic approach to design (Suh, 1984, 1989; Kumar, 1988) and will not be further addressed in this paper. We will present some key experimental results here that have a direct bearing on the process for producing microcellular parts.

All experiments were conducted on DOW XP6065 polystyrene with an average molecular weight of 200,000. Circular disks of 50 mm diameter and 1.6 mm thickness were injection molded from the resin. These disks were saturated with nitrogen in a pressure vessel and foamed in a glycerin bath maintained at 115°C. The microstructure of the resulting foam was studied on a Scanning Electron Microscope (SEM).

To determine the cell density, a micrograph showing 100 to 200 bubbles was obtained and the exact number of bubbles was determined. Figure 1(b) is typical of a micrograph used to determine the cell density. Assuming isotropic distribution of bubbles, the number of cells per cm^3 of foam, N_f, was determined. The average cell diameter D was usually obtained from a second micrograph at a higher magnification. Void fraction v_f in the foam was estimated from

$$v_f = (\pi/6) \ D^3 \ N_f \qquad\qquad\qquad (1)$$

and the number of cells nucleated per cm^3 of the original unfoamed polymer, N_o, was then determined from

$$N_o = N_f/(1-v_f) \qquad\qquad\qquad (2)$$

and is reported here as cell density.

3.0   MICROCELLULAR STRUCTURE AND ITS CONTROL

3.1   GENERAL STRUCTURAL FEATURES

As is evident from Fig. 1, the microcellular foam is extremely homogeneous in comparison to, for example, structural foams. In an effort to quantify the homogeneity of structure, we studied the variations in cell size and cell density in the foamed samples.

3.1.1 Structural Homogeneity

Figure 4 shows average cell size and cell density data from a microcellular foam sample that was saturated with nitrogen at 13.8 MPa (2000 psi) and foamed at 115°C. Overlapping SEM micrographs were taken approximately 250µm apart along the entire length of one specimen (approximately 7 mm long). The micrographs were taken at the center of the specimen so as to avoid any edge effects.

The average cell size is seen to lie between 6 µm and 11 µm. The average cell size is 8.3 µm with a standard deviation of 1.3 µm. The cell density ranges from 170 to 400 million bubbles/cm^3. The average cell density is 280 million/cm^3 with a standard deviation of 60 million/cm^3. Thus we see that both cell size and cell density measurements lie within a factor of two which, for foams, represents a very homogeneous structure.

A close look at Fig. 4 shows that the local variations in cell size and density appear to be out of phase, i.e., when cell density goes up, the average cell size goes down, and vice-versa. The cell size adjusts to the local variations in cell density in such a way that the void fraction remains constant. This is clearly seen in Fig. 5, where void fraction calculated from the data in Fig. 4 has been plotted. The constant void fraction shows that uniform concentration of gas exists in the plastic, and that uniform temperature distribution has been attained during the foaming cycle.

3.1.2 Structure Through Thickness

Our foaming procedure involves saturating the plastic sample under high pressure, bringing the pressure to atmospheric and then heating the sample to a temperature just above the glass transition temperature. From the time that the sample is removed from the pressure vessel to the time that it is heated, a certain amount of gas diffuses out of the sample. We therefore have a variation in the concentration of gas across the sample thickness, it being lowest at the outer edges and increasing as one travels towards the center of the specimen. This variation in gas concentration causes the number of cells nucleated to vary across the sample thickness.

Figure 6 shows the cell density and the average cell size variation across the thickness of a typical microcellular foam sample. We observe three distinct regions to exist. Region I has no bubbles in it, constituting an unfoamed "skin." In this region, the gas concentration is below the minimum that is required for cell nucleation to occur. In region II, the cell density increases from zero in the skin region to a maximum value that corresponds to the pressure under which the plastic sample was saturated. In this region the average cell size is seen to decrease as we move away from the skin region towards the center of the specimen, with the smallest cell size corresponding to the location of highest cell density. Region III represents the "core" of the foam. The cell size and density in this region are constant, within the local variation shown in Fig. 5. This is expected since in this region the gas concentration has not been affected by the loss of gas from outer layers.

3.1.3 Control of Skin Thickness

The thickness of the skin region can be controlled by varying the time over which the gas is allowed to diffuse out of the saturated sample (the gas desorption time). In Appendix A we develop a model for predicting the skin thickness based on one-dimensional diffusion of gas from the polymer sample with initially uniform concentration of gas.

To test this model, a number of polystyrene disks were saturated with nitrogen at 13.8 MPa (2000 psi) and foamed after various gas desorption times ranging from 15 minutes to six hours. Figure 7 shows SEM micrographs from the various samples. We can observe a growth in skin thickness as the gas desorption time increases.

The skin thickness data has been plotted in Fig. 8 and compared to the predictions from the model described in Appendix A. There is good agreement between the predicted and observed values except at small desorption times, where the observed thickness is significantly larger than the predicted value. This may be due to a higher diffusion rate at small values of desorption time, as the diffusion rate in polymers in known to be concentration dependent.

3.2 CONTROL OF CELL NUCLEATION DENSITY

Colton and Suh (1987) studied the phenomenon of nucleation in amorphous polymers in the presence of a nucleating agent. The nucleation process is complex and depends on the solubility, concentration, interfacial energy of any additives present, and the concentration of the foaming gas. The basic thermodynamics of bubble nucleation is discussed in Appendix B.

Although the cell nucleation density can be increased by adding nucleating agents, the best means to control the number of cells nucleated is provided by the gas saturation pressure. Figure 9 shows cell nucleation density in polystyrene samples (without any additives or nucleating agents) as a function of the gas saturation pressure. We see that, at least over the range of saturation pressures used in this experiment, the cell nucleation density increases exponentially with the gas saturation pressure. This establishes gas saturation pressure as a viable process parameter to control cell nucleation density.

3.3   CONTROL OF CELL GEOMETRY

In order to control the void fraction (or the density reduction) of the foam, we need to control both the number of cells as well as their size. Further, it may be desired to control the shape of cells, i.e., spherical or honeycomb, etc., as this may have a bearing on the final foam properties of interest.

The rate at which cells grow at the foaming temperature (typically 10 to 20°C above the glass transition temperature of the polymer) is an important process parameter. This rate is a function of the polymer viscosity and surface tension at temperature, the pressure inside the bubble at a given instant of time, and the external environmental pressure under which the growth is allowed to occur. The pressure inside the bubble changes as the bubble grows, and is affected by the rate at which gas molecules diffuse into the bubble from the surrounding polymer matrix.

The rate of bubble growth in the nitrogen-polystyrene system was experimentally studied. A number of polystyrene samples were saturated with nitrogen at 13.8 MPa (2000 psi) and foamed at 115°C at atmospheric pressure for different lengths of time. Figure 10 shows a plot of the average cell diameter as a function of foaming time. It is seen the cell growth rate is highest in the beginning, and decreases as the cells grow and the driving force for growth is depleted. The cells reach maturity after approximately 100 seconds of foaming time, beyond which little further growth in the average cell size is observed. Most importantly, we see that it takes some 20 to 30 seconds to attain an average cell size of approximately 10 µm.

Figure 11 shows two micrographs from samples foamed under different external pressure conditions with strikingly different cell geometry. In Fig. 11(a) an external pressure was maintained during foaming and subsequent cooling cycle. The external pressure limited the cell growth and spherical cells were formed. In Fig. 11(b) the external pressure was released before the cooling cycle began causing the cells to grow and touch each other until a honeycomb structure was formed. Thus the cell shape can be controlled if desired.

4.0   MANUFACTURE OF MICROCELLULAR PARTS

As mentioned in the introduction, our overall goal was to develop a process in which deformation is integrated in the foaming cycle in such a way that both the microcellular structure and the part geometry can be obtained. One way to achieve this would be to obtain the part geometry first—say in a molding operation—and then subject the part to the microcellular process and a secondary deformation process in order to eliminate the shape distortions shown in Fig. 3. Such a process will involve two heating and cooling cycles, and may be prohibitively expensive. We therefore searched for a process in which microstructure and part geometry could be attained within one thermal cycle.

Our goal could be realized if, starting with a sheet of plastic saturated with gas, we could obtain part geometry first, and then grow the microstructure. Since obtaining part geometry will require heating the plastic above the glass transition temperature, bubbles will nucleate and grow during deformation and the microstructure will probably be damaged. If we could somehow suppress the nucleation of bubbles until deformation was completed, and then nucleate and grow the bubbles, we would achieve our objective. Under these conditions, bubble nucleation will have to occur under a certain external pressure necessary to carry out the deformation step. How would the presence of an external pressure affect the number of cells nucleated?

4.1   EFFECT OF EXTERNAL PRESSURE ON CELL NUCLEATION

We investigated the effect of external pressure on cell nucleation density using the experimental setup shown in Fig. 12. Polystyrene disks were saturated with nitrogen at 13.8 MPa (2000 psi) and placed in the mold where they were subjected to a hydrostatic external pressure provided by a nitrogen cylinder. The mold was then heated by the plattens of a hydraulic press to 115°C, held for a few minutes, and then cooled down to the room temperature. The ample was then removed and examined. The external pressure was maintained throughout the heating and cooling cycle. This experiment was repeated at a number of external pressures, always using polystyrene samples pre-saturated at 13.8 MPa (2000 psi).

Figure 14 shows a plot of logarithm of the cell density as a function of the external pressure used in the experiment described above. We see that the external pressure has no effect on the number of cells nucleated. As the external pressure reduces the degree of supersaturation which drives cell nucleation, we expected cell nucleation density to go down as external pressure increased. The experimental results is therefore surprising and counter-intuitive. The implications of this result, namely that the environmental pressure under which nucleation occurs has no effect on the number of cells nucleated, on the theoretical models of nucleation are discussed elsewhere (Kumar, 1988).

4.2 EFFECT OF GAS SOLUBILITY ON CELL NUCLEATION

In subsequent experiments conducted to explain the above result, we found that the solubility of nitrogen in polystyrene drops substantially—some 40 percent—when the temperature is raised from room temperature to 15°C above the glass transition temperature (see Fig. 14). This drop in solubility provides an additional driving force for cell nucleation that is comparable to that provided by gas supersaturation, and at least partially explains why reduction in gas supersaturation (in the presence of an external pressure) does not reduce the number of cells nucleated.

4.3 A PROCESS TO PRODUCE MICROCELLULAR PARTS

The experiments on gas stability and the effect of external pressure demonstrated further that it is *not possible to suppress* cell nucleation by maintaining an external pressure as we originally thought. Therefore our proposed strategy of (starting with a saturated sheet of polystyrene) getting deformation first, and then obtaining nucleation and growth of bubbles cannot work.

The experimental results, however, suggest the following alternative process strategy. Since cells are bound to nucleate as soon as the gas saturated plastic is heated to the glass transition temperature, one could achieve deformation soon after cell nucleation while the cells are still very small, and then proceed to obtain cell growth (and the desired void fraction) at a higher temperature. This should be possible since it takes some 30 seconds for cells to grow to a size on the order of 10 μm (see Fig. 10) while deformation can be accomplished within one to two seconds.

The result that cell nucleation density is unaffected by any external pressure is a welcome one from the processing standpoint. It suggests that cell nucleation and deformation can be uncoupled in the sense that any external pressures associated in carrying out the required deformation will not affect the number of cells nucleated, and thus our ability to control the cell density is not adversely affected.

The processing strategy is schematically shown in Fig. 15. The presaturated polymer sheet is heated to a temperature T_d above the glass transition temperature T_g at which the required deformation is carried out. The part geometry is obtained in a mold which is pre-heated to a higher temperature T_G at which the desired cell growth is achieved.

4.4 PROCESS DEMONSTRATION

To study the feasibility of the processing strategy, a conventional thermoforming machine was equipped with a mold to produce a plastic container, and the bottom half of the mold was connected to a thermolator which could heat the mold to a desired temperature. A commercial grade high impact polystyrene sheet was saturated with nitrogen at 13.8 MPa (2000 psi) and placed in the thermoformer. The sheet was heated in an infrared oven and formed into a container using a heated mold in a plug-assist vacuum forming operation.

A sample from the bottom of the container and one from the wall of the container was studied under the SEM. Figure 16 shows a picture of the container, and Fig. 17 shows the micrographs from the samples. We see that microcellular structure has successfully been produced in the container.

This demonstrates the feasibility of the process-concept to produce microcellular plastic parts.

5.0 CONCLUSIONS

We have presented a process-concept to produce microcellular thermoplastic parts. In this process, the part geometry and the microstructure can be independently controlled, and the process involves one thermal cycle to achieve both the part shape and the microcellular structure. The feasibility of the process-concept was demonstrated by making a microcellular part from high impact polystyrene sheet on a conventional thermoforming machine. Although the process was demonstrated on polystyrene, the basic process idea is applicable to a variety of thermoplastics.

6.0 ACKNOWLEDGMENTS

This work was funded by the MIT-Industry Polymer Processing Program. The present members of this program are Boeing, DuPont, Kraft, Lockheed, and Lord. This support is gratefully acknowledged.

7.0 REFERENCES

Colton, J.S. and Suh, N.P., 1987, "Nucleation of Microcellular Foam: Theory and Practice," *Polymer Eng. Sci.,* Vol. 27, No. 7, pp. 500-503.

Crank, J., 1956, "Mathematics of Diffusion," Oxford (England), Clarendon Press, 1956.

Durrill, P.L. and Griskey, R.G., 1969, "Diffusion and Solution of Gases Into Thermally Softened or Molten Polymers," *AIChE Journal,* pp. 106-110.

Kumar, V. 1988, "Process Synthesis for Manufacturing Microcellular Thermoplastic Parts: A Case Study in Axiomatic Design," PhD Thesis, Mechanical Engineering, Massachusetts Institute of Technology, Cambridge, Massachusetts.

Kumar, V. and Suh, N.P. 1988, "Structure of Microcellular Thermoplastic Foam," Proceedings of SPE ANTEC 88, Atlanta, Georgia, pp. 715-718.

Martini, J.E., Waldman, F.A. and Suh, N.P., 1982, *SPE Technical Papers,* "Production and Analysis of Microcellular Thermoplastic Foams," *XXVIII,* pp. 674-676.

Martini-Vvedensky, J.E., Suh, N.P. and Waldman, F.A., 1984, U.S. Patent #4, 473, 665.

Suh, N.P., 1984, "Development of the Science Base for the Manufacturing Field Through the Axiomatic Approach," *Robotics and Computer Integrated Manufacturing*, Vol. 1, No. 3/4, pp. 397-415.

Suh, N.P., 1989, *The Principles of Design*, Oxford University Press, Oxford, U.K., (to appear).

Waldman, F.A. 1982, The Processing of Microcellular Foam," S.M. Thesis, Mechanical Engineering, Massachusetts Institute of Technology, Cambridge, Massachusetts.

APPENDIX A

A Model for Skin Thickness

If the region $-\ell < x < \ell$ is initially at a uniform concentration C_0 and the surfaces are kept at a concentration C_1 for $t > 0$, then the concentration distribution, symmetric about $x = 0$, is given by

$$\frac{C - C_0}{C_1 - C_0} = 1 - \frac{4}{\pi} \sum_{n=0}^{\infty} \frac{(-1)^n}{2n+1} e^{\left(\frac{-D(2n+1)^2\pi^2 t}{4\ell^2}\right)} \cos(2n + 1) \frac{\pi x}{2\ell} \tag{A1}$$

where C is the concetration at the distance x at time t, and D is the diffusion constant.

Equation (A1) can be written in terms of dimensionless variables, $T = Dt/\ell^2$ and $X = x/\ell$, and plotted over the range of T and X. Graphs of $\frac{C-C_0}{C_1-C_0}$ for various dimensionless times can be found in standard texts (see, for example, Crank, 1956, p. 46).

For our case the polystyrene disk is saturated at the saturation pressure p_s, and then subjected to atmospheric pressure. Assuming that the nitrogen concentration at the surface immediately assumes the equilibrium value corresponding to atmospheric pressure and stays thereafter, Eq. (A1) can be used to predict the nitrogen concentration in the disk as a function of time. In our experiments the surface of the disk is subject to atmospheric pressure after obtaining equilibrium concentration C_0, so that $C_1 = 0$.

If the diffusion coefficient is known, Dt/ℓ^2 can be calculated for different values of desorption time t. We measured the average diffusion coefficient to be 4.2×10^{-8} cm^2/sec. If C^* represents the minimum gas concentration at hwich nucleation will occur, then the intersection of the line corresponding to the dimensionless concentration of $(1 - C^*/C0)$ with the appropriate Dt/ℓ^2 curve establishes the location x/ℓ beyond which there will be no bubble nucleation. Thus $(1 - x/\ell)$ gives the percentage of the original unfoamed thickness that will become the skin of the foam for the particular nitrogen desorption time.

A value of $C^* = 1$ cm^3 (STP)/g was chosen based on the available data (see Martini et al., 1984), and seems to provide satisfactory predictions for skin thickness. For the nitrogen-polystyrene system, $C^* = 1$ cm^3 (STP) per gram of polystyrene corresponds to a saturation pressure of about 25 atmospheres.

In the skin thickness experiments a saturation pressure of 13.8 MPa (2000 psi) was used, corresponding to a gas concentration of 5.44 cm^3 (STP)/g. For these data the calculated thickness values are given in Table A1, and have been plotted as the predicted curve in Fig. 8.

Table A1
Calculated Skin Thickness Values

t, hr.	Dt/ℓ^2	x/ℓ	$1 - x/\ell$	Skin Thickness (μm)
0.25	0.005	0.98	0.02	17.3
0.5	0.01	0.97	0.03	26.0
1.0	0.02	0.95	0.05	43.2
1.5	0.03	0.94	0.06	51.9
2.0	0.04	0.93	0.07	60.6
3.0	0.06	0.925	0.075	64.9
4.0	0.08	0.91	0.09	77.9
5.0	0.1	0.90	0.10	86.5
6.2	0.125	0.89	011	95.2

APPENDIX B

Effect of Saturation Pressure on Cell Nucleation

Bubbles in polymers may nucleate either homogeneously or heterogeneously. The specific mechanism depends upon whether a second phase is present in the polymer due to an insoluble additive or a nucleating agent. Homogeneous nucleation occurs when the dissolved gas molecules come together for a long enough time to produce a stable bubble nucleus. The homogeneous nucleation rate N_{HOM} is given by (Colton and Suh, 1987)

$$N_{HOM} = C_0 f_0 \exp(-\Delta G'/kT) \tag{B1}$$

where

    C_o =   concentration of gas molecules

    f_o =   frequency factor for gas molecules joining the nucleus

    k =   Boltzman's Constant

    T =   temperature in K

    $\Delta G'$ = activation energy

The activation energy for homogeneous nucleation is given by

$$\Delta G' = \frac{16\pi \, \gamma^3}{3(p_s - p_o)^2} \qquad\qquad (B2)$$

where

    γ =   surface energy of the polymer

    p_s =   gas saturation pressure

    p_o =   environmental pressure

The effect of gas saturation pressure on the cell nucleation density can be qualitatively seen from Eqs. (B1) and (B2) respectively. Equation (B2) shows that a higher saturation pressure leads to a lower activation energy barrier leading to a higher nucleation rate as is evident from Eq. (B1). Thus a higher gas saturation pressure leads to a higher cell density.

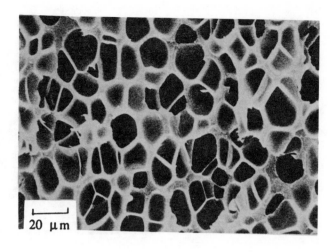

Figure 1. Scanning electron micrograph showing an example of microcellular structure.

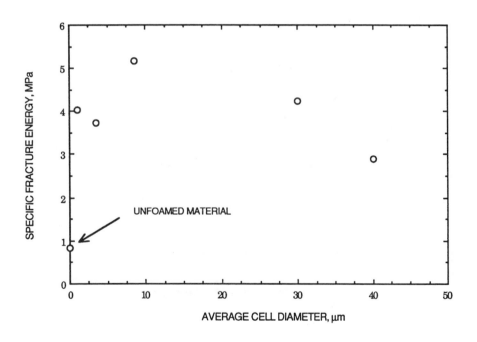

Figure 2. Energy required to fracture microcellular foam as a function of cell size (from Waldman, 1982).

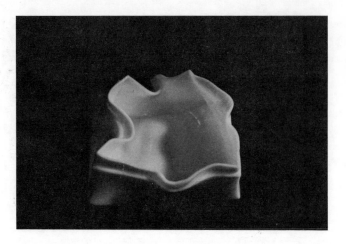

Figure 3. Photograph of an injection molded polystyrene box that was foamed by the microcellular process. Note the gross distortion in the shape of the box.

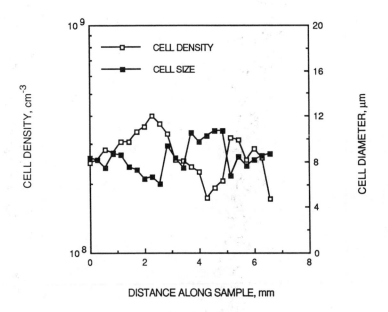

Figure 4. Local variations in microstructure. Note the out of phase character of the cell size and cell density variations.

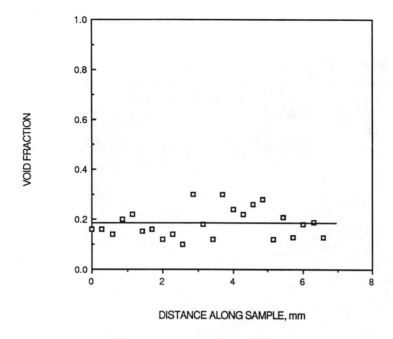

Figure 5. Void fraction along the sample in Figure 4.

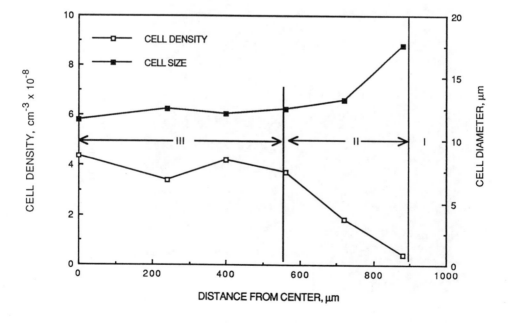

Figure 6. Variation in structure across thickness. The structure is symmetric about the center and has three distinct regions. The data is from one sample.

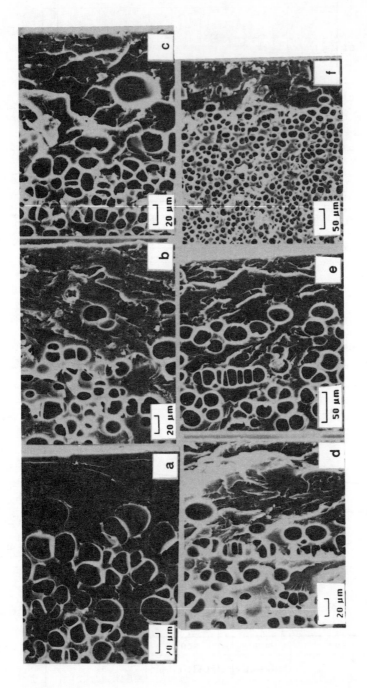

Figure 7. Scanning electron micrographs showing the skin region in six samples (the micrographs are at different magnifications). The nitrogen desorption times are: (a) 0.25 hours, (b) 1.25 hours, (c) 2.25 hours, (d) 4.25 hours, (e) 5.25 hours, and (f) 6.25 hours.

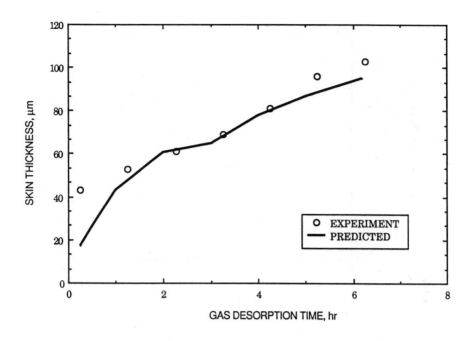

Figure 8. Skin thickness as a function of nitrogen desorption time for polystyrene—comparison of model with experiment.

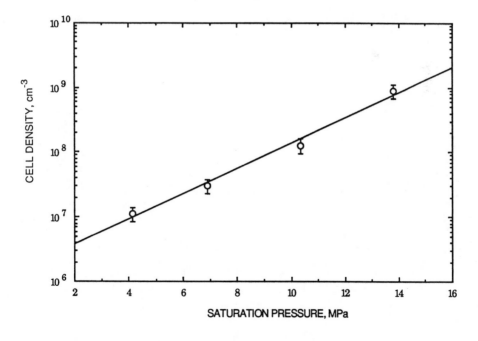

Figure 9. Plot of cell nucleation density in polystyrene as a function of nitrogen saturation pressure.

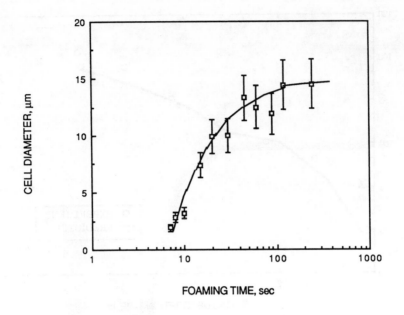

Figure 10. Growth in average cell size as a function of time. All samples were saturated at 13.8 MPa and foamed at 115°C.

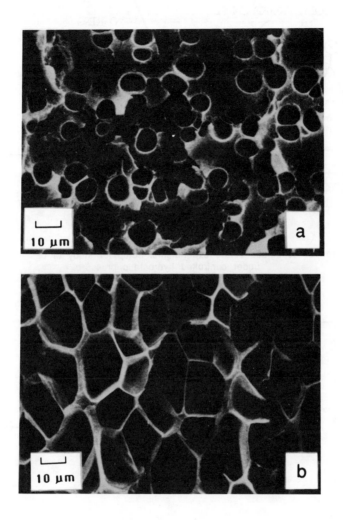

Figure 11. Scanning electron micrographs of polystyrene samples nucleated under external pressure. In (a), the pressure was maintained during the cooling cycle, while in (b), the pressure was released before cooling began.

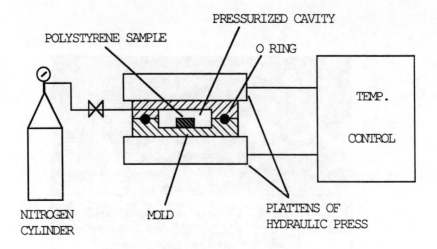

Figure 12. Schematic of the experimental setup for study of cell nucleation under controlled hydrostatic pressure.

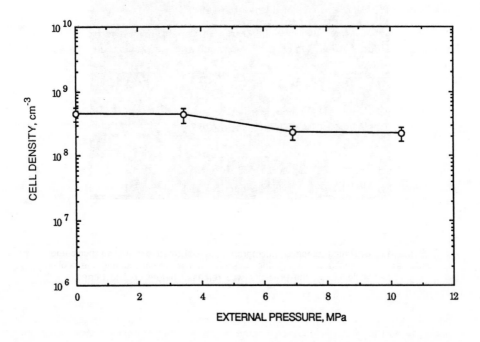

Figure 13. Plot of cell density as a function of external pressure.

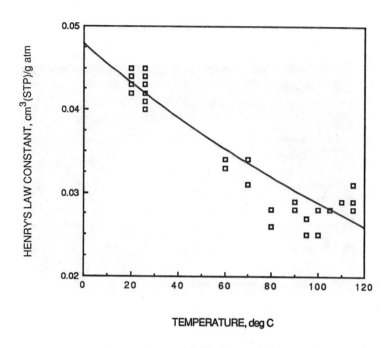

Figure 14. Plot of solubility of nitrogen in polystyrene (expressed as the Henry's Law Constant) as a function of temperature.

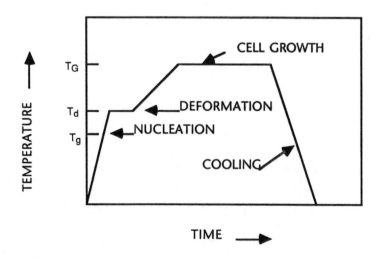

Figure 15. Schematic of the proposed process for producing microcellular parts.

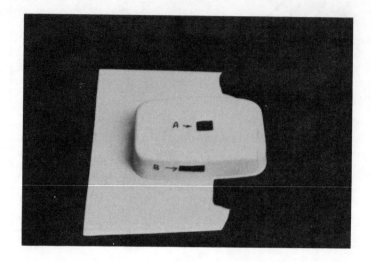

Figure 16. Photograph of a microcellular container made by the process outlined in Figure 15.

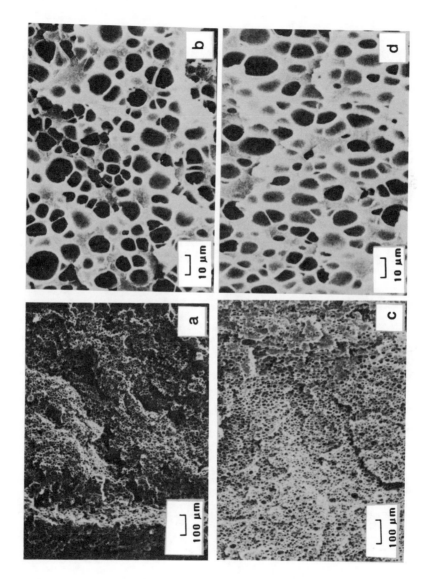

Figure 17. Scanning electron micrographs of samples from the microcellular container: (a) sample from container bottom; (b) close-up of bottom; (c) sample from container wall; and (d) close-up of wall.

A PARAMETRIC STUDY OF BUBBLE GROWTH DURING LOW PRESSURE FOAM MOLDING PROCESS

A. Arefmanesh, S. G. Advani, and E. E. Michaelides
Department of Mechanical Engineering
University of Delaware
Newark, Delaware

ABSTRACT

The bubble size distribution created by the expanding foam plays a key role in controlling the load-bearing and other mechanical properties of the manufactured structural foam part. A numerical method to study the bubble growth and predict the bubble size distribution in polymeric foam is presented. On the microscopic scale, a cell has been used to model the growth of a single bubble. A cell is a closed system composed of a spherical gas bubble and an envelope of polymer with constant mass surrounding the bubble. This cell model accounts for the limited supply of the gas and predicts an equilibrium state for the bubble radius. On the macroscopic scale, the foam has been modeled as a number of cells growing in close proximity to each other. The bubble growth equations for a cell, which are coupled with the field equations for the polymeric fluid, are solved numerically to predict the spatial bubble size distribution and the flow front movement during the expansion process. The effect of processing and material parameters such as the blowing agent concentration, viscosity and diffusion coefficient on the final density reduction and the bubble size distribution is examined.

INTRODUCTION

Polymeric foams are gaining commercial importance as light weight structural materials because of their low density and high impact resistance. They also have good insulation properties because of the entrapped air. Low pressure injection molding is an attractive process to manufacture these foamed engineering plastics because it is possible to mold large parts without increasing the cost as only low injection pressures are required.

Polymeric foams are produced by first dispersing a chemical blowing agent in the polymer melt by pre-blending. Sufficient mass of this mixture to fulfill the weight requirement is injected into a mold cavity filling most of its volume. The mold is continuously cooled by the cooling lines. The trapped gas in the mixture then expands and fills the rest of the mold. Near the walls, as the melt cools rapidly, the growth of bubbles is small due to high viscosity and solidification. The core, which cools slowly, creates the internal cellular structure with large gas bubbles. The bubble size distribution in the fabricated part will determine the physical and mechanical properties of the structure. Hence, it is important to be able to predict and control the growth of bubbles in the mold as the melt front progresses.

The basic phenomena taking place in foam injection molding are the fluid mechanics of the mold filling, nucleation and the non-isothermal growth of bubbles, coalescence of two or more bubbles (interactions among the many growing bubbles of different sizes to form a larger bubble), and solidification. The growth of bubbles is a complex process usually involving interactions between mass, momentum and heat transfer and is governed by many characteristic parameters. It is of prime interest to identify the relative importance of the competing transport phenomena that occur during bubble growth which, will lead to an increased understanding of the processing of these materials.

The present work is a first step towards a more elaborate study on the dynamics of bubble growth. In this work, we examine the growth of many bubbles in close proximity to each other during the mold filling process. A parametric numerical investigation under isothermal conditions has been conducted. The influence of different dimensionless parameters on the growth of spatially distributed bubbles and on the relative reduction in transient bulk density has been predicted. The existence of an axial pressure gradient in the mold due to the bubble growth is demonstrated through numerical experiments.

PREVIOUS WORK

The subject of a single bubble growth in a liquid is well understood from earlier studies on boiling [Hetsroni, 1982] under equilibrium or non-equilibrium conditions. Plesset and Zwick [1954] studied the growth of vapor bubbles in superheated liquids. The subject of bubble growth in viscous liquids, is relatively new. Among the early studies, one can mention the work of Barlow and Langloir [1962]. They studied the problem of diffusion of a gas from a Newtonian liquid into an expanding bubble. Their results show the change of radius of the bubble as a function of time for early stages of the growth (up to two seconds).

Street and co-workers [1971] also studied the growth of bubble in a viscous liquid. In their work, they considered the effect of heat, mass and momentum transfer. They used a non-Newtonian (power-law) constitutive equation to relate the stresses in the fluid to the velocity gradients. They investigated the effect of different parameters such as viscosity, shear thinning, etc. on the bubble growth. More recently, Han and co-workers [1978, 1981] have conducted experimental as well as numerical studies on foam injection molding. In their numerical work, they studied the growth of a single bubble in an infinite sea of liquid. They assumed that the behavior of the single bubble is representative of all the bubbles in the mold. Upadhyay [1985] studied the growth of a single bubble in a viscoelastic fluid under non-isothermal condition.

In the polymeric foam process, the bubbles are separated by short distances as compared to their radii hence a single bubble growing in an infinite sea of liquid is not an adequate model to represent the physical situation. Amon and Denson [1982] used a cell model to account for the depletion of the gas from the polymer melt as the bubble expands. This cell model also takes into consideration that there is much less liquid to be stretched during the bubble expansion surrounded by a thin film of polymer instead of a large mass of liquid. In their work, they combined the equations for bubble growth with the macroscopic flow parameters such as pressure and velocity. However, they assumed that an average bubble can be representative of all the bubbles in the foam. We have relaxed that assumption and solve for growth of many closely spaced bubbles simultaneously.

THEORY

The present analysis is based on the assumption that bubbles are spherical in shape and that they grow symmetrically. Furthermore, we assume that the nucleation is heterogeneous and instantaneous. It has been shown elsewhere [Cole, 1974 and Blander et al., 1975] that homogeneous nucleation is negligible in polymeric foam processes. We also assume that nucleation is uniform throughout the mold cavity. This allows us to nucleate a specific number of bubbles in the initial charge and we study the simultaneous growth of these bubbles as the foam expands. To describe the important features of this system of several bubbles growing in close proximity to one another, we consider the expansion of a bubble surrounded by a film of liquid with finite thickness. The growth of a single bubble surrounded by a film of viscous liquid with finite thickness was first studied by Street and his co-workers [1971]. This concept was later formulated into a cell model by Amon and Denson [1982].

In the present study, we consider a unit cell composed of a spherical bubble surrounded by a finite amount of polymer. Figure 1 shows the schematic representation of a cell. We divide the initial charge into a specific number of cells of equal and constant mass. The number of cells must be equal to the number of nuclei, which is determined from experiments or from thermodynamic considerations. If M is the shot size of the charge and N is the number of bubbles, then the mass (m_{cell}), density (ρ_{cell}) and the volume (V) of a cell can be defined as follows:

$$m_{cell} = \frac{M}{N} \quad ; \quad V = \frac{m_{cell}}{\rho_p} \quad ; \quad \rho_{cell} = \frac{m_{cell}}{V + 4/3\pi R^3} \tag{1}$$

where R is the bubble radius and ρ_p is the density of the polymer.

Using the continuity and momentum equations for the polymer in the cell and neglecting the inertia terms, one can determine the bubble radius as a function of time. For a Newtonian fluid under isothermal conditions, it is given by

$$\frac{dR}{dt} = \frac{1}{4\mu}\left(\{P_g\text{-}P_f\}R\text{-}2\sigma\right) \tag{2}$$

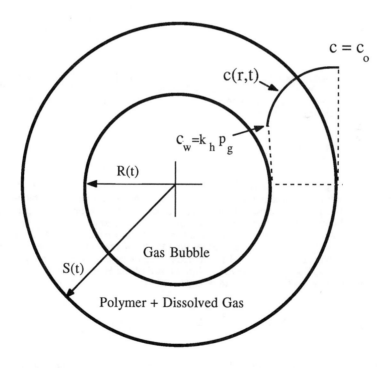

Fig.1: Schematic Diagram of a cell

where μ is the viscosity, σ is the surface tension, P_g is the gas pressure inside the bubble and P_f is the fluid pressure at the outer boundary of the cell. The gas pressure inside the bubble, p_g, and the fluid pressure around the cell, p_f, are needed for the solution of Eq.(2). At this point, we make two assumptions pertaining to the gas inside the bubble. First, we assume that the gas inside the bubble behaves as an ideal gas. Secondly, we assume that the concentration of the dissolved gas at the bubble interface is related to the gas pressure inside the bubble through Henry's law:

$$C_w = K_h P_g \tag{3}$$

where C_w is the gas concentration at the interface of the bubble and the polymer as shown in Fig. 1 and K_h is Henry's law constant. Initially, the gas concentration is uniformly distributed throughout the cell and is equal to c_o. However, as the foam molding process initiates, the growth of the gas bubble will lead to reduction in pressure inside the bubble. Consequently, the concentration of the dissolved gas at the interface will also decreases (Eq.(3)). Hence, at the instant the process initiates $(t=0^+)$ there will exist a concentration profile in the liquid envelope of the cell and the dissolved gas from the polymer will start to diffuse into the bubble. The diffusion process of the dissolved gas in the envelope is governed by the following equation:

$$\frac{\partial c}{\partial t} + v_r \frac{\partial c}{\partial r} = \frac{D}{r^2} \frac{\partial}{\partial r}\left(r^2 \frac{\partial c}{\partial r}\right) \qquad R(t) < r < S(t) \tag{4}$$

where c is the concentration of the blowing agent, D is the diffusion coefficient, R(t) is the radius of the bubble and S(t) is the outer radius of the cell as shown in Fig. 2. Here, v_r is the radial fluid velocity and is determined by conservation of mass of the fluid in the envelope and is given by:

$$v_r = \frac{\dot{R} R^2}{r^2} \tag{5}$$

where r is the radial distance from the center of the bubble and $\dot{R}$ is the rate of change of the bubble radius. Detailed derivation of Eq. (5) is given by Han and Yoo [1981]. To determine the gas pressure inside the bubble, we apply the conservation of mass principle to the gas inside the bubble which results in the following equation:

$$\frac{d}{dt}\left(\frac{P_g}{R_g T} R^3\right) = 3\rho D R^2 \left(\frac{\partial c}{\partial r}\right)_{r=R} \tag{6}$$

where R_g is the ideal gas law constant and T is the thermodynamic temperature. From Eq. (6) one notes that the concentration gradient at the interface relates the gas pressure to the bubble radius. Therefore, we need to know the concentration gradient at the interface before we calculate the pressure inside the bubble. As the concentration profile is restricted to a small diffusion boundary layer inside the cell, we will make use of moment integral method[Rosner et al., 1972] and obtain an approximate solution for the concentration using Eq. (4). In the present study, we restricted to a second order profile for the concentration. The detailed procedure on the use of the moment integral method can be found in [Rosner et al.,1972]. With the use of the Henry's law (Eq. (3)) and the second order concentration profile, Eq. (6) can be expressed as:

$$\frac{d}{dt}\left(\frac{P_g R^3}{R_g T}\right) = \frac{6\rho^2 D K_h^2 (R_g T)[P_{go}-P_g]^2 R^4}{P_g R^3 - P_{go} R_o^3} \tag{7}$$

Equations (2) and (7) form a set of coupled first order ordinary differential equations which describe the microscopic behavior of the cell with the initial conditions as follows:

$$R(t=0) = R_o \tag{8a}$$

$$P_g(t=0) = P_{go} \qquad\qquad (8b)$$

However, these equations that model the microscopic behavior are coupled with the macroscopic flow through the fluid pressure around the cell (P_f).

To find P_f, we need to apply the conservation principles to the macroscopic flow. Figure 2 depicts the schematic of the mold cavity (with the gap height exaggerated). The growth process of the cells give rise to an axial pressure gradient in the mold cavity which drives the fluid. To obtain a detailed representation of the pressure gradient, we model the foam as a system of cells and a compressible medium. Hence, we can apply the continuity condition to this system as follows:

$$\frac{\partial \rho_{cell}}{\partial t} + \underline{\nabla} \cdot (\rho_{cell}\, \underline{V}) = 0 \qquad\qquad (9)$$

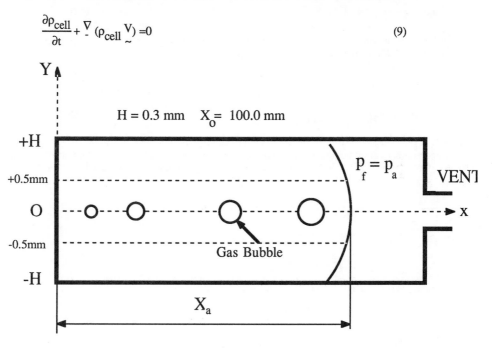

Fig. 2: Schematic diagram of the mold cavity

Since the mold cavity is thin, the usual lubrication approximation is justified. To study the movement and the growth of bubbles as the mold fills, we selected a section of the mold (Fig. 2) at the plane of symmetry for two reasons. First, as we have restricted this analysis to an isothermal case, the temperature field will not vary drastically in the selected zone. Secondly, this allows us to treat the flow as one-dimensional and reduces the governing equation for the fluid pressure in the mold to a second order ordinary differential equation.

With the lubrication approximation and ignoring the inertia terms in the momentum equation, the velocity and axial pressure gradient relation reduces to:

$$v_x = \frac{y^2 - H^2}{2\mu} \frac{dP_f}{dx} \qquad\qquad (10)$$

where we have used the no slip boundary condition at the walls. Substitution of Eq. (10) and the definition of the cell density (Eq. (1)) in the continuity condition (Eq. (9)) results in the following second order ordinary differential equation for the pressure in the mold;

$$\frac{d^2 P_f}{dx^2} + \eta_1 \frac{dP_f}{dx} + \eta_2 \, P_f = \eta_3 \qquad\qquad (11)$$

where

$$\eta_1 = -\frac{4\pi R^2 \frac{\partial R}{\partial x}}{\xi} \quad , \quad \eta_2 = -\frac{2\pi R^3}{H^2 \, \xi} \quad , \quad \eta_3 = -\frac{2\pi R^3 \left(P_g - \frac{2\sigma}{R} \right)}{H^2 \xi}$$

$$\text{and} \ \ \xi = V + \frac{4}{3}\pi R^3 \qquad\qquad (12)$$

To calculate the fluid pressure, the conditions of zero velocity at the gate and atmospheric pressure at the front are used. Mathematically, they may be expressed as

$$\frac{dP_f}{dx} = 0 \qquad \text{at} \qquad x{=}0 \qquad\qquad (13a)$$

$$P_f = P_a \qquad \text{at} \qquad x{=}X_a \, (t) \qquad\qquad (13b)$$

where X_a is the instantaneous position of the melt front. Equation(12) is a boundary value problem. To facilitate numerical implementation, we replace this equation by a set of equivalent initial value problem [Greenberg, 1975]. The equivalent set of initial value problem in this case is as follows:

$$\frac{d^2 p_1}{dx^2} + \eta_1 \frac{dp_1}{dx} + \eta_2 \, p_1 = 0 \qquad\qquad (14)$$

$$\frac{d^2 p_2}{dx^2} + \eta_1 \frac{dp_2}{dx} + \eta_2 \, p_2 = \eta_3 \qquad\qquad (15)$$

Initial conditions for Eqs. (14) and (15) are given by:

$$p_1(x{=}0) = p_a \qquad\qquad (16a)$$

$$\frac{dp_1}{dx}(x{=}0) = 0 \qquad\qquad (16b)$$

$$p_2(x{=}0) = 0 \qquad\qquad (16c)$$

$$\frac{dp_2}{dx}(x{=}0) = 0 \qquad\qquad (16d)$$

After obtaining the solution to p_1 and p_2, the pressure inside the mold cavity can be obtained from the following algebraic relation,

$$p(x) = \left(\frac{p_a - p_2(X_a)}{p_1(X_a)} \right) p_1(x) + p_2(x) \qquad\qquad (17)$$

Equations (2) and (7) along with initial conditions (Eq. (8)), describe the microscopic behavior. They are coupled with Eqs. (14),(15) and (17) and initial conditions(Eq.(16)) which models the macroscopic behavior. This constitutes the system of governing equations to describe the growth of bubbles in the foam molding process.

DIMENSIONLESS FORM OF THE GOVERNING EQUATIONS

We transform the working equations into a dimensionless form to estimate the relative importance of the various characteristic parameters which influence the foam molding process. It is observed that the process has two length scales, X_0 (initial position of the front) and R_0 (the macro and microscales) and two characteristic time scales, $\frac{R_0^2}{D}$, related to the diffusion process and $\frac{P_a}{4\mu}$, which represents the momentum transfer. However, as the characteristic time for diffusion is the dominant one, we will use that as our characteristic value in the dimensionless analysis. Furthermore, we make the forces dimensionless with atmospheric pressure p_a, both in the macroscopic and microscopic equations. Hence, we use the characteristic quantities (R_0, $\frac{R_0^2}{D}$ and p_a) to make the microscopic equations dimensionless and (X_0, $\frac{R_0^2}{D}$ and p_a) to make the macroscopic equations dimensionless. This leads to the following dimensionless variables (denoted with a superscript *):

$$P_g^* = \frac{P_g}{P_a} \tag{18a}$$

$$P_f^* = \frac{P_f}{P_a} \tag{18b}$$

$$R^* = \frac{R}{R_0} \tag{18c}$$

$$t^* = \frac{t\,D}{R_0^2} \tag{18d}$$

$$x^* = \frac{x}{X_0} \tag{18e}$$

$$\xi^* = \frac{v + \frac{4}{3}\pi R^3}{R_0^3} \tag{18f}$$

$$\zeta^* = \frac{p_g R^3 - p_{go} R_0^3}{p_{go} R_0^3} \tag{18g}$$

Eq. (2) and (7) can now be expressed in a dimensionless form as follows:

$$\frac{dR^*}{dt^*} = c1[\, P_{go}^* \, (\frac{1+\zeta^*}{R^{*2}}) - P_f^* R^* - c2\,] \tag{19}$$

$$\frac{d\zeta^*}{dt^*} = 6\left(\frac{\rho_p}{\rho_{go}}\right)^2 \frac{c_0^2}{\zeta^*} \left(\frac{R^{*3} - \zeta^* - 1}{R^*}\right)^2 \tag{20}$$

where, c1 is the dimensionless parameter $\dfrac{R_0^2 \, p_a}{4\mu D}$ which represents the ratio of the characteristic time scale of momentum to the characteristic time scale for diffusion and c2 is the dimensionless surface tension $\dfrac{2\sigma}{p_a R_0}$.

When the triad (X_0, $\dfrac{R_0^2}{D}$ and p_a) was used to make the macroscopic equations for the pressure (14, 15 and 17) dimensionless, the following set of equations was obtained:

$$\frac{d^2 p_1{}^*}{dx^{*2}} + \eta_1{}^* \frac{dp_1{}^*}{dx^*} + \eta_2{}^* p_1{}^* = 0 \tag{21}$$

$$\frac{d^2 p_2{}^*}{dx^{*2}} + \eta_1{}^* \frac{dp_2{}^*}{dx^*} + \eta_2{}^* p_2{}^* = \eta_3{}^* \tag{22}$$

$$P_f{}^*(x^*) = \left(\frac{1 - p_2{}^*(X_a{}^*)}{p_1{}^*(X_a{}^*)} \right) p_1{}^*(x^*) + p_2{}^*(x^*) \tag{23}$$

where coefficients $\eta_1{}^*$, $\eta_2{}^*$ and $\eta_3{}^*$ are the following:

$$\eta_1{}^* = - \frac{4\pi R^{*2} \frac{\partial R^*}{\partial x^*}}{\xi^*} \quad ; \quad \eta_2{}^* = - \frac{2\pi R^{*3}}{\xi^*} \left(\frac{X_0}{H} \right)^2$$

$$\eta_3{}^* = - \frac{2\pi R^{*3}}{\xi^*} \left(\frac{X_0}{H} \right)^2 \left(P_g{}^* - \frac{c2}{R^*} \right) \tag{24}$$

Equations (19-23) form the system of governing equations in dimensionless form. All parametric studies that follow have been conducted by numerically solving these equations with the following initial conditions:

$$R^*(t^* = 0) = 1 \tag{25a}$$

$$\zeta^*(t^* = 0) = 0 \tag{25b}$$

$$p_1{}^*(x^* = 0) = 1 \tag{25c}$$

$$p_2{}^*(x^* = 0) = 0 \tag{25d}$$

$$\frac{dp_1{}^*}{dx^*}(x^* = 0) = 0 \tag{25e}$$

$$\frac{dp_2{}^*}{dx^*}(x^* = 0) = 0 \tag{25f}$$

NUMERICAL IMPLEMENTATION

Our goal is to determine the bubble radius at different positions during the mold filling as a function of time. Eventually, we will use this information to compute the bulk density of the foam. The full system of governing equations consists of two coupled first order ordinary differential equations (Eqs. (19) and (20)) and two coupled second order ordinary differential equations (Eqs. (21) and (22)). A numerical scheme combining second and fourth order Runge-Kutta method was used to solve this system of equations.

The flow domain as shown in Fig.2 is one millimeter thick and wide. The initial length is 10 centimeters and is discretized into a fixed number of nodes, N, which represents the number of bubble nuclei. The center of each cell coincides with the nodal point. The numerical approach permits us to nucleate the bubbles at definite locations in the initial charge domain. However, we assume that the blowing agent is well mixed with the polymer and hence the initial charge has nucleation sites uniformly distributed. Therefore, in this study, initially we space the nodes uniformly along the x-direction with the initial radius, R_0, and the initial concentration, c_0 being equal at every node. The number of bubbles to be nucleated is an input to the program and must be specified before we start the calculations. For each time step, the full system of the governing equations is solved for all the N bubbles simultaneously and the new radii of the bubbles, the gas pressure, fluid pressure and the axial velocity at each nodal point are determined. It is assumed that the bubbles move together with the fluid. Hence, using the axial velocity and the time step, the nodal points (the center of each cell) are convected to their new locations and the calculations are repeated for the next time step. Note that the cells will no longer be equally spaced and there exists a possibility that the bubbles may come within a diameter distance of each other which will lead to coalescence. At the present time, the numerical scheme does not allow coalescence to occur. However, within the present numerical scheme, it is possible to model this physical phenomenon and we plan to add this feature in the future. We stopped our calculations after 10 seconds because our intent is to investigate the relative importance of characteristic parameters during the foam molding process. However, one may continue the calculations until the mold is full.

The numerical simulation was used to investigate the effect of various dimensionless parameters such as c1, which is the ratio of the characteristic time scales for momentum and diffusion, the surface tension parameter, c2, dimensionless initial gas pressure in the bubble and the number of bubbles on the behavior of the system. The selected range of the parameter values for this study were those employed in common industrial foam molding process.

RESULTS AND DISCUSSIONS

To initiate foam molding process, the number of bubbles that nucleate will depend on various thermodynamic properties and the distribution of the blowing agent within the polymer. In this study, we selected a model number of one hundred nuclei since our objective is to demonstrate the usefulness of the cell model and to study the influence of the bubbles growing in close proximity to each other on the overall behavior of the process.

The governing equations were solved numerically, for the case of the transient growth of the coupled system of one hundred cells. The influence of dimensionless parameters, $P_{go}{}^{*}$ and c1 on the rate of growth of the spatially distributed bubbles and on the bulk foam density was examined. For the range of common industrial interest values, the surface tension parameter, c2, did not have a significant influence on the growth process. The initial bubble radius chosen was one micron. It was observed that the growth rate becomes independent of the initial radius after approximately two seconds. Similar observations were made by Upadhyay [1985].

A quantity which is of prime interest in foam molding process is the bulk foam density. We compute this value by averaging the density of all the cells in the domain. Figure 3 shows the effect of changing the dimensionless initial gas pressure on the percent reduction in bulk foam density. Initial gas pressure is directly related to the initial gas concentration through Henry's law. Therefore, increasing the initial gas pressure is equivalent to increasing the initial blowing agent concentration. This leads to faster mass transfer, which becomes the growth rate determining factor and results in higher density reduction.

The dimensionless parameter, c1, exhibits collectively the influence of the diffusion coeffiecient and the viscosity of the fluid. We can vary c1, either by changing D or μ. However, if we change D, we change simultaneously the characteristic time scale for the process. In order to establish a physical understanding of the process and to compare our results with the experimental observations reported in the literature, we plot our results in real time and not in dimensionless time. Hence, the results for the same c1 will be different depending on whether we change the viscosity for the the polymer or if we change the diffusion coefficient, in which case we have changed the respective characteristic time scale.

Figure 4 demonstrates the effect of dimensionless parameter c1 on the percent reduction in bulk foam density. Here, we keep the viscosity constant and change the diffusion coefficient. This allows us to examine the effect of diffusion coefficient on the bulk foam density. Decreasing c1 is equivalent to increasing the diffusion coefficient and this leads to the rapid diffusion of gas into the bubble. Therefore, the bubbles grow faster, resulting in lower foam density. On the other hand, Fig. 5, reveals the influence of changing viscosity of the polymer (which changes c1) on the bulk foam density. Here, the characteristic time scale for the process is constant since the diffusion coeffiecient does not change. Therefore, an increase in c1 reflects the decrease in viscosity which leads to less resistance to the bubble expansion and rapid growth and consequently decrease in the bulk foam density.

To accentuate the importance of spatial location of the bubble, Figs. 6, 7 and 8 show the effect of different parameters on the growth of the bubble nearest to the gate (which we call the first bubble) and the bubble nearest to the front (which we call the last bubble). It is observed that the difference between the radius of the first bubble and the last bubble is directly proportional to the diffusion coefficient and the initial concentration of the blowing agent and varies inversely with the melt viscosity. This model is capable to account for the experimentally observed phenomena of larger bubbles closer to the melt front as compared to those at the gate. The spatial bubble size distribution is attributed to the existence of an axial pressure gradient in the mold which is in turn created due to the dynamics of bubble growth.

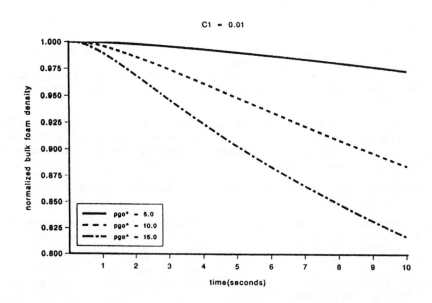

Fig.3: Effect of Initial Gas Pressure on Bulk Foam Density

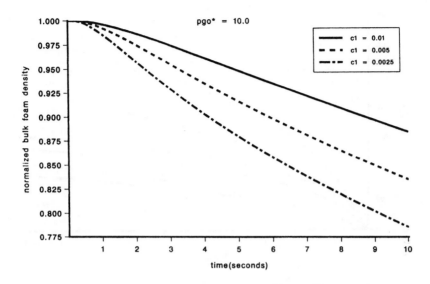

Fig. 4: Effect of Parameter C1 on Bulk Foam Density
(Keeping Viscosity Constant)

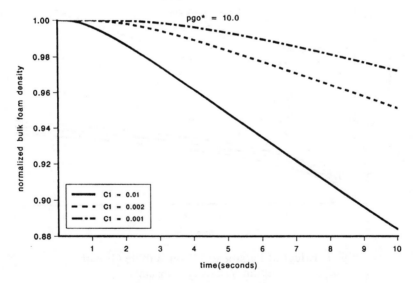

Fig. 5: Effect of Parameter C1 on Bulk Foam Density
(Keeping Diffusion Coefficient Constant)

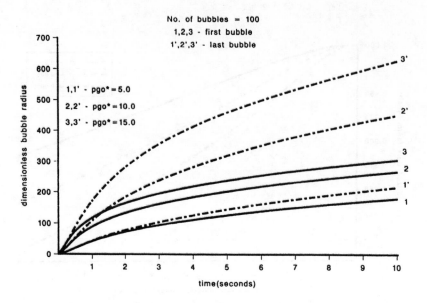

Fig. 6: Effect of Initial Gas Pressure on Bubble Growth

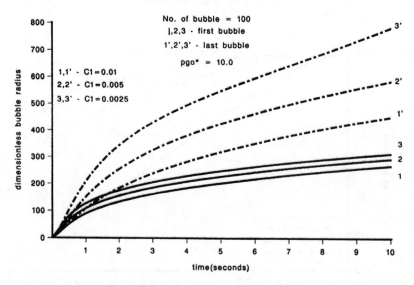

Fig. 7: Effect of Parameter C1 on Bubble Growth
(Keeping Viscosity Constant)

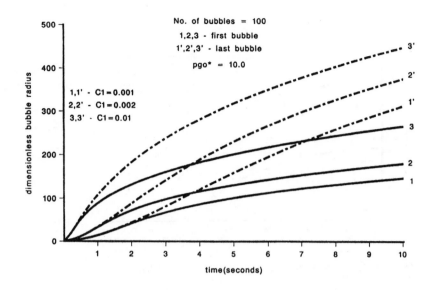

Fig. 8: Effect of Parameter C1 on Bubble Growth
(Keeping Diffusion Coefficient Constant)

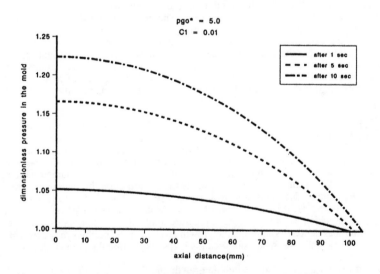

Fig. 9: Axial Pressure Distribution in the Mold

353

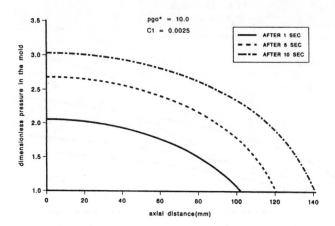

Fig.10: Axial Pressure Distribution in the Mold

(Changing Diffusion Coefficient)

Figures 9 and 10 depict the axial pressure distribution in the mold for different dimensionless parameters after 1, 5 and 10 seconds. The maximum pressure is attained near the gate and the pressure decreases to the atmospheric pressure at the melt front. The gradient of pressure is zero at the gate and it increases to its maximum value near the front. Hence the fluid velocity is much higher near the front. It is due to this high axial pressure gradient, that bubbles closer to the front grow faster than bubbles closer to the gate. Similar observations were made experimentally by Amon[1982]. He observed that the bulk foam density decreases in the axial direction and the rate of decrease was much faster closer to the front, which is the outcome of larger bubbles near the front.

From Fig. 9, it can be seen that lower concentration of the blowing agent and lower diffusion coefficient results in slower bubble growth and ,hence, the axial pressure distribution is uniform in most of the mold cavity, except near the front. Also, due to smaller pressure gradients, the front movement is much slower which results in larger time for mold filling. On the other hand, in Fig. 10, the concentration of the blowing agent is much higher and the characteristic time for diffusion is much smaller, resulting in non-uniform pressure distribution throughout the mold cavity and large pressure gradients due to faster growth rates. Also, note that the front movement is much faster for this case. However, this process will become unstable as larger gradients lead to even faster bubble growth, which in turn create large pressure gradients. In a real industrial process, the concentration of the gas depletes and limits the growth process or the bubbles may come in contact with the mold walls and collapse. Also, solidification helps in arresting the growth process.

Thus, our results suggest that the bubble growth and the pressure distribution are strongly coupled. Hence, it is necessary to consider the spatial variation of the bubble growth phenomena instead of representing the collective growth of all the bubbles in the mold by an average bubble.

CONCLUSION

In this work, we have studied the growth of many bubbles in close proximity to one another, which represents the physical situation in a foam molding process. We investigated the influence of various dimensionless parameter to examine the behavior of the process for the case of one hundred nuclei. The results indicate that there exists a noticeable pressure gradient in the axial direction in the mold, which is created due to the growth of the bubbles. Our model also predicts the existence of a bubble size distribution in the mold and the outcome of larger bubbles at the melt front as compared to the size of bubbles near the gate is consistent with the experimental observations reported in the literature.

REFERENCES

Amon. M.,1982," Theoretical and Experimental Study of Foam Growth Dynamics with Application to Structural Foam Molding," Ph.D. Thesis, Dept. of Chemical Engineering, University of Delaware, Newark, Delaware.

Amon, M., and Denson, C. D., 1984, " A Study of the Dynamics of Foam Growth: Analysis of the Growth of Closely Spaced Spherical Bubbles," *Polymer Engineering and Science*, Vol. 24, pp. 1026-1034.

Amon, M., and Denson, C. D., 1986, " A Study of the Dynamics of Foam Growth: Simplified Analysis and Experimental Results for the Bulk Foam Density in Structural Foam Molding," *Polymer Engineering and Science*, Vol. 26, pp. 255-267.

Barlow, E. J., and Langlois, W. E., 1962, " Diffusion of Gas from a Liquid into an Expanding Bubble," *IBM Journal*, Vol. 6, pp. 329-337.

Blander, M., and Kaltz, J. L., 1975, " Bubble Nucleation in Liquids ," *AICHEJ*, Vol. 21,pp. 833-838.

Cole, R., 1974, " Boiling Nucleation ," *Advances in Heat Transfer*, Vol. 10, pp. 85-166.

Greenberg, M. D., 1978, " Foundation of Applied Mathematics," Prentice- Hall, Inc., Englewood Cliffs, New Jersey.

Han, C. D., and Yoo, H. J., 1981, " Studies on Structural Foam Processing. IV. Bubble Growth During Mold Filling," *Polymer Engineering and Science*, Vol. 21, pp. 518-533.

Hetsroni, G., 1982, " Handbook of Multiphase Systems," McGraw Hill, New York.

Lundberg, J. L., and Mooney, E. J., 1964, " Diffusion of Solubility of Methane in Polyisolutylene," *Journal of Polymer Science*, Vol. 7, pp. 947

Plesset, M. S., and Zwick, S. A., 1954, " The Growth of Vapor Bubbles in Superheated Liquids," *Journal of Applied Physics*, Vol. 25, pp. 493-500.

Rosner, D. E., and Epstein, M., 1972," Effect of Interface Kinetics, Capillarity and Solute Diffusion on Bubble Growth Rates in Highly Supersaturated Liquids," *Chemical Engineering Science*, Vol. 27, pp. 69-88.

Street, J. R., Fricke, A. L., and Reiss, L. P., 1971," Dynamics of Phase Growth in Viscous, Non-Newtonian Liquid," *Ind. Eng. Chem. Fundam.*, Vol. 10, pp. 54-64.

Upadhyay, R. K., 1985, " Study of Bubble Growth in Foam Injection Molding," *Advances in Polymer Technology*, Vol. 5, pp. 55-64.

Villamizar, C. A., and Han, C. D., 1978," Studies on Structural Foam Processing II. Bubble Dynamics in Foam Injection Molding," *Polymer Engineering and Science*, Vol. 18, pp. 669-710.